Structures for Architects

Structures for Architects: Planning, Analysis and Design explains the basics of structural systems to help architects conceive the structural form and analyze and design in a comprehensive manner. The objective is not to explain the structural analysis and design of complex systems in detail but to explain the analysis and design of structural members in a simple and elementary approach. Presenting basic concepts in a simple manner, the book deals with the presentation of structural systems used in the construction of public buildings. Architectural design, urban planning guidelines, basic rules of fire safety, and earthquake-resistant design are discussed in a conceptual manner. Examples included throughout the book shall help students, faculty, and practicing architects understand structural analysis and design in a straightforward manner, and exercise problems are presented to aid in the understanding of the contents presented. This book serves as a useful reference material for both academicians and practicing professionals.

- Strengthens the capacity of architects in fundamental structural analysis and design.
- Presents numerical examples and exercises for self-study in each chapter, including MATLAB® examples.
- Includes the basics for understanding architectural design, urban planning, fire safety, and earthquake-resistant design.
- Examines construction planning and management through CPM and PERT methods to aid in understanding the fundamentals of project management.

Structures for Architects
Planning, Analysis and Design

Jayashree Aniruddha Madav,
Srinivasan Chandrasekaran and Sanjeev Singh

CRC Press
Taylor & Francis Group
Boca Raton London New York

CRC Press is an imprint of the
Taylor & Francis Group, an **informa** business

Designed cover image: Srinivasan Chandrasekaran

First edition published 2025
by CRC Press
2385 NW Executive Center Drive, Suite 320, Boca Raton FL 33431

and by CRC Press
4 Park Square, Milton Park, Abingdon, Oxon, OX14 4RN

CRC Press is an imprint of Taylor & Francis Group, LLC

ISBN: 978-1-032-94698-6 (hbk)
ISBN: 978-1-032-94700-6 (pbk)
ISBN: 978-1-003-58131-4 (ebk)

DOI: 10.1201/9781003581314

Typeset in Times
by Apex CoVantage, LLC

Contents

About the Authors

Jayashree Aniruddha Madav is an urban designer and practicing architect in Mumbai, India. She is currently an Assistant Professor, in the Department of Architecture, Kamla Raheja Vidyanidhi Institute of Architecture and Environmental Studies (KRVIA), Mumbai. She also works as an Associate Director at the Urban Design Research Institute (UDRI), Mumbai. She holds a Bachelor of Architecture and Master of Architecture in Urban Design. She strongly believes that collaborative and participatory methods play a significant role in creating resilient and lively cities. Her master's thesis, titled "An Ecological Approach towards Shaping Urban Peripheries: The Case of Urban Flooding in Chennai City," throws light on ecological and socio-spatial transformations that have led to changing dynamics of peri-urban areas of Chennai City. Over the past five years, she has worked with various communities, local government, institutions and organizations in many projects that hinge on public awareness and participation, community-led development, strategies for communication and outreach, stakeholder engagement and capacity building of local government and communities. She was previously part of the core planning team of the Eastern Waterfront Project at Mumbai Port Trust. She has also been part of the City Investments to Innovate, Integrate and Sustain (CITIIS) program (supported by Ministry of Housing and Urban Affairs (MOHUA), the French Development Agency (AFD), the European Union (EU) and managed by National Institute of Urban Affairs (NIUA)) through projects – Haora Riverfront development, Agartala and Green Mobility Corridor at Hubli-Dharwad while she was working at Urban Design Collective (UDC), Chennai. Currently, she works as an Associate Director at UDRI, and her current research is focused on demystifying the spatial planning process across India. It focuses on understanding the on-ground challenges and best practices adopted in various states as they navigate the spatial planning process guided by state legislations and central schemes. She is involved in the proposals of many prestigious projects of national importance. She has published a number of articles in national newsletters and journals, and is a member of the Professional Council of Architects, India and a Member of Society of Urban Designers, India.

Srinivasan Chandrasekaran is currently a Professor with Higher Academic Grade in the Department of Ocean Engineering, the Indian Institute of Technology, Madras and a Visiting Professor, the Department of Structural Engineering, University of Naples Federico II, Italy. He holds the prestigious fellowship conferred by the Ministry of Italian University Research (a MiUR Fellow). He commands a rich experience in teaching, research, and industrial consultancy of about 30 years. He is also a Visiting Professor, Faculty of International Maritime Studies, Kasetsart University, Sri Racha Campus, Thailand. He has supervised many sponsored research projects and offshore consultancy assignments, both in India and abroad. His active research areas include the dynamic analysis and design of offshore structures, the structural health monitoring of ocean structures, risk and reliability, the fire resistant design of structures, and use of functionally graded materials (FGM). He was a visiting fellow under the Ministry of Italian University Research's invitation to the University of Naples Federico II for two years. During his stay in Italy, he researched the advanced nonlinear analysis of buildings under earthquake loads and other impact loads with experimental validation on full-scale models. He has authored around 170 research papers in peer reviewed international journals and refereed conferences organized by professional societies worldwide. He has authored 22 textbooks with various publishers of international repute. He is an active member of several professional bodies and societies, both in India and abroad. He has also conducted around 30 engineering courses on structural engineering for the National Program on Technology-Enhanced Learning (NPTEL), Government of India. He is a vibrant speaker and has delivered many keynote addresses at international conferences, workshops, and seminars organized in India and abroad. He holds life membership of many national and international professional societies.

Sanjeev Singh is a full professor in the Department of Architecture, SPA, Bhopal. He holds a doctoral degree from the National University, Singapore and a Master of Environmental Studies (by Research) from University of Melbourne, Australia with an Honors Grade. He is an active researcher and dedicated faculty, teaching architectural design, environmental planning and vernacular architecture. He has authored many research articles in peer-reviewed international journals and has presented his research works at many conferences organized by the professional societies of architecture and planning groups in India and abroad. He has also contributed to many book chapters, namely *Corporate Environmental Responsibility in Singapore and Malaysia: The Potential and Limits to Voluntary Initiatives* (edited by Peter Utting), *The Greening of Business in Developing Countries, Stakeholder Influence and Business Behavior*, and *Sustainable Development and Wisdom of Vernacular Architecture*. He has chaired many national level committees in planning and is member of many international committees on architecture and planning. He is member of many professional bodies including the Council of Architects, India.

Preface

Structures for Architects: Planning, Analysis and Design addresses the importance of geometric form in structural design, from an architect's perspective. It provides a comprehensive overview of structural forms conceived by architects, beginning with the site selection guidelines and region-wise classification of geometrical forms and their construction methods, followed by modern structural forms. It includes planning guidelines against earthquake loads, fire and blast loads, and a few code guidelines. The emphasis is that any architectural form ideation is not just an exercise that has a given philosophical narrative or an aesthetically conscious approach but also a valuable creation. This could not be achieved without proper structural design, making the form of ideation a reality. Book chapters deal with the overviews of loads acting on buildings. Beginning with the classification of loads, various types, including gravity, wind, impact and blast, earthquake, and fire loads are presented. Each load type is discussed in detail, highlighting its unique characteristics and effects on buildings. Materials used for construction, namely, steel and concrete, are presented with their structural properties and variations at high temperatures.

A successful architectural design integrates structural considerations from the beginning, rather than treating structure as an afterthought. This integration allows for innovative designs that push the boundaries of traditional construction methods. The form–function–structure relationship is presented implicitly, making it convenient for architects to follow, less mathematics but intensive explanations. Detailed presentation of the methods of structural analysis, developed from the first principles, are presented. It introduces structural analysis with the aid of the MATLAB® program to facilitate a fast and accurate method of analysis. The conceptual analysis is developed from the idealizations of the architectural form of buildings, and one-to-one mapping is explained. The design procedures of steel and the reinforced cement concrete section (RCC), as per the current design code, are discussed. Details of bolted connections are discussed with example problems. Design code in current prevalence is used to solve the design problem examples. The book also discusses the basics of construction planning using CPM and PERT, making it useful for practicing engineers, architects, students, and researchers in architecture, urban planning, and civil engineering.

- Simple explanation with a lot of illustrations.
- Problem-solving using the MATLAB program to improve speed and accuracy.
- Architectural form blended with structural design is the unique feature of this book.
- Guidelines for planning and design addressing earthquake and fire loads are added advantages to the user.
- Basics of construction planning and management through CPM and PERT help understand the fundamental project management methods.

The book is not a substitute for the classical structural analysis and design of members, as only the preliminary level of the contents is dealt with. The objective is to develop sufficient capacity building for architects and help them understand a relevant and structurally viable geometric form in the conceptual stage itself. In this exercise, the authors wish to thank the Center for Outreach and Digital Education (CODE), Indian Institute of Technology Madras, for the administrative support extended under the Book Writing Scheme. They also wish to express their gratitude to the director, School of Planning and Architecture, Bhopal, for the support and encouragement offered in publishing this book.

The authors sincerely acknowledge their teachers' valuable input, which imparted courage and confidence while expanding their classroom notes into the contents of the book. As practicing architects and well-experienced faculty, the authors felt the gap in knowledge transfer between architects and structural engineers. Therefore, the primary objective is to fill up that gap, and hope the readers will find it useful and truly simple.

Jayashree Aniruddha Madav
Srinivasan Chandrasekaran
Sanjeev Singh

Forewords

The book, titled *Structures for Architects: Planning, Analysis and Design*, is a good attempt by the authors to bridge the gap of the architect–structural engineer interface. The concepts of structural analysis and design, being complex, were dealt with in detail in the civil engineering curriculum, but neglected in architecture. As a result, innovative architectural forms conceived by architects get disrupted by the structures' intervention, and vice versa. This book will strengthen the capacity building of architects and help them gain confidence toward a committed achievement of the chosen building form. As stated by the authors, the book is not an attempt toward advanced analysis and design, for which there are too many, but to illustrate the basics in a simple perspective. For instance, many other books authored by one of the authors address highly advanced topics in structural analysis and design, and they are quite popular in India and abroad.

With my wide experience of teaching civil and structural engineering and consulting major construction and design projects, I felt a strong need for such a book. I am very glad to see the outline of this book, mapped to the demand. The approach of explaining the transformation of different architectural forms of different eras, from Roman to Gothic and modern architecture, helps align the thought process of conceiving new building forms. Subsequent bridging of the forms to idealized analysis and then design of elements is a nice approach authors followed, which is appreciated.

In the recent past, more attention has been drawn to earthquake analysis and the design of structures with a demand to understand ductile detailing. This is because many buildings do not comply with the revised code guidelines under earthquake loads. Unless engineers and practicing architects are updated with the simple and basic rules of seismic-resistant design, engineered structures can neither be constructed nor retrofitted. However, the book does not explain any advanced design examples of earthquake resistance but focuses on the methodology and approach for the design. I feel the contents of this book are vital for the architecture curriculum and useful for practicing engineers. I am happy to recommend this book as an important source of reference to leading architectural schools in India and abroad. I also congratulate the authors for their contributions toward capacity building of infrastructure growth.

Arvind Kumar Jain, FIE, AM ASCE, M ISOPE, F ISET
Professor and Former Head (Superannuated)
Department of Civil Engineering
Indian Institute of Technology Delhi
Haus Khas, New Delhi, India

* * * * *

In architecture and engineering, integrating strong structures with elegant designs is the foundation of great creations. *Structures for Architects: Planning, Analysis and Design* emerges as a much-needed bridge between the disciplines of architecture and structural engineering, offering an integrated perspective that is both practical and visionary. This book serves as a guide for architects, engineers, and researchers motivated to create structures that are not only functional and safe but also deeply aligned with the artistic and philosophical principles of design.

The authors, recognized experts in civil engineering and architecture disciplines, have carefully compiled a wealth of knowledge that starts with the origins of architectural forms and progresses to their realization through detailed planning, analysis and design. By addressing site selection, geometrical classifications, and construction methods, the book sets the stage for understanding the intricacies of modern structural forms. It delves into the dynamic interactions between materials and loads, offering insights into how environmental and operational factors such as earthquakes, fire, and blast loads influence design decisions. The comprehensive inclusion of spectrum plots and material properties under varied conditions underscores its technical depth.

A unique feature of this book is its accessibility to both architects and engineers. With an emphasis on conceptual understanding, the chapters navigate complex topics—such as structural analysis using the stiffness method and design procedures for steel and RCC members—through clear explanations and practical examples. The use of computer tools such as MATLAB® for analysis highlights the integration of modern tools to enhance precision and efficiency in structural modelling.

The text highlights the important connection between form, function, and structure, emphasizing that structural design should be a key part of architectural planning. The emphasis on planning tools like CPM and PERT adds an additional dimension, enabling readers to grasp the fundamentals of project management alongside technical expertise.

Illustrated with simplicity and clarity, this book is a valuable resource for practicing professionals and a foundational guide for students and researchers. Its multidisciplinary approach fosters a collaborative ethos, encouraging architects and engineers to engage in a dialogue that enriches their practice and expands the possibilities of design.

Structures for Architects: Planning, Analysis and Design is not just a book; it is an invitation to explore the limitless potential of architecture when harmonized with complex engineering principles. For anyone passionate about shaping the built environment with ingenuity and precision, this book is an essential companion.

Rouzbeh Abbassi
Professor and Director of Research
School of Engineering, Macquarie University, Australia

* * * * *

Understanding structures in developing architectural planning and design is essential to ensure safety and stability, enhancing functional efficiency, innovative design, sustainability, environmental response, and economic viability of any building project. Therefore, for any state-of-the-art architectural design, the structure is a silent hero. The structure is more than an assembly of the skeletal framework; it interlaces science and art, coalescing form and function in a compatible relationship. The most promising materials used in architectural structures are reinforced concrete, structural steel, wood, bricks, and stones. These materials possess multiple properties that ensure quality and high aesthetics and are thus widely used. Apart from providing safety and durability, architectural structures also influence the form and functionality of buildings. Different structural solutions produce different shapes, spaces, and functions for the construction. Architects must balance technical requirements with aesthetics, creating safe and efficient structures and visually appealing and unique in form. The architectural structure also expresses each era's architectural style, trends, and ideas. Innovative structural materials, techniques, and technologies always accompany architectural development. Creative approaches to structure have invented many new design possibilities, creating unique and iconic structures.

In this book, *Structures for Architects: Planning, Analysis and Design*, written by Srinivasan Chandrasekaran, Jayashree Aniruddha Madav, and Sanjeev Singh, the chapters are designed to address the importance of the geometric form in structural design from an architect's perspective. It provides a comprehensive overview of structural forms conceived by architects, beginning with site selection guidelines and region-wise classification of geometrical forms and their construction methods, followed by modern structural forms. It includes planning guidelines for earthquake loads, fire and blast loads, and a few code guidelines. The emphasis is that any architectural form ideation is not just an exercise with a given philosophical narrative or an aesthetically conscious approach but also a valuable creation. This could not be achieved without proper structural design, making the form of ideation a reality.

Book chapters deal with the overviews of loads acting on buildings. Beginning with the classification of loads, various types, including gravity, wind, impact and blast, earthquake, and fire loads, are presented. Each load type is discussed in detail, highlighting its unique characteristics and effects on buildings. Materials used for construction, namely, steel and concrete, are presented with their structural properties and variations at high temperatures.

A successful architectural design integrates structural considerations from the beginning rather than treating structure as an afterthought. This integration allows for innovative designs that push the boundaries of traditional construction methods. The form–function–structure relationship is presented implicitly, making it convenient for architects to follow less mathematics but more intensive explanations. Detailed presentations of the methods of structural analysis, developed from the first principles, are presented. It introduces structural analysis with the aid of the MATLAB® program to facilitate a fast and accurate analysis method. The conceptual analysis is developed from the idealizations of the architectural form of buildings, and one-to-one mapping is explained. The design procedures of steel and the reinforced cement concrete section (RCC), as per the current design code, are discussed. Details of bolted connections are discussed with example problems. The book also discusses the basics of construction planning using CPM and PERT, making it useful for practicing engineers, architects, students, and researchers in architecture, urban planning, and civil engineering.

Nasir Shafiq, ASCE, ACI, PEC, BEM, CSM
PhD, Structural Engineering and Materials (Leeds, UK)
MSc, Structures (AIT Bangkok)
BE, Civil (NEDUET Karachi Pakistan)
Professor, Civil (Structural) Engineering
Universiti Teknologi PETRONAS
32610 Bandar Seri Iskandar, Perak

1 Structural Forms Conceived by Architects

1.1 INTRODUCTION

Buildings are not constructed in solitude; they are intended to accommodate, support, and inspire various human activities formed by their time's social, economic, and political demands. Buildings must overcome limits in natural and constructed surroundings while capitalizing on development potential. As a result, when designing and erecting a structure, carefully considering the contextual factors on the site is critical (Sucuoğlu and Akkar, 2014; Torunbalci and Ozpalanlar, 2008). A site's microclimate, geography, and natural ecosystems impact early design decisions. Sustainable design considers the location's natural qualities to increase human comfort, preserve energy, and use resources efficiently (Chandrasekaran et al., 2021, 2024). This strategy guarantees that buildings are in harmony with the landscape, considering the sun's path, wind patterns, and water movement on the site (Buonocunto et al., 2023; Verderame et al., 2014). Figure 1.1 shows details of bamboo housing, a common type of construction in the rural sector. The components of bamboo housing are shown in Figure 1.2. Various house components are shown in Figures 1.3–1.5. Arches are the most popular building forms of load-bearing structures in the Roman style. While stone masonry or course-rubble ones are considered to be better for gravity loads, their construction process is also tedious and shows a large variation in strength based on their construction practice. Figures 1.6–1.8 show details of stone masonry construction, which is still the common form of load-bearing structures in rural areas.

Beyond environmental factors, regulatory frameworks such as zoning rules influence how and where buildings are built. These restrictions consider existing land-use practices, specify permissible uses for the site, and set limits on building size, shape, and placement. Site circumstances influence building design, construction material, construction type and method, and maintenance, which place demands on transportation networks, utilities, and other infrastructure services (ASTM, 2020). Calculating how much development a place can support without overloading these systems, overconsuming energy, or creating environmental damage is a significant difficulty. In this context, sustainability becomes a prominent element, necessitating a careful balance between development and the environment, ensuring that structures suit human requirements while reducing their environmental impact.

1.2 SITE ANALYSIS

Site analysis investigates the contextual forces that shape a building, such as its location, spaces, organization, external design, and interaction with the surrounding

DOI: 10.1201/9781003581314-1

FIGURE 1.1 Bamboo housing details.

FIGURE 1.2 Bamboo housing components.

environment. The approach starts with gathering thorough physical information about the site. Key steps include the following information:

- Map the site's shape and extent using legal boundaries.
- Identify mandatory setbacks, easements, and rights-of-way for the site.
    - Estimate space requirements for the project, site amenities, and future expansions.
    - Analyze terrain and soil characteristics to identify ideal construction and outdoor activities places.

FIGURE 1.3 Components of bamboo roofing.

FIGURE 1.4 Flooring and foundation details of bamboo housing.

FIGURE 1.5 Cross section of a wall through the foundation of bamboo housing.

FIGURE 1.6 Arches: common form of load-bearing structure in Roman age.

- Identify steep and moderate slopes that may not be suitable for development.
    - Rainwater management: Identify appropriate drainage zones and map existing patterns.
    - Determine the water table level, and identify places prone to surface water flow, flooding, or erosion.

FIGURE 1.7 Stone masonry: common form of load-bearing structure in a rural area.

FIGURE 1.8 Section details of stone masonry.

- Conserve natural vegetation by identifying existing trees and native species.
    - For conservation, map natural water features, such as wetlands, streams, floodplains, and shorelines.
    - Analyze climatic conditions, including sun path, wind direction, and rainfall forecast, to influence design decisions.
    - Analyze climatic parameters, such as sun path, wind direction, and rainfall forecast, to help inform design decisions.
    - Consider how the site's topography and surrounding structures affect solar access, wind flow, and potential glare.
- Determine the feasibility of using solar radiation as an energy source.
- Determine access routes from public roadways or adjacent transit stops.

- Investigate possible pedestrian and vehicular routes, from access points to building entrances.
- Check the availability of critical services, such as water, sewage, gas, power, telephones, and fire hydrants.

- Ensure you are close to municipal services, like police and fire protection.
- Determine both acceptable and unwanted views of the site.
- Determine the probable sources of noise and traffic congestion at the site.
- Determine how well the site's proposed use complements surrounding land uses.
- Create a map of artistic or historical resources that should be conserved.
    - Consider how the surrounding area's scale and character may influence the building's design.
    - Consider the proximity to public, commercial, medical, and recreational amenities.

1.3 TOPOGRAPHY AND LANDFORM

Topography refers to the arrangement of natural surface features on a plot of land, which is crucial in determining how and where to build a site. Conducting a thorough topographic survey enables architects and urban planners to design structures harmonizing with the natural landscape, reducing the need for significant excavation or alterations. This understanding also helps optimize the placement of buildings, roads, and drainage systems, ensuring that the site's natural features are effectively utilized and preserved. Architects often utilize site sections or plans that include contour lines, slopes, elevation variations, and geological characteristics to understand how a building design interacts with the land's topography.

Contour lines are imaginary lines that connect points at the same elevation above a specific reference point or datum. Each line traces the shape of the terrain at a given height. These lines are continuous and never intersect, except when they meet a vertical surface in a plan view. The contour interval is the vertical difference in elevation between two adjacent contour lines on a topographic map or site plan. The interval is chosen based on the scale of the drawing, the size of the site, and the slope's characteristics. Topographical data provides essential insights, such as pinpointing high and low elevations, analyzing water drainage patterns, and assessing soil stability and areas prone to erosion.

1.4 GEOMETRIC FORMS

The concept of geometric forms conceived by architects delves into architecture's creative and functional expressions, where form and structure intertwine to bring buildings to life. A building's geometric form is not merely an aesthetic choice but a response to numerous factors, such as functionality, material properties, environmental context, and technological advances. Over the centuries, architects have developed and conceived various structural forms that have transformed how we interact with spaces, addressing the needs of their time while pushing the boundaries of design and engineering.

1.4.1 Historical Perspective: Ancient Civilizations

Ancient architecture laid the foundation for understanding geometric/structural forms. Civilizations like the Egyptians and Greeks conceived structures such as pyramids and temples based on the post-and-lintel system. Figure 1.9 shows the typical Egyptian and Greek architectural forms. As shown in the figure, the Greek Parthenon is a prime example of the post-and-lintel system, designed to withstand the elements and offer monumental aesthetic appeal.

Roman contributions included the invention of arches and vaults, which were incorporated into massive structures like the Pantheon and aqueducts. Figure 1.10 shows a typical Roman architectural form, depicting arches and vaults in the building forms. The arch became a breakthrough in structural forms as it distributes weight more efficiently.

Stonehenge, Wiltshire, UK **Parthenon, Greece**

FIGURE 1.9 Egyptian and Greek architectural forms.

Aqueducts, Rome, Italy **Pantheon, Rome, Italy**

FIGURE 1.10 Roman architecture forms.

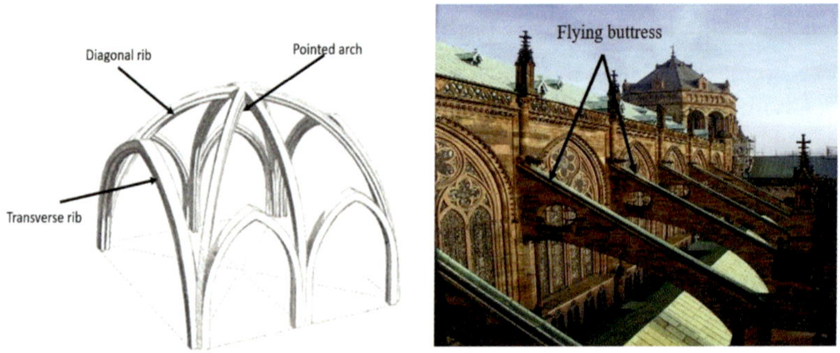

Ribbed vault with pointed arch

FIGURE 1.11 Gothic form.

Baroque architecture

FIGURE 1.12 Florence's Santa Maria del Fiore, Italy.

In the Middle Ages, Gothic architecture revolutionized structural design by developing the ribbed vault, pointed arch, and flying buttress. Figure 1.11 shows the classical Gothic forms, with ribbed vaults and flying buttresses. These structural innovative forms allowed for taller, lighter buildings with expansive stained-glass windows, exemplified by cathedrals like Notre Dame. Gothic forms were organic and dynamic, balancing the structural need for stability with aesthetic aspirations for height and lightness.

During the Renaissance, architects revisited classical forms but introduced new complexities in geometry. See, for example, Florence's Santa Maria del Fiore by Brunelleschi (Figure 1.12). Domes became iconic forms, combining art with structural ingenuity. The Baroque architecture further pushed the limits of structural forms with its dramatic use of space and movement, favoring fluid, complex geometries.

1.5 BUILDING FORMS AND ARCHITECTURE

In India, the rich diversity of landscapes—encompassing mountains, plains, coastal areas, deserts, and even tropical forests—demands various architectural approaches, structural systems, and foundation techniques. These geographic diversity, climatic variations, cultural influences, and traditional knowledge have shaped a unique architectural heritage across the country. Each region's architectural styles and construction techniques are inherently linked to the landscape, addressing local environmental challenges while utilizing region-specific materials.

1.5.1 Hilly and Mountainous Regions (e.g., Himachal Pradesh, Uttarakhand)

The primary architectural challenge in mountainous areas like the Himalayas is building on steep slopes and dealing with seismic activity (Barrell et al., 2013, 2011). Structures in these regions are often made from locally available stone, timber, and mud, offering both durability and thermal insulation in cold climates.

1.5.1.1 Building Types

Traditional timber houses, such as those seen in the Kath-Kuni architecture (Figure 1.13) of Himachal Pradesh, are constructed using a combination of stone and timber. This method offers excellent stability against earthquakes, with flexible jointing systems that allow the structures to absorb seismic forces effectively. Modern eco-friendly cottages and resorts have also embraced these principles in recent years, blending seamlessly into the mountainous landscapes. These structures often use sustainable materials, like timber and bamboo, offering lightweight, energy-efficient

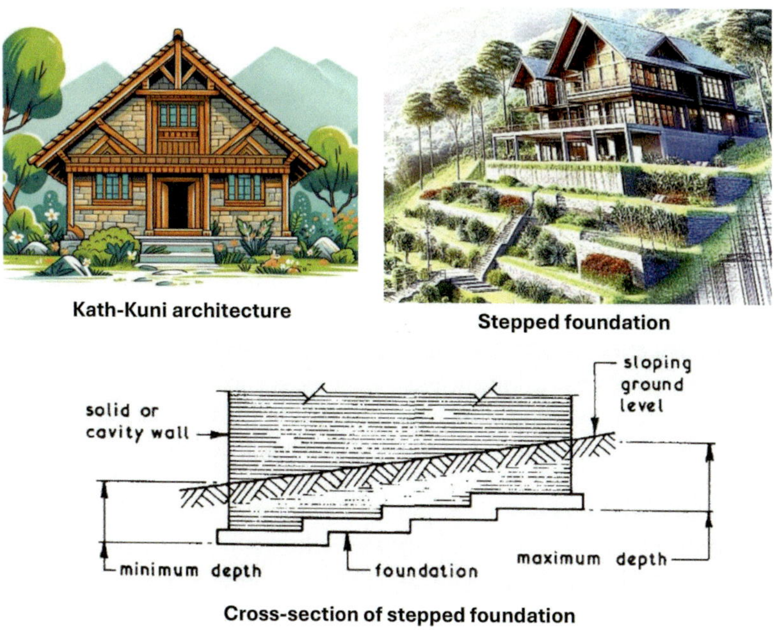

FIGURE 1.13 Kath-Kuni architecture.

alternatives that harmonize with the environment while maintaining the region's architectural heritage.

1.5.1.2 Foundation Details

Stepped foundations shown in the figure are commonly used in sloped terrains, following the natural gradient of the land to provide stability and minimize erosion risks. These foundations help distribute the building's load evenly by aligning with the terrain, ensuring structural integrity on uneven ground. In contrast, raft and pile foundations are employed in unstable or soft soils. Raft foundations spread the load over a large area to prevent sinking. Pile foundations anchor the building deep into the ground or bedrock, offering more excellent stability in challenging soil conditions, such as wetlands or flood-prone regions.

1.5.1.3 Architecture

Vernacular architecture in mountainous and high-rainfall regions is distinguished by sloping roofs designed to effectively manage heavy rainfall and snow. These roofs prevent water and snow accumulation, reducing the risk of structural damage. These constructions often use local materials, such as stone and timber, providing durability and thermal insulation. Pitched roofs, in particular, are crucial in these areas, as their steep angles allow snow to slide off easily, preventing excessive weight from building up and ensuring the longevity and safety of the structure.

1.5.2 Plains and Riverine Regions (e.g., Northern Plains, Ganges Basin)

The plains of India, such as the Indo-Gangetic plains, have vast expanses of fertile land, moderate climates, and abundant water resources, making them suitable for large agricultural settlements and urban growth. Architectural styles here have evolved to reflect the rural economy and urban development.

1.5.2.1 Building Types

Cities in these regions showcase a wide range of architectural styles, from traditional brick-and-mortar houses to modern high-rise buildings that dominate urban centers. In rural areas, homes are often constructed using locally available materials, like mud, bamboo, and thatch, offering sustainable and climate-responsive designs. Traditional courtyard houses, or **havelis**, found in cities like Lucknow and Varanasi, are designed for hot climates with inner courtyards that promote natural ventilation and provide shaded, cool spaces. Meanwhile, modern residences and apartment buildings in urban areas are designed to maximize space, incorporating contemporary materials and design principles to meet the needs of growing populations.

1.5.2.2 Foundation Details

Shallow foundations are commonly used in the firm soils of the plains, where the ground provides adequate support for structures. Strip and spread footings are typical in these areas, distributing the building's load evenly across the soil and providing a stable base for walls and other structural elements. For column-based structures, such as apartment buildings, isolated footings (Figure 1.14) are often employed,

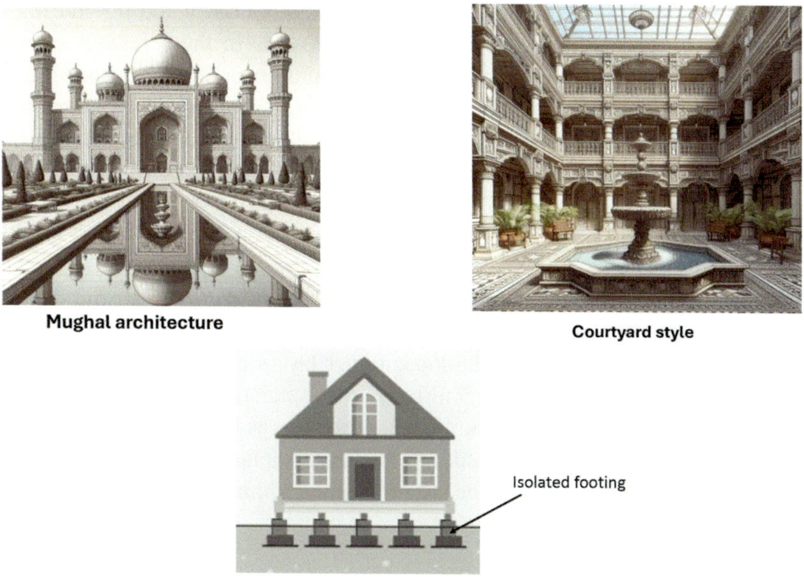

FIGURE 1.14 Mughal architecture.

supporting individual columns by spreading the load over a larger area to prevent settlement. These foundation types are efficient and cost-effective, well-suited to the stable soil conditions in plains and urban developments.

1.5.2.3 Architecture

Colonial and Mughal architectural influences shown in Figure 1.14 are evident in the use of large arches, high ceilings, and expansive verandas, all designed to cope with the high temperatures typical in many regions of India. These elements not only add grandeur but also facilitate better air circulation, helping keep interiors cool. Courtyard design, a key feature in both styles, is particularly effective in hot climates, as it allows for natural ventilation and cooling by creating shaded, open-air spaces that promote airflow and reduce the heat inside the building. This blend of aesthetic appeal and practical climate adaptation has left a lasting impact on the architectural landscape.

1.5.3 Coastal Regions (e.g., Kerala, Tamil Nadu, Goa)

India's extensive coastline, spanning thousands of kilometers along the Arabian Sea, the Bay of Bengal, and the Indian Ocean, presents unique architectural challenges due to high humidity, salt-laden winds, and the threat of cyclones. The architecture here focuses on resisting harsh weather conditions while maximizing airflow and sea views.

1.5.3.1 Building Types

Stilt houses are a standard architectural solution in coastal and flood-prone areas, with homes elevated aboveground to protect against flooding and rising sea levels.

These elevated structures ensure that living spaces remain dry and safe during seasonal floods or storm surges. Beach villas and resorts in these regions often use locally sourced materials, like bamboo, thatch, and coconut timber, blending sustainability with the natural environment. In contrast, concrete-based modern homes are built to withstand the humid climate and the high saline content of coastal soils, providing durability and resistance to corrosion while maintaining a contemporary aesthetic.

1.5.3.2 Foundation Details

Deep foundations, particularly pile foundations, ensure structural stability in areas with loose sandy soil, such as coastal regions. These foundations anchor buildings deep into the ground, bypassing the loose upper layers of soil and reaching firmer strata, thus providing a solid base. Similarly, raft foundations are employed in areas prone to flooding or where soil conditions are uneven, as they help prevent differential settlement by spreading the building's load across a larger surface area, ensuring uniform support and minimizing the risk of structural damage.

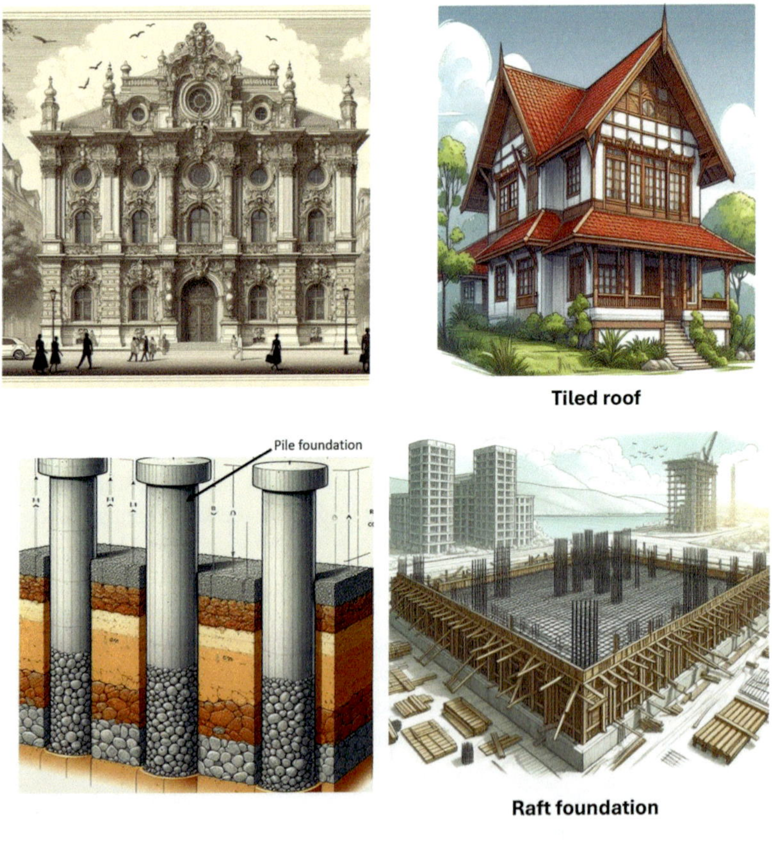

FIGURE 1.15 Portugal architecture.

1.5.3.3 Architecture

In coastal regions with heavy monsoonal rains, tiled roofs (Figure 1.15) made from locally sourced clay are a common feature, designed to shed water and withstand the intense weather efficiently. Homes often incorporate open verandas and balconies to promote cross-ventilation, which helps reduce humidity levels and keep interiors cool. The architecture in coastal towns like Goa showcases strong colonial and Portuguese influences (Figure 1.15), with buildings, painted in bright colors, featuring wooden beams, large windows, and decorative details. These design elements enhance aesthetic appeal and address the region's climatic challenges, blending form with function. The plains of India, such as the Indo-Gangetic plains, have vast expanses of fertile land, moderate climates, and abundant water resources, making them suitable for large agricultural settlements and urban growth. Architectural styles here have evolved to reflect the rural economy and urban development.

1.5.4 Desert Regions (e.g., Rajasthan, Gujarat)

The Thar Desert in Rajasthan presents an extreme environment characterized by arid conditions, scarce water, and extreme temperatures. Traditional desert architecture is designed to mitigate these harsh conditions, focusing on thermal mass and natural cooling techniques.

1.5.4.1 Building Types

Traditional havelis and sandstone forts in arid regions are built to withstand extreme heat, featuring thick walls that provide excellent insulation, keeping interiors cool during the day and warm at night. In rural desert areas, circular mud huts (*bhungas*) are ingeniously designed to be wind-resistant, with rounded shapes deflecting solid winds. In contrast, mud construction naturally regulates interior temperatures. In modern architecture, solar passive homes are gaining popularity. They are designed to minimize heat gain by incorporating energy-efficient materials, strategic window placements, and shading devices to reduce the need for artificial cooling, all while harmonizing with the local climate.

1.5.4.2 Foundation Details

Strip and shallow foundations are commonly used in regions with firm sandy soils, as the soil provides sufficient support for these less-intrusive foundation types. These foundations distribute the building's load effectively across the ground, making them a cost-effective choice in stable soil conditions. In areas where the soil needs reinforcement, stabilized soil foundations are implemented, where local soil is mixed with cement or lime to increase its strength and load-bearing capacity. This method enhances the durability of the foundation, ensuring the structure remains stable, even in less-than-ideal soil conditions.

1.5.4.3 Architecture

In hot desert climates, buildings have thick walls and small windows to minimize heat absorption and protect interiors from the harsh environment, including sandstorms. Traditional features like *Jharokhas* (overhanging balconies), as shown in Figure 1.16, provide shade and natural cooling by allowing air to circulate while

FIGURE 1.16 *Jharokha* architecture.

reducing direct sunlight exposure. Additionally, courtyards are the key architectural elements, promoting cross-ventilation and aiding in heat dissipation. It aids in creating a cooler space within the enclosure. These design strategies are integral to maintaining comfortable living conditions in extreme desert climates.

1.5.5 Wetland and Flood-Prone Areas (e.g., Assam, West Bengal)

In regions like Assam and West Bengal, prone to heavy rainfall, flooding, and wetlands, architecture has evolved to adapt to these unique environmental challenges. Structures in these areas are designed to withstand periodic flooding, high humidity, and damp soil conditions. The focus is elevating the living spaces above flood levels and ensuring proper drainage and ventilation to combat moisture and waterlogging.

1.5.5.1 Building Types

In flood-prone regions like Assam, traditional stilted bamboo houses, known as *Changghars*, are elevated on stilts to protect against rising water levels during floods, ensuring that living spaces remain dry. Innovative floating homes are being adopted in areas experiencing more frequent and severe flooding. These homes, built on pontoons, can rise and fall with the water levels, providing a resilient solution to the challenges posed by unpredictable and rising floodwaters. Both designs reflect the adaptability of local architecture to environmental conditions.

1.5.5.2 Foundation Details

Pile foundations are essential for securely anchoring buildings in unstable soils in wet, marshy regions, providing stability and elevation to protect structures from

Stilted bamboo houses Foundation on pontoons

FIGURE 1.17 Floating houses.

floodwaters. These deep foundations ensure the building remains intact even in saturated ground conditions. For more innovative solutions, floating foundations are used to construct floating homes (Figure 1.17), allowing these structures to rise and fall with floodwaters. This design offers a flexible and adaptive approach to living in flood-prone areas, ensuring homes remain safe and habitable during extreme weather events.

1.5.5.3 Architecture

In flood-prone and wetland areas, lightweight structures made from materials like bamboo, thatch, and other locally sourced resources are commonly used to reduce the load on foundations, making the buildings more adaptable to shifting, unstable soils. Additionally, raised verandas and platforms are integral to the design, ensuring that living spaces remain elevated above flood levels, protecting rising waters while maintaining comfort and accessibility. These architectural strategies balance traditional building practices and the need for resilience in challenging environments.

1.5.6 Tropical Rainforest Regions (e.g., Northeast India, Western Ghats)

In tropical regions like the Western Ghats and Northeast India, the architecture is influenced by high humidity, heavy rainfall, and dense vegetation. The emphasis is on lightweight structures that withstand torrential rains and humid conditions.

1.5.6.1 Building Types

Traditional wooden houses in regions with heavy rainfall are constructed using bamboo and wood, often featuring sloping roofs to efficiently manage water runoff and prevent water damage. These structures are practical and sustainable, utilizing local materials suited to the climate. Additionally, eco-lodges and treehouses have become popular in forested areas, particularly for tourism, offering a unique experience that blends seamlessly with the natural environment. These eco-friendly designs prioritize minimal environmental impact while providing shelter that harmonizes with the surrounding landscape.

1.5.6.2 Foundation Details

In wetland areas, stilt foundations are commonly used to elevate buildings above the ground, protecting them from waterlogging and local wildlife. This design allows air circulation beneath the structure, helping keep it dry and stable. Pile foundations are often employed for regions with soft, saturated soils, where deep piles are driven into the ground to provide a solid base and prevent the building from sinking. Both foundation types are essential for ensuring the durability and safety of structures in challenging, moisture-prone environments.

1.5.6.3 Architecture

In humid and rain-prone regions, thatched roofs with broad overhangs, as shown in Figure 1.18, are a traditional solution, using natural materials like palm leaves or bamboo to protect from heavy rainfall while promoting water runoff effectively. These roofs are not only sustainable but also offer excellent insulation. Additionally, ventilated walls are a key feature in such climates, designed to allow air circulation and reduce moisture buildup inside the home. This natural ventilation helps maintain a cooler, drier indoor environment, making the structure more comfortable and resistant to the challenges posed by high humidity.

1.6 MODERN AND CONTEMPORARY FORMS

The evolution of structural forms in architecture reflects advancements in materials and construction techniques and changes in cultural, environmental, and functional

FIGURE 1.18 Thatched roof and overhang houses.

needs. From ancient pyramids to modern parametric designs, architects continue to innovate, balancing the aesthetics of form with structure demands. Each era has its defining structural forms, shaped by the technology and ideas of the time, and the future of architecture will likely see even more complex and sustainable forms emerge through the use of digital design tools and new materials. The design of tall structures has seen significant advancements over the years, driven by the need to create efficient and resilient buildings in the face of increasing urbanization and environmental challenges. Modern engineering has introduced innovative structural systems, allowing taller buildings and improving their stability, performance, and sustainability. Critical structural systems, such as tube-in-tube, outrigger, and bio-inspired structures, have emerged as leading solutions for constructing skyscrapers and high-rise buildings.

Tube-in-tube structures enhance lateral load resistance by combining an inner tube, typically formed by the core of the building, with an outer tube made up of the exterior framework. This dual-system approach provides stability against wind and seismic forces while optimizing material usage. On the other hand, outrigger structures utilize horizontal beams that link the central core to the exterior columns, distributing the lateral forces and reducing the building's sway, making it possible to construct taller, slender buildings. Additionally, the field of architecture has increasingly turned to bioinspired structures, which draw inspiration from natural forms and systems to achieve both efficiency and adaptability. These designs often mimic the load distribution and flexibility seen in nature, resulting in buildings that are more energy-efficient, sustainable, and resilient to environmental stresses.

Together, these structural systems reflect the evolution of engineering and architecture in response to the growing demands for taller, more efficient, and resilient buildings in modern cities, blending form, function, and sustainability in groundbreaking ways.

1.6.1 Tube-in-Tube Structures

Tube-in-tube structures are an efficient solution for tall buildings (Figure 1.19), where the exterior and interior are primary structural elements to resist lateral forces, like wind and seismic activities. This design maximizes the building's strength while minimizing material usage, making it highly efficient for skyscrapers.

Design Overview: The tube-in-tube structure consists of an *outer tube* formed by closely spaced exterior columns and beams and an *inner tube* created by the elevator core or shear walls. Both tubes work together to resist lateral loads and distribute them evenly throughout the building.

Lateral Load Resistance: The exterior tube takes on most of the wind forces, while the inner tube supports the building against additional lateral loads, such as seismic activity. The dual system allows for a more robust, stiffer structure.

Efficiency and Cost-Effectiveness: Because the load is distributed between two systems, less material is needed than in a conventional solid core structure, making it a cost-effective solution for tall buildings.

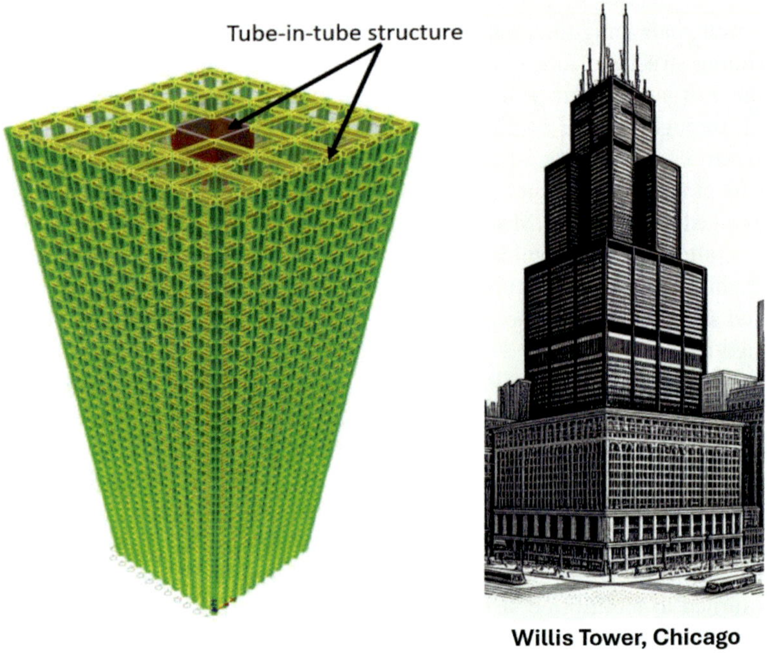

FIGURE 1.19 Tube-in-tube tall structure.

Applications: Tube-in-tube designs are widely used in high-rise office buildings and residential towers, including iconic structures like the Sears Tower (now Willis Tower) in Chicago.

1.6.2 Outrigger Structures

Outrigger structures represent another advancement in the design of tall buildings (Figure 1.20), where horizontal beams or trusses—called outriggers—are connected between the central core and the exterior columns. This system enhances the building's resistance to wind and seismic forces by extending the load-bearing capacity outward.

Design Overview: In an outrigger structure, *outriggers* are horizontal structural elements that extend from the central core of the building to the perimeter columns, essentially linking the core with the outer frame. These outriggers act like rigid arms that prevent the core from bending under lateral loads.

Lateral Stiffness: The outriggers significantly increase the lateral stiffness of the building by engaging the perimeter columns, which absorb a portion of the wind or seismic forces. This reduces the bending and swaying of the central core, allowing for taller, slender structures.

FIGURE 1.20 View of outrigger structures.

Reduction of Drift: One of the primary advantages of the outrigger system is that it reduces lateral drift—the movement of the building under wind loads—by distributing forces throughout the structure.

Applications: Outrigger systems are standard in supertall buildings, such as the Burj Khalifa in Dubai (Figure 1.20) and the Shanghai Tower in China, where the building height requires additional stabilization.

1.6.3 Bioinspired Structures (Form-Dominant)

Bioinspired structures take inspiration from nature, applying biological principles and organic forms to architecture and engineering. These structures mimic natural systems' efficiency, adaptability, and resilience, leading to innovative and sustainable design solutions, especially for tall buildings.

Design Overview: Bioinspired structures often draw inspiration from natural forms, such as the exoskeletons of insects, the skeletons of trees, or the cellular structures of plants and animals. These natural systems demonstrate efficient load distribution, flexibility, and adaptability, which can be applied to modern architecture.

Energy Efficiency and Sustainability: Bioinspired structures are often designed to be more energy-efficient by incorporating passive ventilation, heat regulation, and lightweight yet strong materials. Nature-inspired forms also reduce material waste, making the construction process more sustainable.

The Eden Project, UK

Cybertecture Egg, Mumbai

Taipai tower, Taiwan

FIGURE 1.21 Bioinspired structural forms.

Load Distribution and Resilience: Like in nature, where structures evolve to carry loads and adapt to environmental stresses efficiently, bioinspired designs are often used to enhance the building's ability to withstand lateral loads, such as wind or seismic activity. The natural forms, such as branching or cellular networks, allow for better load distribution throughout the structure.

A few examples of bioinspired forms are shown in Figure 1.21. The Eden Project in the UK, which mimics the structure of soap bubbles for its geodesic dome; the Beijing National Stadium (Bird's Nest), whose lattice structure is inspired by a bird's nest; and Cybertecture Egg, Mumbai, India, are few examples of how bioinspiration can be applied to architecture. In skyscrapers, bioinspired designs may resemble trees with branching supports or the sponge-like skeletons of sea organisms that can handle tension and compression, allowing the structure to flex without breaking. Bioinspired skyscrapers, like the Taipei 101 in Taiwan, employ earthquake-resistant designs inspired by natural structures' flexibility and resilience. Other examples include towers inspired by termite mounds promoting natural ventilation and cooling.

1.7 EARTHQUAKE-RESISTANT PLANNING AND DESIGN

Seismic activities across the globe pose challenges in the planning, design, and construction of buildings in general. India's geographical positioning makes it highly susceptible to seismic activities, mainly due to the collision of the Indian plate with the Eurasian plate. This has led to the establishment of various guidelines and building codes to mitigate earthquake risks. As a result, code guidelines and governance are updated, leaving many old buildings unfit to be qualified under the revised design norms (Chandrasekaran et al., 2008a, 2008b, 2009a, 2009b). It is quite interesting and important to note that instead of changing the books, and code provisions, it is better to cultivate thoughts, methods, and techniques to improve the strength of the existing buildings. Engineers and architects execute the construction based on the knowledge gained by education and experience, which is a time-bound phenomenon (Chandrasekaran and Roy, 2006; Chandrasekaran and Tripati, 2005). The Bhuj earthquake in India, the recent earthquake in Istanbul, Turkey, and many others have created a scenario such that earthquakes are no longer problems associated with academia but a concern of the common man. Therefore, both the strength and monitoring aspects of buildings are gaining importance in the construction domain (Chandrasekaran et al., 2010, 2008; Jia, 2016). Earthquake engineering refers to "fighting with nature with the available resources," which is not an easy task even for a qualified engineer/architect (Chopra, 2003; NDMA, 2016; Johnston and White, 2018; Mittal, 2017; Murty et al., 2012).

Earthquakes are low-probability but high-consequence events (Ealangi, 2010a, 2010b; IBC, 2015). Let us try to understand the difference between earthquake disaster and earthquake hazard. For example, driving a car in the Grand Trunk Road can be a hazard. But considering the fact that the GT Road lies on an active fault, more possibilities of earthquakes can result in meeting an accident while driving a car; this can be a disaster. Even water level rising in a dam can result in an earthquake, with the Koyna earthquake, Maharashtra, 1967, as an example. It was a devastating event, referred to as a reservoir-triggered seismic activity that killed about 200 people and injured many. Although the probability of such events is very low (about 2–3%), they are a cause of serious concern for landscape architects in particular.

Earthquakes result from sudden shaking. Earlier, the Earth was homogeneous, making the heaviest objects spin about its center. Later, homogeneity is divided into seven tectonic plates; one of the longest is on the Himalayan belt (about 3000 Km long) and poses high seismic vulnerability to UP and Bihar states, as they sit on their soft sediments. Rings of fire are the epicenter where earthquakes occur around the world. In simple terms, earthquakes are likely to occur wherever mountains or hilly landscapes are present. Earthquakes are expressed in terms of three parameters, namely, frequency, amplitude, and time duration.

A few terms that are important in seismology are as follows:

- A *fault* lies in between two tectonic plates, or it can be even inside one of the plates.
- *Slip* is the residual displacement between the tectonic plates.
- *Focus* is the first point that gives up, causing an earthquake.
- An *epicenter* is located directly above the focus on the Earth's surface.
- *Focal depth* is the radial distance between the focus and the epicenter.

Design codes, as practiced in many countries, have developed seismic zoning maps, which are crucial for determining the construction practices necessary for buildings in different zones. The recommendations outlined in the code (IS 1893, 2016) are based on several key factors to ensure an earthquake-resistant construction, reflecting its vulnerability to seismic activity. The fundamental guidelines suggested by the code help assess seismic loads on various structures and outline earthquake-resistant design principles. One part is focused on categorizing India into four seismic zones, based on the level of seismic risk. It helps determine the appropriate design parameters based on regional seismicity. It specifies methods for calculating design for seismic forces and the necessary structural response to these forces. A few details with reference to seismic zoning in India are discussed in what follow.

1.7.1 Zone II: Low Seismic Risk

This zone covers much of Peninsular India, including parts of Madhya Pradesh, Karnataka, and Tamil Nadu. The salient design practices favorable for this zone are summarized here.

1.7.1.1 Unreinforced Masonry Buildings

- In rural areas, unreinforced masonry (URM) buildings, such as brick or stone, are still typical. While these are more vulnerable to earthquakes, the region's low seismic risk has led to the continued use of such traditional construction methods.
- In some areas, retrofitting is done by adding reinforcements like roof ties and wall bands to improve earthquake resistance.

1.7.1.2 RCC Frame Structures

- In urban areas, RCC frame structures are commonly used. Although seismic considerations are less stringent here, buildings still follow the basic principles of earthquake-resistant design outlined in IS 1893.

1.7.1.3 Rehabilitation of Historical Buildings

- In cities like Bangalore and Chennai, efforts are made to retrofit and strengthen historic structures and heritage buildings to prevent damage in an earthquake.

1.7.2 Zone III: Moderate Seismic Risk

This zone includes regions like Kerala, Goa, Maharashtra, and parts of West Bengal. The salient design practices favorable for this zone are summarized here.

1.7.2.1 Confined Masonry Construction

- In areas like Kerala and Goa, confined masonry is commonly used. This construction technique uses reinforced concrete (RC) frames to enclose brick or stone walls. This combination improves the structure's ability to resist earthquakes.
- Buildings have horizontal and vertical reinforcement bands at regular intervals to enhance their seismic resistance.

1.7.2.2 Retrofitting and Rehabilitation

- Many old masonry structures in urban areas, especially historical buildings, undergo retrofitting to meet modern earthquake-resistant standards. Techniques like adding tie beams and using fiber-reinforced polymer (FRP) strips to strengthen the walls are typical.

1.7.2.3 Pile Foundations in Coastal Regions

- In coastal regions like Goa and Kerala, buildings are often constructed on pile foundations to prevent damage from soil liquefaction during earthquakes, which can destabilize structures.

1.7.2.4 Low-Rise RCC Buildings

- In semi-urban and rural areas, low-rise RCC frame buildings with earthquake-resistant detailing are typical. These structures typically include lateral bracing and reinforced joints to resist seismic loads.

1.7.3 Zone IV: High Seismic Risk

This zone covers parts of Northern Bihar, Delhi, the Indo-Gangetic plains, and Gujarat. The salient design practices favorable for this zone are summarized here.

1.7.3.1 Reinforced Concrete (RC) Buildings

- RC frame buildings follow IS 1893 and IS 13920 for ductility and lateral load resistance.

1.7.3.2 Reinforced Brick Masonry (RBM)

- It is widely used in semi-urban and rural areas. Reinforcement bands (usually steel bars or concrete beams) anchoring walls to the roof structure are provided at various levels in the masonry structure to improve earthquake resistance.

1.7.3.3 Foundation Design

- Special care is given to foundation design, as soft soil conditions (common in the Indo-Gangetic plains) amplify earthquake effects. Pile foundations and raft foundations are commonly used to counter these effects.

1.7.3.4 Seismic Retrofitting

- In cities like Delhi and Patna, retrofitting older buildings using column jacketing or by adding steel bracing is critical to improve their seismic performance. Traditional brick masonry buildings are retrofitted using techniques like reinforcing the walls with mesh or steel bars.

1.7.3.5 Seismic Joints

- Larger buildings incorporate seismic joints to allow movement during an earthquake and avoid pounding between adjacent structure sections.

1.7.3.6 Base Isolation for Critical Infrastructure

- Base isolation systems are used in critical buildings, such as hospitals and emergency operation centers, in high-risk regions. The base isolation system minimizes the impact of ground shaking on the structure.

1.7.4 Zone V: Very High Seismic Risk

This zone includes parts of Northeast India, Jammu and Kashmir, Himachal Pradesh, Uttarakhand, and the Andaman and Nicobar Islands. The region is highly vulnerable to major earthquakes, and the building designs prioritize resilience and safety. The salient design practices favorable for this zone are summarized here.

1.7.4.1 Reinforced Concrete (RC) Frame Buildings

- RC frame buildings with shear walls are the most common in urban areas. The shear walls are designed to absorb seismic loads, making the structure more resilient.
- Ductile detailing (IS 13920) is mandatory to ensure that beams and columns deform in a controlled manner during earthquakes without collapsing.

1.7.4.2 Timber Structures

- In regions like Northeast India, where timber is abundant, traditional timber houses are famous. Timber's flexibility makes it suitable for earthquake resistance, as it absorbs seismic energy without collapsing.
- Ikra houses, in traditional timber-framed construction style, are prevalent. These are lightweight, which reduces seismic forces, and the materials can flex during an earthquake.

1.7.4.3 Steel Structures

- In critical areas, like hospitals, bridges, and large commercial buildings, steel frames are used, as they provide flexibility and high tensile strength, making them better suited for earthquake resistance.

1.7.4.4 Lightweight Prefabricated Buildings

- In disaster-prone areas like Uttarakhand, lightweight prefabricated buildings are gaining popularity. These structures are easy to assemble, and their low mass reduces the seismic forces acting on them.

1.7.4.5 Earthquake-Resistant Traditional Buildings

- Koti Banal architecture in Uttarakhand uses stone and timber in a unique seismic-resistant manner. Timber bands are placed between stone layers to absorb seismic forces.

1.7.4.6 Base Isolation Systems

- These are used in critical structures, like hospitals and emergency service buildings, to reduce ground acceleration.

1.7.4.7 Retrofitting

- Many existing buildings, including schools, hospitals, and historical structures, undergo retrofitting to improve earthquake resistance.

1.8 RECOMMENDATIONS COMMON TO ALL SEISMIC ZONES

Energy absorbed by the construction material, while moving into the non-linear zone of stress–strain deformation, is the key savior in earthquake-resistant design (Bureau of Indian Standards, 2015, 2016; Chandrasekaran, 2013a, 2015b, 2015a, 2015c). When buildings are designed for seismic resistance, it is meant to take care of the extra reserve energy due to its plasticity, allowing permanent deformation and no collapse (CEN, 2002, 2004, 2005). It is characterized in the design by computing the ductility demand, which is the ratio of maximum deformation to the yield displacement. One of the underlying philosophies of seismic-resistant design is the capacity design concept (Chandrasekaran and Gaurav, 2008). Structures to resist the most severe ground shaking without damage are very expensive to build (Chandrasekaran et al., 2006). As most severe seismic events are infrequent, it is not necessary to build structures to meet those demands.

An architect looking for the primary objectives of seismic-resistant design shall focus on the following, in which structures should be able to resist thus:

- Minor and frequent shaking with no damage
- Moderate shaking with no structural damage but some non-structural damage permitted
- Severe and infrequent shaking with structural damage but no collapse

The last objective emphasizes saving lives and property inside and adjoining the structure. Hence, in seismic-resistant design, the level of earthquake protection needs to be pre-decided at the design stage (Chandrasekaran et al., 2008a; Chandrasekaran, 2016a, 2016b). It depends on the consequences of damage. For example, essential services, namely, water mains, electric wires, and transmission towers, hospitals, primary health centers, and communication systems, must remain in service post-earthquake (Chandrasekaran, 2017, 2019b). Hence, higher levels of protection are applied to these structures in comparison to the others.

Furthermore, certain structures may not be lifelines, but their damage could result in mass destruction. For example, dams and nuclear plants. Hence, they also enjoy the degree of maximum protection under earthquakes. The following are a few critical recommendations common to all buildings across all seismic zones:

- **Ductile Detailing in Reinforced Concrete Structures:** Across all seismic zones, ductile detailing (as per IS 13920) is applied to ensure that buildings can deform under seismic loads without collapsing. This involves reinforcing the beam column joints and ensuring the structure has sufficient flexibility to dissipate earthquake energy (Chandrasekaran et al., 2011, 2010).
- **Horizontal Bands in Masonry Construction:** Horizontal bands made of reinforced concrete or steel bars are provided at critical levels (such as the

plinth, lintel, and roof levels) in masonry buildings. These bands tie the structure together and prevent the collapse of walls during earthquakes.

- **Base Isolation and Energy Dissipation Devices:** Base isolation systems and energy dissipation devices like viscous dampers are used in high-rise buildings and critical structures in seismically active regions. These devices reduce the amount of seismic energy transferred to the building.
- **Seismic Gaps:** Seismic gaps are provided between adjacent buildings to prevent them from hitting each other during an earthquake, which could lead to structural damage. This is particularly important in densely populated urban areas.
- **Bracing Systems and Shear Walls:** Braced frames and shear walls are used extensively in earthquake-resistant buildings to provide lateral stability and resist seismic forces. These systems are especially critical in tall and slender buildings in high-risk zones.

1.9 CRITICAL PRINCIPLES FOR EARTHQUAKE-RESISTANT DESIGN

- **Structural Design:** Buildings should incorporate reinforced concrete frames, shear walls, and bracing systems to enhance their ability to withstand seismic forces.
- **Foundation Design:** Deep and well-anchored foundations, such as pile foundations, are essential to ensure stability during seismic events. The foundation must effectively transfer the building's load to the ground.
- **Material Selection:** High-quality materials like reinforced concrete and steel are preferred for their strength and flexibility, which help absorb seismic energy. Avoiding brittle materials is crucial to prevent catastrophic failures during an earthquake.
- **Retrofitting:** Older buildings that do not meet current standards can be retrofitted to improve their seismic performance. This includes reinforcing structures and ensuring compliance with modern building codes. Active and passive control methods are also used to mitigate the serious consequences of earthquakes on old buildings (Chandrasekaran, 2016a, 2016b; Chandrasekaran et al., 2009a, 2009b; Zhang and Ali, 2021).
- **Innovative Technologies:** Adopting advanced technologies such as base isolation systems and dampers can significantly reduce the impact of seismic forces on buildings (Constantinou and Symans, 1992; El Ouni et al., 2022; Fenz and Constantinou, 2008; Inaudi and Kelly, 1993; Nuzzo et al., 2018; Juhn et al., 1992; Politopoulos, 2008; Renzi and Serino, 2004; Sakamoto, 1993; Eem and Hahm, 2019; Skinner et al., 1993).

The four important verticals of earthquake-resistant design are strength, stiffness, ductility, and building configuration. More details on these factors are not within the scope of this chapter, but readers can refer to other published literature (Chandrasekaran et al., 2024; Guzman Pujols and Ryan, 2020; Hernried and Lei, 1995; Li et al., 2014; Losanno et al., 2019; Chen et al., 2021).

1.9.1 Construction Challenges

Despite the existence of these guidelines and standards, several challenges hinder effective implementation:

- **Enforcement of Building Codes:** There is often poor enforcement of existing building codes, leading to the construction of vulnerable structures.
- **Awareness and Education:** There is a need for greater awareness among architects, engineers, and builders about earthquake-resistant construction practices.
- **Economic Constraints:** Many builders may face economic limitations that prevent the adoption of advanced technologies and materials

1.10 MANDATORY DESIGN FEATURES

Building standards (NBC, 2016) provide comprehensive guidelines to ensure safety, functionality, and sustainability standards. The key aim is to mitigate risks associated with natural disasters, including earthquakes, and to guide homeowners, engineers, and contractors in building resilient structures. Various building codes and standards mandate several design features to ensure earthquake-resistant construction. These features are essential for enhancing the structural integrity of buildings in seismic zones. A few major design features are discussed in what follows. The primary goal is to ensure life safety. Hence, structures may be permitted to suffer minor damage but should not collapse. It enables occupants to evacuate (or to be evacuated) safely during seismic events.

Site Selection: The emphasis is on the importance of choosing a suitable construction site, particularly in earthquake-prone areas. Poor site selection can amplify earthquake damage due to soil instability, liquefaction, or landslides.

Soil Conditions: A proper geotechnical investigation should be conducted to assess soil stability. Soil liquefaction occurs during solid earthquake shaking, when poorly graded, fine-grained, cohesionless soil with a high water table is subjected to prolonged seismic activity, causing heavy objects to sink and lighter objects to rise. If liquefaction is predicted at a site, soil strengthening or pile foundations may be used, but choosing sites where liquefaction is unlikely is preferable to minimize damage. Hill slopes of soft soils steeper than 30° are prone to landslides, especially in moist conditions or areas with perennial rainfall. For safer construction, it's crucial to consult geotechnical experts and consider lightweight structures on hill slopes.

When constructing on hill slopes, start from the bottom to the top, ensuring uphill buildings are lighter than those downhill. The horizontal distance between structures should be at least three times the difference in their founding levels to prevent landslides and ensure stability. Ensure the foundation system is stiff, especially on steep slopes, where structural walls or braced moment frames are necessary. Additionally, it protects against landslides, maintains sufficient distance between buildings to avoid foundation damage, and ensures proper drainage to prevent water damage to the foundation. Reference should be made to the seismic zoning map, which categorizes the country based on the intensity of seismic activity. Structures should be designed accordingly for the local seismic zone.

Architectural Form: One of the fundamental aspects of earthquake-resistant design is regularity in the building's architectural form. Buildings with regular, symmetrical shapes perform better during earthquakes than those with irregular shapes. Symmetry and balance should be ensured while designing overhangs, cantilevers, or irregular floor plans, as these create stress concentrations and torsional movements during an earthquake. Buildings with symmetrical and regular plan shapes (like square or rectangular) perform better in earthquakes, as they translate uniformly and avoid twisting during seismic shaking. Non-rectangular shapes, like T, L, or C plans, are structurally weak but can be improved with corner strengthening or by separating them into rectangular units with flexible junctions. The plinth height of a building should be raised from the surrounding ground or road levels to prevent dampness and flooding. A minimum plinth height is prescribed (for example, 450 mm in Indian code) from the natural ground level, road level, or high flood level, whichever is higher. This helps protect the building from moisture and potential flooding. Door and window openings should be kept small, positioned well away from the wall junctions, preferably centrally located, to avoid weakening walls. Standard door and window sizes should be followed to ensure modular construction. A continuous lintel band over openings enhances earthquake resistance. It is better to avoid using masonry arches in earthquake-prone areas, as they cannot handle lateral forces during seismic shaking.

It is preferable to utilize the lowest story created by the sloping ground for storage or cattle housing, as it reduces the mass of the upper floors, making the building lighter. Limit openings in this space to access doors, and ensure any openings are properly stiffened for structural stability. Soft stories (like open-ground floors for parking) are strongly discouraged unless they are specially designed and strengthened with bracing systems or shear walls. These open spaces make the building vulnerable to collapse under seismic events.

Structural Safety: Structural safety is a cornerstone of earthquake-resistant building design. Buildings should be constructed using materials and techniques to effectively absorb and dissipate seismic energy. Reinforced concrete (RC) frames with ductile detailing (IS 13920) are critical for withstanding seismic forces. Ductile detailing ensures that the building deforms in a controlled manner during an earthquake without sudden failure. Tall buildings in higher seismic zones should be designed with shear walls built into the structural framework. They help resist lateral seismic forces and provide additional strength to the building. For steel structures, cross-bracing is used to increase the lateral stiffness and help resist shaking forces arising from earthquakes.

Ductility and Flexibility: Materials like reinforced cement concrete (RCC) and steel, which can deform without breaking, are crucial. Ductility allows structures to absorb and dissipate energy during seismic events, reducing the likelihood of catastrophic failure. The "strong column, weak beam" design principle ensures columns directly transfer the seismic forces to the foundation without causing the beams to fail first. Adequate reinforcement in columns, beams, and walls is essential. Vertical and horizontal reinforcement bars should be placed at critical points, such as corners and junctions, to improve load distribution and prevent shear failure.

Base Isolation: Implementing base isolators, which act as shock absorbers between the building and the ground, can significantly reduce the transmission of seismic forces to the structure. These isolators allow the building to move independently of ground motion, minimizing damage.

Shear Walls and Bracing: Incorporating shear walls in the design provides lateral strength and stiffness to resist horizontal seismic forces. These walls should be strategically placed to ensure they are effective during an earthquake. Diagonal bracing, between columns and beams, enhances the lateral load resistance of the structure, preventing excessive swaying during seismic activity.

Foundation Design: Foundations must be designed based on the seismic zone, with isolated or raft foundations recommended for areas with high seismic activity. Proper anchorage of the structure to the foundation is critical to prevent uplift and sliding during an earthquake.

Openings and Wall Design: It is advisable to limit the number and size of wall openings to maintain structural integrity. Large openings can weaken walls and lead to failure during seismic events. Providing continuous horizontal bands of reinforcement at various wall levels to tie the structure together enhances its ability to withstand lateral forces.

Irregularities: Buildings should preferably have regular shapes and symmetrical designs to avoid torsional irregularities. Otherwise, it can lead to an uneven distribution of forces during an earthquake. It is also advisable to avoid large overhangs and cantilevers.

Quality Control and Materials: It plays an important role in seismic-resistant design. Ensure high-quality materials that meet the standards specified in the design codes. Regular inspections during construction to check/assess the quality of materials are vital to maintain quality control.

Retrofitting of Structures: Retrofitting methods, such as column jacketing, adding shear walls, bracing, or base isolators, can improve the earthquake resistance of older structures, when they do not meet the revised design standards, in particular. For heritage structures with cultural significance, non-invasive techniques, like fiber-reinforced polymer (FRP) wraps, can strengthen structural elements without altering the building's architectural form. In addition to structural elements, non-structural components, like facades, ceilings, and partition walls, must be securely fixed to prevent them from collapsing during an earthquake. Anchoring systems, including heavy equipment, such as HVAC systems, water tanks, and elevators, must be adequately anchored to prevent them from becoming hazards during an earthquake.

Masonry buildings depict a few major deficiencies under earthquake loads. Architects shall address these while retrofitting them. They are as follows:

- Masonry buildings possess poor in-plane shear strength and stiffness.
- The diaphragm shear capacity and the ability to transfer forces are very less.
- They possess very poor out-of-plane stability when the walls are slender, in particular.
- They possess poor structural integrity, as the entire building does not behave as a single rigid body due to a lack of proper connections between the components.

When applied to seismic strengthening of buildings, the philosophy of earthquake-resistant design is to initially calculate the maximum elastic seismic forces (the demand) and then reduce this demand to the design level. The reduction is done to account for ductility and overstrength. The capacity design concept is adopted more successfully in the seismic strengthening of (old) buildings. In each structural system, identify the sequence of failure (who fails first?). It will be important to understand the load transfer path of a typical framed structure. Loads acting on the superstructure, comprising gravity and lateral loads, act on the roof, which distributes it to the supporting beams. The beams carry loads from the adjacent bays and distribute them to the supporting columns, which are further disbursed to the ground through foundations. This strength hierarchy helps in understanding the consequences of failure: Slab failure shall lead to a local failure of that space (only that room). A beam failure shall involve adjacent spans, which is slightly more severe than the former. Failure of a column shall cause serious consequences, as it supports critical loads from the roof and the beams.

Failure of foundation is the most serious concern, in addition to the failure arising from soil liquefaction. Therefore, while strengthening is carried out, ensure ductile elements yield before the failure of brittle elements. A better design would be based on sequencing the failure strategy. As quoted very often, the strength of the chain depends on the strength of its weakest link. If the statement is applied to the forces (as demand), then the ductility (capacity) of the chain depends on the ductility of the weakest link. Hence, the weakest link (member) in the failure hierarchy is more ductile, which helps sustain the load distribution path until the failure chain is complete.

1.11 CONSTRUCTION PROJECT PLANNING

A construction project comprises various jobs, activities, functions, and interrelated tasks that are required to be completed within the project duration. Each construction project serves a specific set of objectives, starting at a defined point, and project duration, keeping into account the return on investment, and the completion of the building in a fully accomplished manner. To successfully execute a construction project, two fundamental resources are essential, namely, material resources and workforce. It is very important to note that the overall responsibility of the project management is under the auspices of an architect (or the company/firm). He is held responsible in case of any delay in completing the project, for whatever reason.

Hence, it is necessary to understand the basics of project planning and management features. Despite having abundant material resources, construction productivity and effective project management become vital. On the other hand, nations with fewer natural resources have excelled in construction output due to their workforce's talent, skills, experience, and expertise. The availability, quality, and effective use of human resources are critical factors in ensuring successful project delivery.

In recent years, while technological and engineering advancements have accelerated, there has yet to be an equivalent improvement in managing workforce relations and construction processes. In other words, our ability to manage human resources and team dynamics in construction has not advanced at the same rate as technology. With improvements in project management and human interaction, one can fully harness the potential of modern construction technologies, which emphasizes the vital role of project (construction) management. While technology focuses on planning and arranging material resources and equipment, project management ensures the optimal use of both material resources and workforce, making it indispensable for the success of any construction project. Effective project management ensures timely completion, optimal resource utilization, and cost-effective execution. It can significantly enhance productivity by strategically integrating technological advancements and human factors.

Every construction project, regardless of scale, has three primary objectives:

- Timely completion, which helps minimize project duration without compromising quality
- Resource efficiency, which helps avoid delays, utilize available workforce and other resources judiciously
- Cost optimization, which aids in the timely completion of the project within the budget constraints

Project management in construction typically involves three distinct phases:

1. *Planning.* This initial phase encompasses the following activities:
   a. Defining project scope, goals, and deliverables
   b. Identifying key stakeholders and their roles
   c. Developing a detailed project schedule
   d. Allocating necessary resources (materials, equipment, personnel)
   e. Creating a risk management plan
2. *Scheduling.* This phase involves the following activities:
   a. Breaking down the project into smaller, more manageable tasks
   b. Establishing dependencies between tasks
   c. Creating a visual representation of the project timeline (e.g., Gantt chart)
   d. Assigning resources to tasks
   e. Monitoring and controlling project progress
3. *Controlling and reviewing.* This ongoing phase focuses on the following items:
   a. Tracking project performance against the established plan
   b. Identifying and addressing potential issues or deviations
   c. Making necessary adjustments to the plan as needed

Project Planning is the cornerstone of effective project management in the construction industry. It involves a systematic approach to define project objectives, identify necessary tasks, estimate resource requirements, and develop a timeline for successful completion. Key elements of project planning are discussed in the following:

- **Defining Objectives:** Clearly outlining the desired outcomes and goals of the project.
- **Task Identification:** Breaking down the project into smaller, more manageable tasks or activities.
- **Resource Estimation:** Determining the necessary materials, equipment, and workforce to accomplish each task.
- **Cost and Duration Estimation:** Predicting the costs and time frames associated with each activity.

Importance of Planning

- *Provides direction.* A well-defined plan guides the project team toward a common goal.
- *Unifies framework.* It creates a cohesive structure for the project, ensuring all stakeholders are working toward the same objectives.
- *Reveals opportunities and threats.* Planning helps identify potential risks and challenges, allowing for proactive measures.
- *Establishes performance standards.* It sets benchmarks for measuring progress and success.

Planning Process

- *Plan formulation.* Creating a detailed plan outlining the required steps, resources, and timelines.
- *Strategy development.* Considering company policies, procedures, and rules to develop effective strategies.
- *Resource allocation.* Assigning appropriate resources to each activity based on requirements.

1.12 METHODS OF PLANNING AND PROGRAMMING

Complex research and development projects can be managed more effectively if project managers can plan and control the schedules and costs required to meet their technical performance goals. When planning a project, several critical questions arise:

- How will the work be executed? What resources are necessary?
- How long will the project take?
- What will be the total cost?

Modern project management techniques provide solutions to all these questions. Architects or project management consultants, at all levels, require the implementation of enhanced techniques throughout a project's life cycle to (1) clearly define the tasks to be carried out; (2) develop more accurate schedules and cost estimates based on the resources allocated for the tasks; (3) determine where resources should be allocated to meet the time, cost, and technical performance objectives; and (4) identify areas at risk of delays or cost overruns early enough to allow for timely corrective measures. Some of the critical tools and techniques used in project management are bar charts, milestone charts, and network diagrams.

1.12.1 Bar Charts and Milestone Charts

Henry Gantt bar charts, proposed in the early 20th century, help visualize project schedules by dividing them into minor, manageable activities. While these activities are plotted along the abscissa, the duration of each activity is plotted along their ordinate. Subsequently, bar charts are updated into milestone charts, highlighting significant events that mark the start or completion of a task. The bars in the latter are divided into segments, each representing critical events during the project. Each event has a crucial timestamp, which is to be managed critically to achieve the targeted progress.

1.12.2 Network Methods

Network diagrams are improved versions of milestone charts, drawn based on the principle that any project, regardless of complexity, is executed through a series of well-defined steps. For instance, when constructing a foundation, the project might involve sequential steps: (1) marking the layout, (2) excavation, (3) placing sideboards, and (4) pouring concrete. Network techniques graphically illustrate these steps as a network diagram. Management methods have gained more widespread use and recognition for project planning, scheduling, and monitoring than these network techniques. Various names exist for these methods, including PERT, CPM, UNETICS, LESS, TOPS, and SCANS. However, most of these systems originate from two foundational methods, namely, PERT and CPM (Punmia and Khandelwal, 2002).

PERT (program evaluation and review technique) uses a network diagram consisting of events that must be established to achieve project objectives. An *event* refers to a specific time when a particular part of the plan is completed, marking a milestone without consuming any resources. PERT employs event-oriented network diagrams, where arrows connect successive events. For example, in the construction of a bridge, events could include "piling completed," "foundation set," "superstructure installed," and "final inspection completed." This event-oriented approach in the network diagram arose from the need to track project progress through clearly defined management milestones. PERT system is typically used for projects or operations that are non-repetitive or where precise time estimates for activities cannot be determined. Experience offers limited guidance in such projects, and management must rely on estimates rather than historical data. These projects are called once-through operations, meaning, the work has never been done before. An example of this would be a spacecraft launch, where the complexity of technical challenges makes time estimation difficult. In such research and development projects, the time estimates may be based on assumptions or educated guesses. PERT is ideal for managing such projects.

CPM (critical path method) is another project management technique that divides the project into distinct tasks or operations, known as activities. These activities represent operations that require time and resources for completion. The points where activities intersect are referred to as *events*. CPM networks are typically described as activity-oriented diagrams, with each activity represented by an arrow and the sequence of arrows indicating the order in which these tasks are performed. For instance, in the construction of a building, the activities might include tasks such as "laying the foundation," "erecting the structural framework," "installing plumbing and electrical systems," and "completing interior finishes." This excerpt provides an overview of the critical path method (CPM) used in project management. It emphasizes breaking down complex projects into individual activities and visualizing their dependencies through network diagrams. CPM helps ensure efficient project execution by identifying the critical path—the most extended sequence of dependent activities—and determining the project's overall duration. Table 1.1 compares CPM and PERT management techniques.

TABLE 1.1
Comparison of CPM and PERT Network Techniques

Description	PERT	CPM
Diagram type	Event-oriented network diagrams	Activity-oriented network diagrams
Focus	Emphasis on events (milestones)	Emphasis on activities (tasks)
Time estimates	Uncertain and variable; used for probabilistic scenarios	Reasonably accurate and can be determined with precision
Cost optimization	Assumes a direct correlation between time and cost, minimizing time minimizes cost	Prioritized, but time and cost not directly proportional
Controlling factor	Time the primary controlling factor; reducing time critical to minimizing costs	Cost the primary controlling factor; project duration based on cost optimization
Type of projects	Ideal for projects with high uncertainty, such as research and development projects	Suited for projects with well-established processes, such as construction and manufacturing
Application areas	Frequently applied in R&D projects, where time frames are unpredictable	Commonly used in construction, manufacturing, and other projects where time and cost are predictable
Project characteristics	Used for projects with uncertainties in development, design, and construction	Applied to projects that use stable technology and well-defined tasks
Flexibility	High flexibility in managing uncertain timelines and tasks	Lower flexibility; more suitable for projects with fixed schedules and budgets

1.13 ELEMENTS OF NETWORK

Network technique is recognized as one of the most advanced tools in project management. Any project can be systematically divided into specific, clearly defined tasks or jobs, referred to as activities. The start or completion of each activity marks an event within the project. A *network*, therefore, is a graphical representation that illustrates the flow of activities and events arranged logically and sequentially. In this diagram, arrows show activities, and events are typically represented by circles, as shown in Figure 1.22.

As seen in the preceding table, a PERT network is event-focused, while a CPM network emphasizes the activities. CPM and PERT are project management techniques that rely on graphical and diagrammatic representations. These methods assist management in planning, scheduling, and controlling project operations.

1.14 COMPONENTS OF PERT/CPM NETWORK

An event, also known as a *node*, marks the start and completion of an activity in a project (ref. Figure 1.23). It represents a specific point in time but only consumes resources like time, money, or energy. Events are usually depicted as circles in network diagrams. The starting event of an activity is referred to as the "tail event" (ith event), while the event marking the activity's completion is called the "head event" (jth event). Events serve as milestones within the project, signifying the transition between activities, but do not represent any work or effort in themselves.

Key Aspects of Event Representation

- *Node or circle.* Events are usually represented by a circle (or node) in network diagrams. Each node signifies a distinct point in time where one or more activities start or finish.

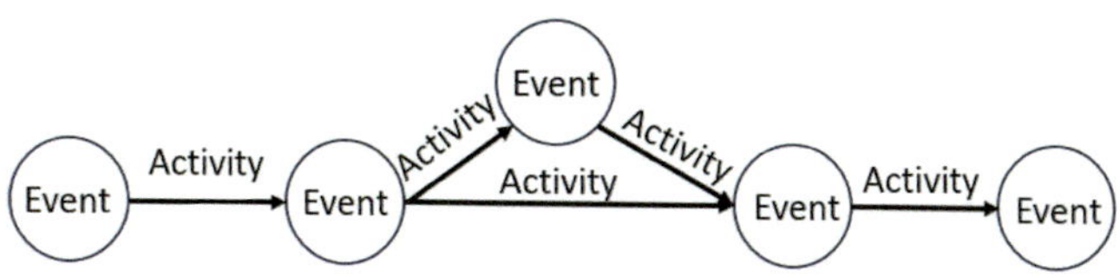

FIGURE 1.22 Network diagram.

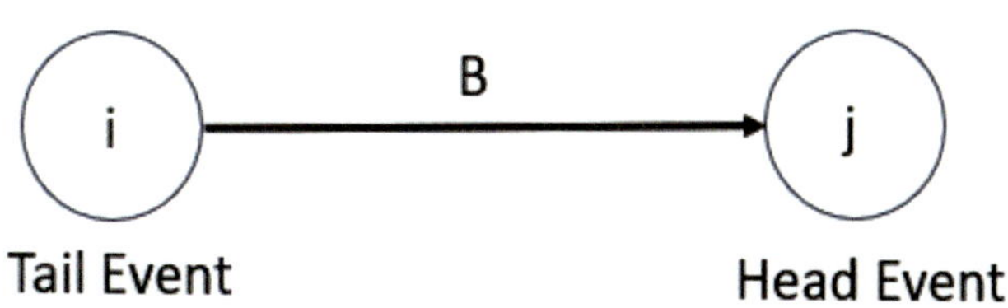

FIGURE 1.23 Representation of events.

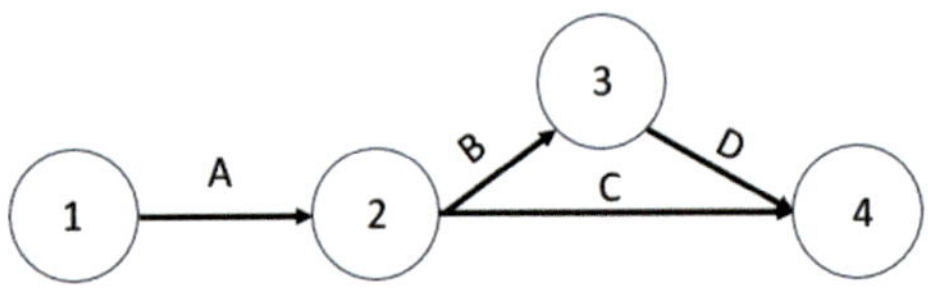

FIGURE 1.24 Numbering of events.

- *Numbering events.* Events are assigned unique numbers (Figure 1.24) to help distinguish them. The starting event (node) is usually numbered as 1, and subsequent events are numbered sequentially as the project progresses.

1.14.1 Examples of Events in Construction

1. *Foundation excavation completed.* Marks the completion of digging for the foundation.
2. *Concrete pour completed.* This indicates that the foundation has been filled with concrete.
3. *Scaffolding setup finished.* The scaffolding has been erected and is ready for use.
4. *Walls built to full height.* Indicates that bricklaying or wall construction is completed.
5. *Roof installed.* This signifies that the roof structure is entirely in place.
6. *Wiring completed.* Marks the end of electrical installations.
7. *Plumbing tested.* Indicates that the plumbing system is installed and has passed testing.
8. *Painting finished.* All painting work has been completed.
9. *Windows and doors fitted.* Marks the installation of doors and windows.
10. *Project handover.* The final event when the building is complete and ready for use.

In project networks, events can serve as a convergence point (merge event) or a divergence point (burst event) for multiple activities (Figure 1.25). A *merge event* occurs when two or more activities converge and complete the same event, meaning, the project's next phase can only begin once all preceding activities are finished. Conversely, a *burst event* happens when one event triggers the start of two or more subsequent activities that can be executed in parallel. These events are crucial in a project's logical sequencing and flow, ensuring that dependencies and parallel tasks are correctly managed for smooth project execution.

An *activity* represents a specific task or action within a project that requires the consumption of resources, such as time, money, or effort, to complete. Each activity contributes to the project's progress and typically has a defined start and end point. In network diagrams, an activity is represented by an arrow, with the length of the arrow often reflecting the time required to complete the task. Additionally, activities can be dependent (requiring prior tasks to be completed) or independent, allowing for parallel execution. In Figure 1.26, "B" is an activity. In network diagrams, activities

FIGURE 1.25 Merge and burst events.

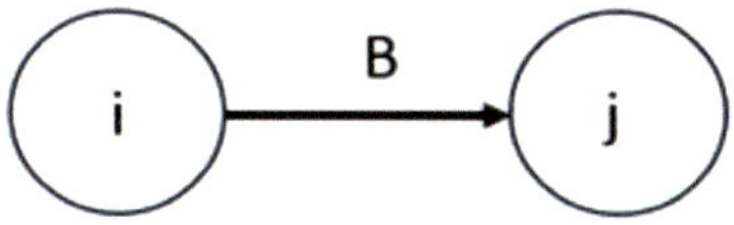

FIGURE 1.26 Representation of activity.

are represented by arrows or directed lines, pointing from the tail event (where the activity starts) to the head event (where the activity ends). This shows the flow and sequence of the project's tasks. Activities are labelled with a short description of the task being performed (e.g., "Excavate foundation," "Install roof"). Additionally, some diagrams may include the duration of the activity on or near the arrow. Activity B can be represented as (i–j).

1.14.2 Examples of Activities in Construction

1. *Excavation of foundation.* The process of digging the foundation trench.
2. *Pouring concrete.* The task of filling the foundation with concrete.
3. *Erecting scaffolding.* Setting up temporary structures for workers to access high areas.
4. *Bricklaying.* Building the walls by laying bricks.
5. *Roof installation.* Placing the roof structure and materials.
6. *Electrical wiring.* Installing electrical wiring throughout the building.
7. *Plumbing installation.* Installing pipes for water and drainage systems.
8. *Painting.* Applying paint to interior and exterior surfaces.
9. *Window and door fitting.* Installing doors and windows in the building.
10. *Flooring.* Laying tiles, carpets, or hardwood flooring.
11. *Interior finishing.* Tasks like plastering, installing lighting fixtures, or cabinetry.

Activities before a specific event are referred to as preceding activities, while activities after a given event are known as succeeding activities. Preceding activities must be completed before the event, and project dependencies must be set up. Similarly, succeeding activities can begin once the event has been completed. These relationships between preceding and succeeding activities are critical in defining the project's flow, ensuring proper sequencing, and avoiding delays due to task dependencies. Additionally, they help identify essential paths and potential bottlenecks in project execution. Activity "A" precedes activities "B" and "C" (Figure 1.27).

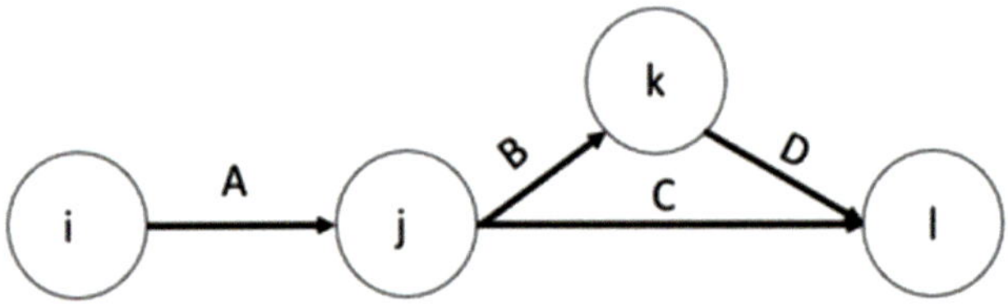

FIGURE 1.27 Preceding and succeeding activities.

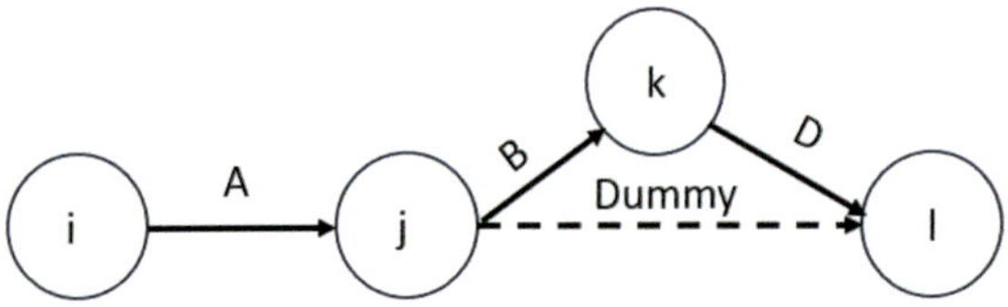

FIGURE 1.28 Dummy activity.

A *dummy activity* is an artificial or imaginary activity, represented by the dashed arrow, that consumes no resources, such as time, money, or labor, and solely maintains logical relationships between events in a project network (Figure 1.28). Dummy activities are typically used to show dependencies between tasks, when no actual work is performed. They ensure the correct sequencing of activities without altering the overall project duration. In network diagrams, dummy activities are represented by dotted or dashed lines. Their primary function is to preserve the logical integrity of the project, mainly when dealing with parallel tasks or complex task dependencies.

Dummy activities serve two primary purposes within a network: grammatical and logical purpose.

Grammatical Purpose: A dummy activity is often introduced to ensure that each activity has a unique identification, especially when two activities share the same starting and ending points. For instance, if two activities, P and Q, begin at node 1 and end at node 2 (Figure 1.29a), confusion arises, as there's no clear distinction between them. This lack of uniqueness can lead to errors when performing network computations. As shown in the revised network diagram (Figure 1.29b), a dummy activity is inserted to resolve this. By introducing a dummy activity, identifying activities remains distinct, ensuring clarity in representation. For example, instead of using two arrows with the same start and end points, a dummy link helps maintain grammatical accuracy in the network and avoids computational errors.

Logical Purpose: Dummy activities are not only necessary to clarify logical dependencies but also play a crucial role in maintaining the logical flow of the network. For example, consider two activities, D and E, ending at the same node. B and C are predecessors to D, while C and A are predecessors to E (Figure 1.30a). With a dummy activity, the network representation

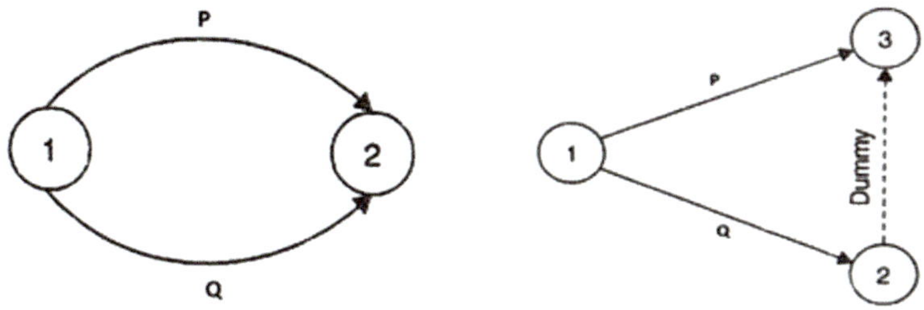

FIGURE 1.29 Grammatical purpose of activities.

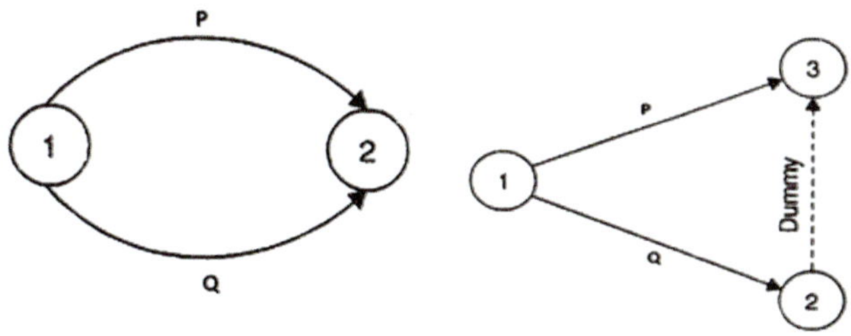

FIGURE 1.30 Logical purpose of activities.

can lead to clarity, as activity C appears to have dual identities, which is impossible. By introducing a dummy activity, the network ensures that each activity and event is uniquely identifiable, preserving the logical flow of the network (Figure 1.30b). By leveraging the power of dummy activities, project networks are refined, eliminating grammatical and logical errors. This precision enables accurate scheduling and efficient resource management, underscoring the benefits of using dummy activities in project networks.

Some conventions of the network diagram are shown in the following Figures (a), (b), (c), and (d):

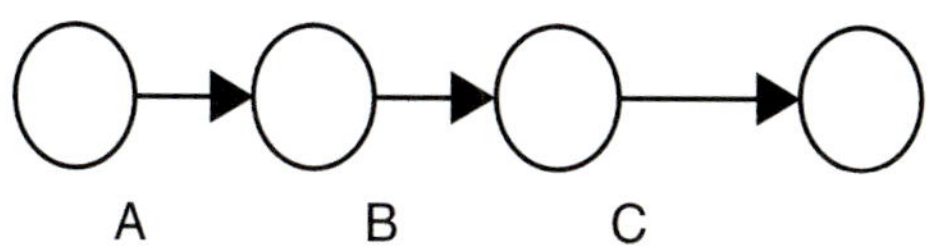

Activity B can be performed only after Activity A, and C can be performed only after Activity B.

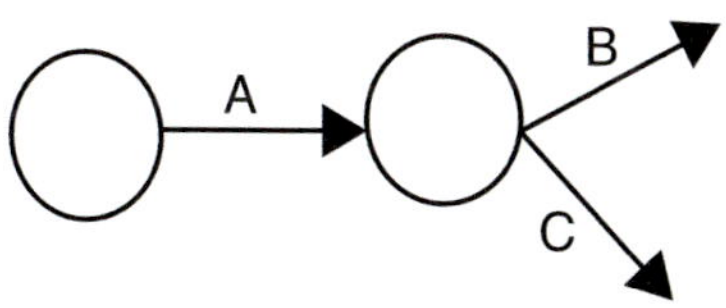

Activities B and C can start simultaneously only after A is completed.

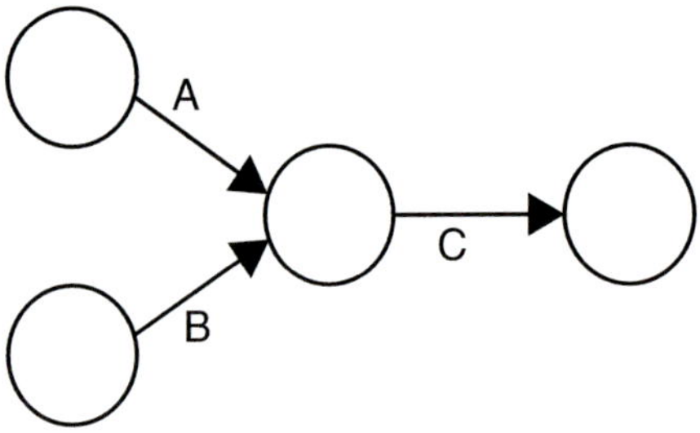

Activities A and B must be completed before Activity C is started.

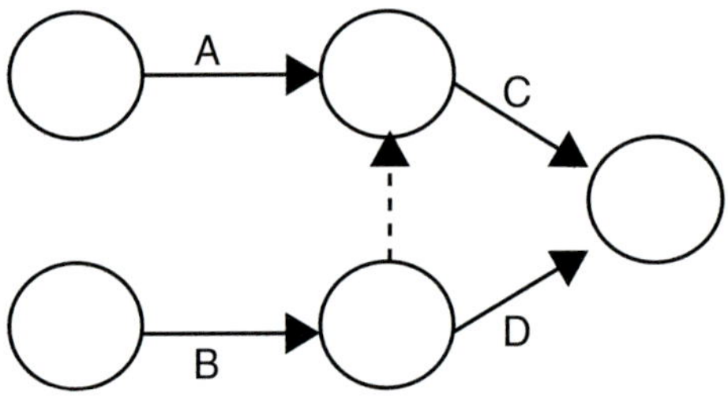

Activity C must start only after Activities A and B are completed. But Activity D can start after Activity B is completed.

1.15 ERRORS IN CONSTRUCTING A NETWORK

- When two activities originate from the same starting event, they should not share the same ending event. As Figure 1.31a illustrates, a dummy activity must be introduced to prevent this.
- Looping errors, which create cycles of repeated activity performance, must be avoided in the network. This can lead to an endless loop, as shown in Figure 1.31b.
- As the following diagram demonstrates, a network should have a single starting and ending point, ensuring a clear and logical flow of activities.
- Arrows should consistently point from left to right to avoid confusion and maintain the proper direction of activities in the network.

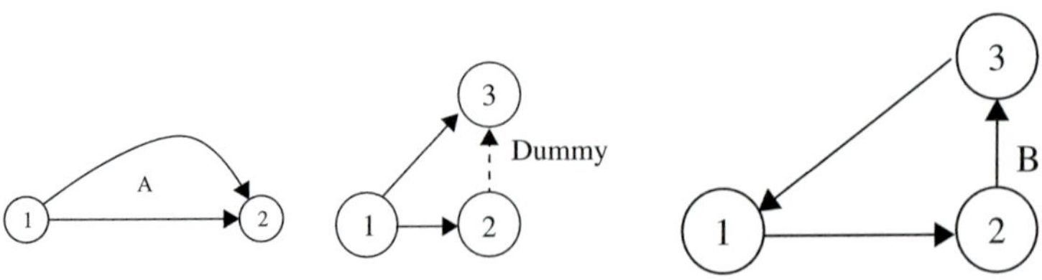

FIGURE 1.31 Network errors: (a) Dummy activity to prevent error. (b) Looping error.

1.15.1 Example Problems

1.15.1.1 Problem 1

You are managing the construction of a residential building. Following is a list of tasks involved in the project. Separate the activities and events from this list.

1. Project started.
2. Dig foundation trenches.
3. Lay the foundation concrete.
4. Foundation completed.
5. Erect columns.
6. Complete wall construction.
7. Wall constructed.
8. Place roofing structure.
9. Install electrical and plumbing systems.
10. Apply plaster and paint walls.
11. Finish flooring work.
12. Flooring work completed.
13. Install doors and windows.
14. Project ended.

1.15.1.1.1 Solution

1.15.1.1.1.1 Activities

1. *Dig foundation trenches.* Requires time, labor, and resources.
2. *Lay the foundation concrete.* Involves using materials and equipment.
3. *Erect columns.* Requires resources like steel and labor.
4. *Complete wall construction.* Involves brickwork or blockwork.
5. *Place roofing structure.* Requires roofing materials and installation.
6. *Install electrical and plumbing systems.* Needs labor and technical equipment.
7. *Apply plaster and paint walls.* Labor-intensive activity.
8. *Finish flooring work.* Includes tiling or flooring.
9. *Install doors and windows.* Involves carpentry and installation.

1.15.1.1.2 Events

1. *Project started.* Marks the beginning of the construction process.
2. *Foundation completed (after "Lay the foundation concrete").* Signals the foundation is done.
3. *Wall constructed (after "Complete wall construction").* Marks that the walls are done.
4. *Floor work completed (after "Finish flooring work").* Signals the project is nearing the end.
5. *Project ended.* Marks the conclusion of the project.

1.15.1.2 Problem 2

Prepare a table showing the immediate predecessor activities and events, predecessor activities and events, immediate successor activities and events, and successor activities and events for various activities and events of the following network diagram.

1.15.1.2.1 Solution

Create a simple network diagram based on a construction project with the following activities and events:

- Activity A: Excavation (Events 1 to 2)
- Activity B: Foundation Laying (Events 2 to 3)
- Activity C: Erecting Columns (Events 3 to 4)
- Activity D: Placing Roof (Events 4 to 5)
- Activity E: Electrical Installation (Events 4 to 6)
- Activity F: Finishing (Events 5 to 7)

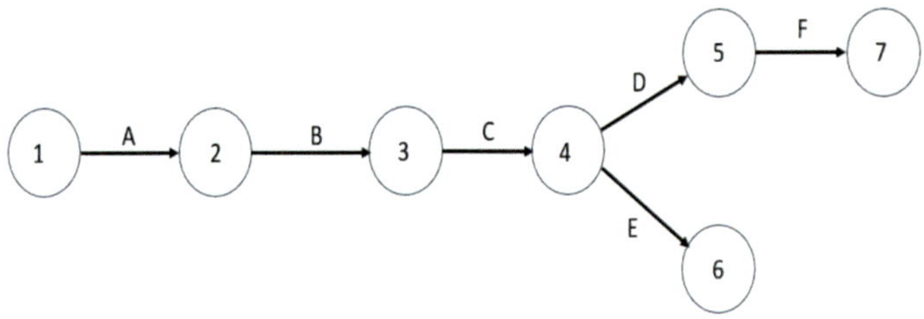

	Immediate Predecessor	Predecessor	Immediate Successor	Successor
Activities				
A	–	–	B	B, C, D, E, F
B	A	A	C	C, D, E, F
C	B	A, B	D, E	D, E, F
D	C	A, B, C	F	F
E	C	A, B, C	–	F
F	D	A, B, C, D, E	–	–
Events				
1	–	–	2	2, 3, 4, 5, 6, 7
2	1	1	3	3, 4, 5, 6, 7
3	2	1, 2	4	4, 5, 6, 7
4	3	1, 2, 3	5, 6	5, 6, 7
5	4	1, 2, 3, 4	7	7
6	4	1, 2, 3, 4	–	7
7	5	1, 2, 3, 4, 5, 6	–	–

1.16 PERT TIME ESTIMATES

Program evaluation and review technique (PERT) focuses heavily on estimating the time required to complete activities in a project. Since PERT is commonly used in projects with high uncertainty, like research and development, time estimates are treated as probabilistic rather than deterministic. In PERT, each activity is assigned three different time estimates:

- *Optimistic time* (t_o). The shortest time to complete the activity, assuming everything proceeds better than expected. It assumes there are no delays or issues.
- *Most likely time* (t_m). The time the activity is expected to take under normal conditions, based on realistic assumptions. This is the most probable time to complete the task.
- *Pessimistic time* (t_p). The longest time the activity might take, assuming significant delays or problems occur.

Expected time (t_e) is the weighted average of the three time estimates and is calculated as follows:

$$t_e = \frac{t_o + 4t_m + t_p}{6} \tag{1.1}$$

Please note that the preceding equation emphasizes on the "most likely time (t_m)," as it is weighted four times, while the optimistic and pessimistic times are considered once each. PERT also calculates the variance of the time estimates to measure the uncertainty associated with each activity. It is given by:

$$\text{Variance} = \left[\frac{t_p - t_o}{6}\right]^2 \tag{1.2}$$

This helps determine the degree of risk or uncertainty involved in each activity.

The project's total duration in PERT is determined by summing the expected times (t_e) along the critical path. The critical path is the longest path through the network of activities and represents the shortest time to complete the project. To evaluate the variability of the project completion time, the standard deviation of the critical path is calculated as follows:

$$\begin{matrix}\text{Standard deviation} \\ \text{of critical path}\end{matrix} = \sqrt{\sum\left(\text{Vaiances of the critical path activities}\right)} \tag{1.3}$$

With the expected project completion time and the standard deviation known, PERT allows project managers to calculate the probability of completing the project by a specific deadline. This is done using the Z-score formula, as follows:

$$Z = \frac{\text{Target time} - \text{Expected time}}{\text{Standard Deviation of critical path}} \tag{1.4}$$

Eq. (1.4) is used to determine how many standard deviations away the target completion time is from the expected time. It helps derive the probability of meeting the deadline using a standard regular (normal distribution) distribution table. The following points are important:

- PERT uses probabilistic time estimates, which makes it suitable for projects with high uncertainty.
- The expected time provides a weighted average of the optimistic, most likely, and pessimistic estimates.
- Variance helps measure the uncertainty of each activity's duration.
- PERT allows managers to compute the probability of project completion within a specified time, which is helpful for planning and risk management.

1.16.1 Problem 3: Estimating Activity Duration

A construction project has an activity, foundation laying, with the following time estimates: optimistic time (t_o), 4 days; most likely time (t_m), 6 days; pessimistic time (t_p), 10 days. Calculate this activity's expected time (t_e) and variance.

1.16.1.1 Solution

Expected time (t_e) =

$$t_e = \frac{t_o + 4t_m + t_p}{6}$$

$$t_e = \frac{4 + 4(6) + 10}{6} = 6.33 \text{ days}$$

$$\text{Variance} = \left[\frac{t_p - t_o}{6}\right]^2 = \left[\frac{10 - 4}{6}\right]^2 = 1$$

1.16.2 Problem 4: Probability of Project Completion

Consider a construction project where the wall painting activity has the following data: optimistic time (t_o), 4 days; most likely time (t_m), 7 days; and pessimistic time (t_p), 10 days. The project manager wants to know the probability of completing the activity within 8 days.

1.16.2.1 Solution

$$t_e = \frac{t_o + 4t_m + t_p}{6} = \frac{4 + 4(7) + 10}{6} = 7 \text{ days}$$

$$\text{Variance} = \left[\frac{t_p - t_o}{6}\right]^2 = \left[\frac{10 - 4}{6}\right]^2 = 1$$

$$\text{Standard deviation} = \sqrt{1} = 1$$

Z-score to calculate the probability of completing within 8 days:

$$Z = \frac{\text{Target time} - \text{Expected time}}{\text{Standard Deviation}} = \frac{8 - 7}{1} = 1$$

Using the Z-table, the probability corresponding to a Z-score of 1 is approximately 0.8413, or 84.13%. There is an 84.13% probability of completing the wall painting activity within 8 days.

1.16.3 Problem 5: Critical Path Time Estimate

For a building construction project, the following activities along the critical path have time estimates as shown. Calculate the total expected time and variance of the critical path:

Activity	Optimistic Time (t_o)	Most Likely Time (t_m)	Pessimistic Time (t_p)
Excavation	2 days	4 days	6 days
Foundation	5 days	7 days	9 days
Framing	3 days	5 days	7 days

1.16.3.1 Solution

For excavation:

$$t_e = \frac{2+4(4)+6}{6} = 4 \text{ days}$$

$$\text{Variance} = \left[\frac{6-2}{6}\right]^2 = 0.444$$

For foundation:

$$t_e = \frac{5+4(7)+7}{6} = 5 \text{ days}$$

$$\text{Variance} = \left[\frac{7-3}{6}\right]^2 = 0.444$$

For framing:

$$t_e = \frac{3+4(5)+9}{6} = 7 \text{ days}$$

$$\text{Variance} = \left[\frac{9-5}{6}\right]^2 = 0.444$$

Total expected time for the critical path $= 4+5+7 = 16$ days.

Total variance for the critical path $= 0.444+0.444+0.444 = 1.332$.

The standard deviation of the critical path $= \sqrt{1.332} = 1.154$ days.

The total expected time for the critical path is 16 days, with a standard deviation of 1.154 days.

1.16.4 Problem 6

A construction company is building a residential house. The following table provides the activities and their time estimates (in weeks). Use the PERT method to determine the critical path and expected project completion time.

Activity	Predecessor	Optimistic Time (t_o)	Most Likely Time (t_m)	Pessimistic Time (t_p)
A	None	2	4	6
B	A	3	5	7
C	A	2	3	8
D	B	1	2	3
E	C	4	6	10
F	D, E	2	3	4

1.16.4.1 Solution

Step 1: Calculate the expected time for each activity.

$$t_e(\mathrm{A}) = \frac{2+4(4)+6}{6} = 4\,\text{weeks}$$

$$t_e(\mathrm{B}) = \frac{3+4(5)+7}{6} = 5\,\text{weeks}$$

$$t_e(\mathrm{C}) = \frac{2+4(3)+8}{6} = 3.67\ \text{weeks}$$

$$t_e(\mathrm{D}) = \frac{1+4(2)+3}{6} = 2\,\text{weeks}$$

$$t_e(\mathrm{E}) = \frac{4+4(6)+10}{6} = 6\,\text{weeks}$$

$$t_e(\mathrm{F}) = \frac{2+4(3)+4}{6} = 3\,\text{weeks}$$

Step 2: Draw the network diagram. Draw the PERT network, showing each activity as a node and arrows indicating precedence relationships.

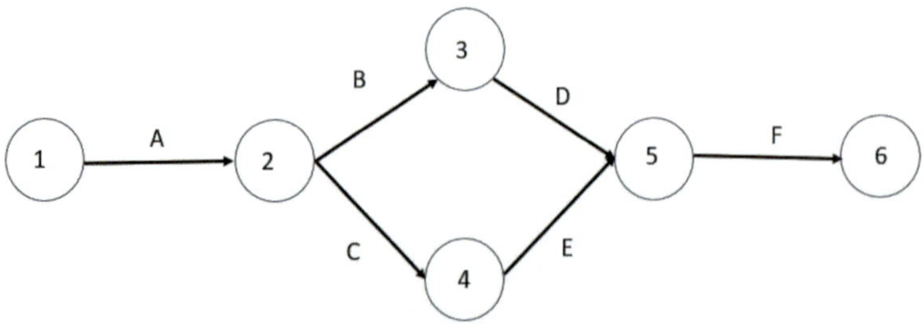

Step 3: Determine critical path and project duration. Find all possible paths, and sum the expected times.

- Path 1: A → B → D → F

$$\text{Total time} = 4 + 5 + 2 + 3 = 144 + 5 + 2 + \ 3 \ = \ 144 + 5 + 2 + 3 = 14 \text{ weeks}$$

- Path 2: A → C → E → F

$$\text{Total time} = 4 + 3.67 + 6 + 3 = 16.674 + 3.67 + 6 + 3 = 16.674 + 3.67 + 6 + 3 = 16.67 \text{ weeks}$$

Critical Path: A → C → E → F
Project Completion Time: 16.67 weeks

1.16.5 Problem 7

The following activities and time estimates are provided for constructing a school building. Using PERT, calculate the expected project completion time and the probability of finishing the project within 24 weeks.

Activity	Predecessor	Optimistic Time (O)	Most Likely Time (M)	Pessimistic Time (P)
A	None	3	5	8
B	A	2	4	6
C	A	4	6	9
D	B	1	3	5
E	C	3	5	7
F	D, E	2	4	6

1.16.5.1 Solution

Step 1: Calculate the expected time for each activity.

$$\text{Expected time } (\text{t}_\text{e}) = \frac{t_o + 4t_m + t_p}{6}$$

$$t_e(\text{A}) = \frac{3 + 4(5) + 8}{6} = 5.33 \text{ weeks}$$

$$t_e(\text{B}) = \frac{2 + 4(4) + 6}{6} = 4 \text{ weeks}$$

$$t_e(\text{C}) = \frac{4 + 4(6) + 9}{6} = 6.17 \text{ weeks}$$

$$t_e(\text{D}) = \frac{1 + 4(3) + 5}{6} = 3 \text{ weeks}$$

$$t_e(\mathrm{E}) = \frac{3+4(5)+7}{6} = 5\,\text{weeks}$$

$$t_e(\mathrm{F}) = \frac{2+4(4)+6}{6} = 4\,\text{weeks}$$

Step 2: Calculate variance for each activity:

$$\text{Variance} = \left[\frac{t_p - t_o}{6}\right]^2$$

Variance (A) = 0.69
Variance (B) = 0.44
Variance (C) = 0.69
Variance (D) = 0.44
Variance (E) = 0.44
Variance (F) = 0.44

Step 3: Calculate expected completion time and critical path:

Critical Path: A → C → E → F
Total Expected Time = 5.33 + 6.17 + 5 + 4 = 20.5 weeks
Total Variance = 0.69 + 0.69 + 0.44 + 0.44 = 2.26

Step 4: Calculate probability of completion within 24 weeks:

$$Z = \frac{\text{Target time} - \text{Expected time}}{\text{Standard Deviation}} = \frac{24 - 20.5}{\sqrt{2.26}} = 2.33$$

Using a Z-table, the probability of completing the project within 24 weeks is approximately 0.99, or 99%.

1.17 BASIC SCHEDULING COMPUTATIONS

The following nomenclature is followed in the basic scheduling of events:

1. **(i, j).** An activity connecting two events, with ***i*** as the starting (tail) event and ***j*** as the ending (head) event.
2. **E_i.** Earliest occurrence time of event ***i***. It is the earliest event ***i*** can occur, based on the project schedule up to that point.
3. **L_j.** Latest allowable occurrence time of event ***j***. This is the latest time that event ***j*** can occur without delaying the entire project.
4. **D_i.** Estimated activity completion time ***(i, j)***. This represents the estimated duration or time to complete an activity ***(i, j)***.

5. **$(Es)_{ij}$**. Earliest starting time of activity ***(i, j)***. The earliest time activity ***(i, j)*** can begin. It depends on the earliest occurrence of event ***i*** and the completion of its preceding activities.
6. **$(Ef)_{ij}$**. Earliest finishing time of activity ***(i, j)***. The earliest time activity ***(i, j)*** can be completed. It is calculated as follows:

$$(Ef)_{ij} = (Es)_{ij} + D_{ij}$$

7. **$(Ls)_{ij}$**. Latest activity starting time ***(i, j)***. The latest time activity ***(i, j)*** can start without delaying the project completion. It is calculated as follows:

$$(Ls)_{ij} = (Lf)_{ij} - D_{ij}$$

8. **$(Lf)_{ij}$**. Latest finishing time of activity ***(i, j)***. The latest time activity ***(i, j)*** must finish without delaying the overall project. It depends on the latest occurrence time of event ***j***.

1.17.1 Steps for Scheduling

1. *Calculate the earliest occurrence times (forward pass).* Begin at the start event (usually node 1), and calculate each event's earliest occurrence time E_i by moving forward through the network diagram. For each activity (i, j), calculate the earliest start time $(Es)_{ij}$ as the earliest occurrence time of event i:

$$(Es)_{ij} = Ei$$

   - Then, calculate the earliest finish time $(Ef)_{ij}$ for each activity:

$$(Ef)_{ij} = (Es)_{ij} + D_{ij}$$

   - Update the earliest occurrence time E_j of event j as the maximum of the earliest finish times of all activities leading to event j.
2. *Calculate the latest occurrence times (backward pass).* Start from the last event, and calculate the latest occurrence time L_j for each event by moving backward through the network diagram. For each activity (i, j), calculate the latest finish time $(Lf)_{ij}$ as the latest allowable occurrence time of event j:

$$(Lf)_{ij} = L_j$$

   - Then, calculate the latest start time $(Ls)_{ij}$ for each activity:

$$(Ls)_{ij} = (Lf)_{ij} - D_{ij}$$

   - Update the latest occurrence time L_i of event i as the minimum of the latest start times of all activities leading out of event i.
3. *Identify the critical path.* The critical path consists of activities where $(Es)_{ij} = (Ls)_{ij}$ and $(Ef)_{ij} = (Lf)_{ij}$, that is, the activities with zero float or slack. These activities cannot be delayed without delaying project completion time.

1.17.2 Problem 8

A construction project consists of the following activities, with given durations and dependencies. Schedule the project, identify the critical path, and indicate whether there can be any delay in the critical path events.

Activity (i, j)	Duration D_{ij} (days)	Immediate Predecessors
A (1, 2)	4	None
B (1, 3)	6	None
C (2, 4)	3	A
D (3, 4)	5	B
E (4, 5)	2	C, D

1.17.2.1 Solution

Step 1: Draw the network diagram. The network consists of five events and five activities. The events are numbered 1 through 5. Following is the basic layout of the network:

- Activities A and B originate from event 1.
- Activities C and D converge at event 4.
- Activity E connects event 4 to the final event 5.

Now, one can prepare the earliest and latest times for the events.

Step 2: Perform a forward pass to compute the earliest times. The forward pass calculates the earliest start (*Es*) and earliest finish (*Ef*) times.

1.17.2.1.1 For Activity A (1, 2)

- $(Es)_A = E_1 = 0$
- $(Ef)_A = (Es)_A + D_A = 0 + 4 = 4$

Thus, $E_2 = 4$ (earliest occurrence time of event 2).

1.17.2.1.2 For Activity B (1, 3)

- $(Es)_B = E_1 = 0$
- $(Ef)_B = (Es)_B + D_B = 0 + 6 = 6$

Thus, $E_3 = 6$ (earliest occurrence time of event 3).

1.17.2.1.3 For Activity C (2, 4)

- $(Es)_C = E_2 = 4$
- $(Ef)_C = (Es)_C + D_C = 4 + 3 = 7$

Thus, E_4 depends on the max of $(Ef)_C$ and $(Ef)_D$.

1.17.2.1.4 For Activity D (3, 4)

- $(Es)_D = E_3 = 6$
- $(Ef)_D = (Es)_D + D_D = 6 + 5 = 11$

Thus, $E_4 = 11$ (since it's the max of $(Ef)_C$ and $(Ef)_D$).

1.17.2.1.5 For Activity E (4, 5)

- $(Es)_E = E_4 = 11$
- $(Ef)_E = (Es)_E + D_E = 11 + 2 = 13$

Thus, $E_5 = 13$ (earliest completion time of the project).

Step 3: Perform a backward pass to compute the latest times. We calculate the latest finish (*Lf*) and latest start (*Ls*) times by performing a backward pass.

1.17.2.1.6 For Activity E (4, 5)

- $(Lf)_E = E_5 = 13$
- $(Ls)_E = (Lf)_E - D_E = 13 - 2 = 11$

Thus, $L_4 = 11$ (latest occurrence time of event 4).

1.17.2.1.7 For Activity C (2, 4)

- $(Lf)_C = L_4 = 11$
- $(Ls)_C = (Lf)_C - D_C = 11 - 3 = 8$

Thus, $L_2 = 8$.

1.17.2.1.8 For Activity D (3, 4)

- $(Lf)_D = L_4 = 11$
- $(Ls)_D = (Lf)_D - D_D = 11 - 5 = 6$

Thus, $L_3 = 6$.

1.17.2.1.9 For Activity A (1, 2)

- $(Lf)_A = L_2 = 8$
- $(Ls)_A = (Lf)_A - D_A = 8 - 4 = 4$

1.17.2.1.10 For Activity B (1, 3)

- $(Lf)_B = L_3 = 6$
- $(Ls)_B = (Lf)_B - D_B = 6 - 6 = 0$

Thus, $L_1 = 0$.

Step 4: Identify critical path. The critical path consists of activities where (Es) = (Ls) and (Ef) = (Lf). For these activities, there is no float or slack.

1.17.2.1.11 Check Each Activity

- For **Activity A:** $(Es)_A = 0$, $(Ls)_A = 4 \rightarrow$ Not on the critical path.
- For **Activity B:** $(Es)_B = 0$, $(Ls)_B = 0 \rightarrow$ Critical.
- For **Activity C:** $(Es)_C = 4$, $(Ls)_C = 8 \rightarrow$ Not on the critical path.
- For **Activity D:** $(Es)_D = 6$, $(Ls)_D = 6 \rightarrow$ Critical.
- For **Activity E:** $(Es)_E = 11$, $(Ls)_E = 11 \rightarrow$ Critical.

Thus, the **critical path is B → D → E**. The project will take 13 days to complete, and any delay in these activities will delay the entire project.

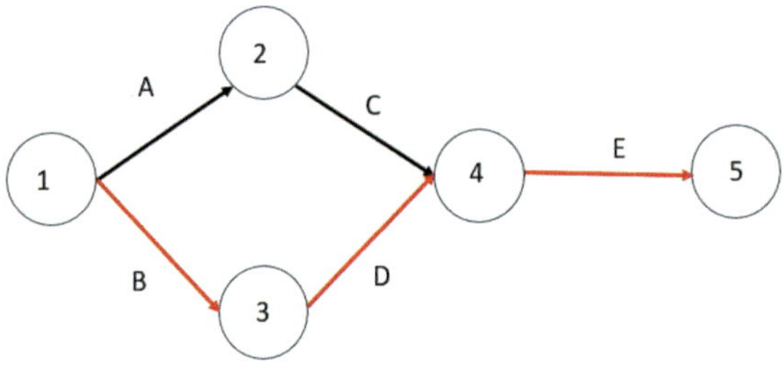

Final Table of Results

Activity (i, j)	D_{ij}	$(Es)_{ij}$	$(Ef)_{ij}$	$(Ls)_{ij}$	$(Lf)_{ij}$	Total Float
A (1, 2)	4	0	4	4	8	4
B (1, 3)	6	0	6	0	6	0
C (2, 4)	3	4	7	8	11	4
D (3, 4)	5	6	11	6	11	0
E (4, 5)	2	11	13	11	13	0

This problem demonstrates the basic scheduling computations for a construction project. We've calculated each activity's earliest and latest start/finish times, determined the critical path (B → D → E), and found that the project will take 13 days to complete.

2 Form Formation and Geometry in Architectural Thinking

2.1 INTRODUCTION

Architectural practice has always seen structural framework as a prerequisite to architectural form, and the source of such forms has been the physical world around us. Architects have been visualizing complex forms by using geometrical elements and design principles to create a form language for organizing architectural spaces with a philosophical narrative that is aesthetically intriguing and economically viable. Ideally, architects are responsible for the aesthetics of any structure they design; without a structural engineer, that form could not be achieved. If an engineer has knowledge of basic aesthetics or he or she is aesthetically conscious, then both architect and engineer together could innovate on new structural forms. The Platonic idea of architectural forms has a multiplicity of meanings that are embedded in metaphysical, aesthetical, epistemological, logical, and ethical meanings. In the Renaissance, and after that, the epistemological meaning of form became the most important one, where relating the mind to reality was emphasized. Architectural forms were perceived and conceived so that they could be easily apprehended and could be grasped structurally. As a result, geometric forms became a formative idea in form formation (Evans, 1995).

Humans always perceived essential characteristics of natural elements through abstraction and symbolic notations inscribed in geometry, such as the sun, moon, mountains, etc. "The source of form is thoroughly embedded in the physical world itself and owes nothing to an extra-sensory divine overseer" (Gelernter, 1988). The magnanimity of mountains could be the perception behind the pyramids. Geometric figures as such have also been used in shapes of buildings and decorative motifs in ancient civilizations but were not demonstrative enough, until the Greeks effectively used them using polygons, circles, and other three-dimensional geometric shapes. Thales, Pythagoras, Plato, and Euclid are philosophers who laid the foundation of geometry that continues to date. Modern geometry continues to be on the foundations laid by Greek geometry but is not confined to it. Elliptic and hyperbolic geometry were new developments in modern geometry that revolutionized form formation in architectural thinking and made the conceptualization of structural framework a possibility. This chapter tries to develop an understanding of the conceptualization of form formation and geometry as a formative idea in architectural thinking. It explores the process of form formation using geometry and visual grammar/design principles.

DOI: 10.1201/9781003581314-2

If the structural engineer is aware of aesthetics and can understand the conceptualization of the form by an architect, he or she can understand the language of form formation and the structural components that can be feasible to design and would help architects develop structural forms that are creative and innovative (Fergusson, 2020).

2.2 IDEATION OF FORMS IN ARCHITECTURE

> Architecture, although it depends on Mathematics, nevertheless it is a flattering Art, that by no means wants to be distasteful: even if many of its rules follow the dictates [of Mathematics] when its observed demonstrations result to be offensive for the sight, it changes them, abandons them and eventually contradicts the same; so that, to know what an Architect should observe, it will not be unfruitful to see the real scope of Architecture, as well as its way of proceeding.
>
> **(Guarini, 1937)**

Geometry can be seen as a formative idea in architectural thinking. Before discussing geometry and architecture, we first explore other creative disciplines, like music/dance or the performing arts, visual arts, and linguistics, and the parallels drawn between them and architecture. Architecture, music, and linguistics share a clear cultural lineage, and the creative language embedded in all three is not just about the composition of its basic elements but is also their combination and arrangement with each other. These creative pursuits go beyond the creative and constructive manipulation of the basic elements, but also the meaning behind them and the impact it has on the people. Design perspective and its transformational process when compared with linguistics, and transformational generative grammar by Chomsky (through a stylistic approach), helps to understand and appreciate a given text through phrase-structure rules, transformational rules, and morphophonemic rules. Architecture with elements of design and visual grammar makes any architectural creation a subject of objective interpretations. When comparing architecture to music also, such parallels could be seen. The notion of sound and sense in music compositions is like what we have seen in architecture and its interpretation. Indeed, architecture as "frozen music" can also be equated to the art of harmonic and rhythmic order that is performing arts, especially music.

The alphabets (as in English), or varnmala (as in Hindi), are at the very foundation of any language; similarly, the Sargam/musical notes are at the foundation of any musical creation. In architecture, the elements of design, such as points, lines, shape, plane, color, texture, and geometrical figures, are the fundamentals for any form formation process. These have been discussed in Euclidean geometry, which studies these elements through a logical system based on ten assumptions, namely, common notions and postulates. These elements, through analytic geometry, elaborate on the three-dimensional cartesian system (the x-, y-, and z-axis) and facilitate any architect in a three-dimensional environment, where the structural elements of the form could be visualized. Fractal, non-Euclidean, and descriptive geometries also play a key role in the ideation of forms and their structural calculations. Knowledge of geometry not only empowers us to conceive forms with ease but also makes us precisely understand them. It gives us a set of ready forms that can be subject to visual manipulations in many ways.

Architecture is not just about built forms; the functional dimension of architecture cannot be ignored. Any architectural creation cannot ignore the set of activities that shall be going on in the building or around it. It is an ensemble that encloses the given human activity through place, space, and other elements and facilitates the completion of the set of tasks to be conducted within the given volume. This given volume could also be seen as a physical space, a behavioral space, and an experiential space. The experiential space is where a user or any observer perceives and experiences while he or she moves along within that space. As a result, not only is the functional or behavioral quality of spaces important, but their experiential value also generates emotions. The feeling of pleasantness or unpleasantness (emotional qualities) in any architectural creation represents the distinction between the artistic and factual, just like the distinction between a phrase and poetry.

Architecture is a creative discipline that involves a thinking process and includes idea generation, evaluation of the idea generated, developing task strategies for the idea generated, information searching and its synthesis, organizing the ideas, developing an architectural concept, developing the conceptual thought into an architectural product, its implementation and monitoring. Every architect must go through the process of ideation of form, which is cognitive and a conscious mental activity. This comes through reading, remembering, thinking, reasoning, understanding, visualizing, and then expressing and needs an elevated level of mental process. An architect, for such ideation of forms, needs to convert the information gathered from the aforementioned mental activities into knowledge which then helps him/her in visualization, which leads to creating an architectural ensemble. This requires skills like memory, judgment, and problem-solving, leading to imagination and perception. To understand the ideation of forms, it is necessary to understand the fundamentals of the processes that lead to architectural conceptualization. The next section discusses the elements of design that are at the foundation of understanding the process and how geometry plays a key role in the ideation of forms.

2.3 ELEMENTS OF DESIGN AND GEOMETRY IN FORM FORMATION

Architecture can provide a symbolic meaning through its built form, which generates an ambience, an aesthetical quality which cannot be measured and analyzed easily. A step-by-step approach could help us understand the form formation process. The most effective factors that shape any architectural identity are, first, the design concept; second, the visualization of shape and form; third, technological innovation; fourth, the material used for construction; and finally, its relationship with the surroundings. The *conceptualization* of all forms is dependent on how an architect visualizes the design in his or her mind before it is realized into a physical shape. "Form as the final architectural product is shaped by the visual cues that produce symbols as a powerful tool in identifying a specific architectural trend". It starts with an idea that reflects an architect's feelings, thinking process, and visualization of design that creates an architectural identity. This could be through an abstraction of visual elements that are built upon the conceptual elements of form formation. Abstraction of visual elements, once realized, could reflect upon the intangible experiences that the user or visitor feels while moving through or along those finished spaces and forms. The symbols, as mentioned by Baper, could be pure geometrical entities or could be a result of a combination of given geometrical entities.

Design is an ensemble of elements of design through a process of visualization that is in line with the purpose for which it is meant. The design elements provide a starting point for the ideation of form, as the combination of these elements, as described by theories and philosophies in architecture, becomes the foundation for any architect to define his language of form formation either through "Function follows form" or through "Form follows function" as popular known phrases in architecture. These can also be rephrased from the ideation of forms perspective as "Whole is the sum of parts" and "Whole is greater than the sum of its parts." Every architect, although aware of the design elements and the philosophy established, sometimes acts in defiance of these established rules to create forms through his interpretation of these elements. That is the reason we can identify or read the form formation pattern language that any architect practices.

The basic elements as earlier described can be manipulated through a given framework of organizational philosophies or principles, also known as design principles or visual grammar, generating an aesthetically pleasing and structurally feasible architectural form. These elements can also be seen as vocabulary for any design language that reflects aesthetical prowess through form formation. Geometric shapes, angular planes, and solids, either singly or in combination, create forms that could be visually appealing. Architects extend a grid pattern of geometrical shapes into a three-dimensional entity reflecting space for a certain designated activity and symbol or a sign with a meaning behind it. So all architectural creations have three components: space that defines an activity, the sign or symbol that has an aesthetical meaning, and the structural feasibility through the load distribution path, along with the strength of the material.

Ideation of architectural forms and spaces consists of two fundamental elements: firstly, the conceptual elements and, secondly, the visual elements. The visual elements of architecture can be assumed to be corresponding to words of the language and the relations of these elements to the grammar. As a result, distinctive styles of architecture could also be considered as different languages. Architecture could also be seen as a fundamental form of expression and communication reflected through its tangibility. Conceptual elements consist of points, lines, planes, and volume, and visual elements consist of shape size, color, texture, etc. The line could be seen as an element reflecting direction, endpoints, and boundary. Usually in design, the elements are visual and are generally three-dimensional, but these three-dimensional volumes are defined by vertex (point), edges (line), and planes (segment). *Point* in geometry has also been described as the prime generator of form. The line then becomes the second generator, the plane is the third generator, and solids the fourth generator.

Geometry in mathematics is all about the relationship between lines and surfaces and so has been seen as a key element in design, visual arts, and architecture. In architecture design, apart from points, lines, planes, and solids, there are elements like color, texture, and material. The use of colors in architectural thinking has mainly been due to its psychological aspects and the human response to it. It can be a medium to define a building's identity and could enhance the spatial ambience of any space where users can perceive and feel, creating different moods. The use of color, and its impact, is also influenced by another element, namely, texture. The varying texture of the finished surfaces with the same color would have different effects when that surface is subjected to natural or artificial light. Texture in architecture is visualized in two ways, either by its visual or by its touch. Texture adds to the visual weight of any surface and could be rough/smooth, hard/soft, dull/shiny, opaque/transparent, metallic, etc. Materials in architecture have evolved along with the technologies that have produced them. In contemporary practice, architects have been experimenting with new materials and have gone as far as to test their limits. The structural design component of the material and its decorative or aesthetic design component are now being integrated to create interesting forms.

Geometry is all about shape and form, which are two-dimensional and three-dimensional entities. Circles, squares, and triangles are standard and regular geometric shapes, whereas shapes like ovals, rectangles, octagons, hexagons, pentagons, parallelograms, etc. are further derivatives of regular geometrical shapes. These geometrical shapes also help in defining the conceptual elements in architectural design, for example, windows, arches, doors, and decorative elements of a building, which are an integral part of any architectural creation. More complicated shapes, like free-flow shapes, organic, non-standard, or any irregular shape, also are the result of derivations from the derivatives of regular geometrical shapes. Architects have always used these derivations and the derivations from these derivatives, respectively, in their novel visualization and have created forms that, although complex, have regular geometries at their foundations. The digital tools for form formation 3D modelling and structural analytical tools have made deriving forms that are more complex and easier. Architecture is all about surfaces that enclose a given volume and can be divided into two types. Firstly, ruled surfaces can be derived by rotating or moving a straight line in a given direction or track, for example, cylindrical plane or conical plane, and secondly, non-ruled surfaces are where the surface cannot be formed by a straight line moving along a trajectory, for example, ellipsoid, paraboloid, or hyperboloid. When understanding the ideation of forms in architecture, both Euclidean and non-Euclidean geometry play a role starting from the forms based on basic geometry to hyperbolic geometry, which leads to complex forms that are loved by architects.

Geometry could be seen as the starting point of any architectural design process, where shape and form are at the very foundation. As we explore Euclidean geometry, we see that points, lines, planes, and geometrical figures are fundamental to any form that evolves from that. Point is the primary generator of form, line is the second generator, plane is the third generator, and three-dimensional solids are the fourth generator. The fourth generator, namely, solids, occupies space and consists of all points, lines, and planes (extensions of lines or surfaces). Analytic geometry investigates a flexible way of understanding and manipulating curved lines. Conic forms, or forms derived from cones, and shell structures, ruled surfaces or doubly ruled

FIGURE 2.1 Form formation through basic geometry.

surfaces, could be conceptualized through analytic geometry. Fractal geometry has been an approach that was a departure from traditional Euclidean as well as analytic geometry. It deals with free-flowing forms, just like those that exist in nature. Non-Euclidean geometry has also brought the concept of spherical and hyperbolic geometry, which help architects visualize much more complex forms. Figure 2.1 shows the conceptualization of form through basic geometry.

Let us explore the building forms that have been dictated by basic geometry, for example, the Kadthal Meditation Pyramid in Hyderabad, which is one of the largest in India, where meditation and spiritual workshops are organized. This is one of the simplest geometrical buildings which has a simple square plan. There are many examples in the world of such structures, like the ancient pyramids, the Louvre in Paris, the Infosys Pyramid building at its headquarters in Bangalore, etc. (see Figure 2.1). Such basic geometrical structure has also been the backbone of structures in vernacular/traditional architecture, for example, the Bhunga house in Gujrat; tipi or tepee of Native Americans; trulli houses in Apulia, Italy; etc. (see Figure 2.2). Euclidean and analytic geometry dictate such three-dimensional formations, guided by basic or simple forms, where straight lines have been used, and their extensions, into planes, either straight, as in pyramids, or curved as, in cones. Figure 2.3 shows the sketch of St. Peter's Basilica, Rome (Jasieńko et al., 2021).

With the development of digital tools, non-Euclidean geometry also became a part of architectural thinking that enabled architects to develop free-flowing forms that were structurally sound through highly advanced software where spatial structures have been modelled in the hyperbolic, elliptic, or fractal geometry. Non-Euclidean

FIGURE 2.2 Formation in vernacular architecture.

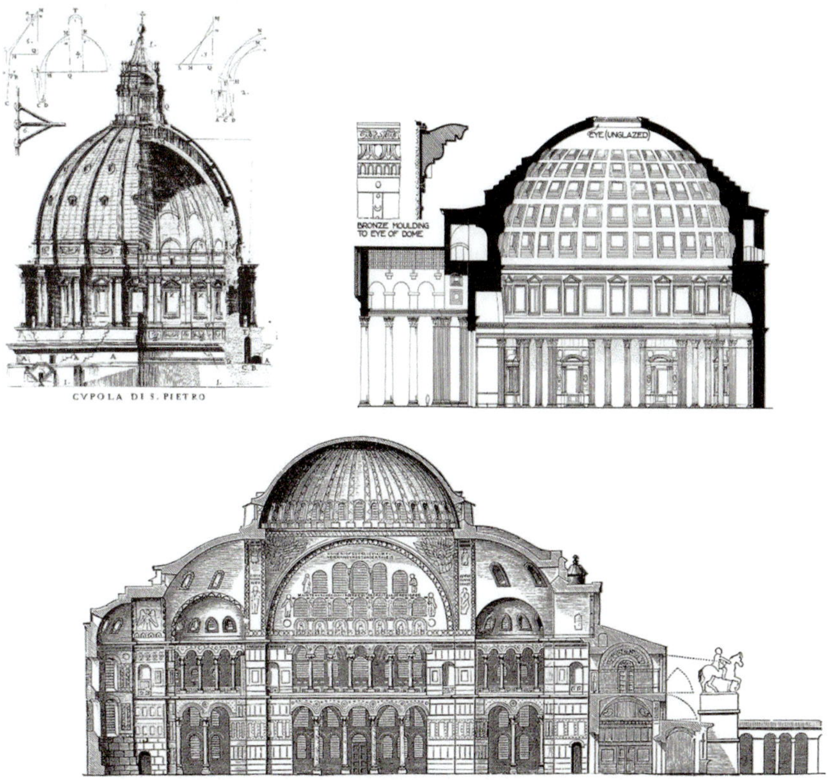

FIGURE 2.3 St. Peter's Basilica, Rome, Italy.

FIGURE 2.4 Built form based on elliptic geometry.

geometry helps architects visualize more complex forms where designs can be curved, deformed, or otherwise distorted. One such geometry is elliptic geometry, also called geometry of spherical or curved surfaces. The dome of the Pantheon in Rome, Hagia Sophia, and St. Peter's Basilica are some early examples of elliptic geometry (Figure 2.4). The Gherkin by Norman Foster, Shell House by Kotaro Ide, Zaha Hadid's Galaxy SOHO, and Heydar Aliyev Centre, Baku, Azerbaijan, are a few of the many examples of the modern world (Figure 2.4).

The basis of hyperbolic geometry is a hyperbolic plane that has a negative curve that ignores Euclid's axioms and Euclid's parallel postulate. Hyperboloid structures and hyperbolic paraboloid structures could be visualized and modelled through this form of geometry. Vladimir Shukhov, a Russian engineer, was the first to build hyperboloid structures. Sukhov Towers in Moscow was designed as a broadcasting tower and is an early example of a hyperboloid structure (Figure 2.5). It is a classical example of a broadcasting tower designed of hyperbolic geometrical structure. In hyperbolic paraboloid, lines on a saddle-shaped surface are never straight, because the surface is curved. In hyperbolic geometry, or Lobachevskian geometry, triangles have angles that add up to less than 180°.

Hyperbolic geometry is where parallel postulates of Euclidean geometry do not apply. In this geometry, parallel lines move away from each other, and there are no similar triangles. It is based on the hyperbolic plane, which has a consistent negative

FIGURE 2.5 Sukhov Towers, Moscow.

FIGURE 2.6 Hyperbolic paraboloid structural form.

curve. In Euclidean geometry, parallel lines never meet, and the angles of a triangle sum to exactly 180°. However, in hyperbolic geometry, parallel lines can intersect, and the sum of angles in a triangle is always less than 180°. Xing Hai Concert Hall, Guangzhou, China, and San Vincente de Paul Chapel in Mexico City are good examples of hyperbolic paraboloid structures (Figure 2.6). The roofs of both structures have four sides which are in straight lines.

FIGURE 2.7 Hyperbolic paraboloid structural form.

In the same typology, but with a bit more complexity, the horizontal projection of the roof is close to a rectangle, but with an inward depression or a negative curve in the middle. Chapel Lomas de Cuernavaca, built in 1958 by Flix Candela, is a very suitable example of such a roof formation. These structures could attain even more complexity by cutting the roof of such types and combining them, making it an even more complex hyperbolic paraboloid structure. For example, Los Manantiales Restaurant in Xochimilco, Mexico City, designed by the same architect, is a more complex hyperbolic paraboloid roof building consisting of four hyperbolic paraboloids joined together (Figure 2.7).

Hyperbolic paraboloid is ruled space surface which gives the roof an aesthetically sound surface and is used in architectural design to achieve a free-flowing and flexible formation. Due to the complexity of curved surfaces, many architects love to navigate them but, at the same time, feel not confident to navigate such structures. Hyperbolic paraboloid structures become more complex when the roof projection or plan is not in straight lines, say, rectangle but is oval. The London Velodrome for the 2012 Olympics is an incredibly good example where the roof, which is a hyperbolic paraboloid structure in steel frame, has been placed on a 360° glazed concourse ventilation (Figure 2.8).

FIGURE 2.8 London Velodrome: hyperbolic paraboloid-shaped steel frame.

The smooth and regular shapes of classical geometry have helped humans measure and understand the world around them but have failed to describe shapes whose uneven contours mimicked the irregularities found in nature. Fractal geometry is a geometry where elements or components that constitute its form appear to be similar but are of varied sizes or scales, as in the case of features that occur in natural forms. Fractal geometry can also be seen as an extension of the classical geometry. A middle ground between the extreme order of Euclidean geometry and the geometric chaos of complexities (Mandelbrot, 1982). It reaches at the core of design composition, allowing any architect to express an understanding of the construct of nature. Branching of trees, veins that form a leaf structure, mountain ridges that seem to be uneven and at varying scales, rivers, vegetable forms are all examples of fractals that exist in nature.

The concept of fractals was first introduced by Polish French mathematician Benoit Mandelbrot in a much more complex way and in its applications in geometry. He described Euclidean geometry as cold and dry, as it failed to account the natural forms, like the "shape of a cloud, mountain, coastline, or tree. Clouds are not spheres, mountains are not cones, coastlines are not circles, and bark is not smooth, nor does lightning travel in a straight line" (Mandelbrot, 1982). There are a number of architectural examples where fractal geometry plays a vital role. The form and structure of historic Hindu temples in India can be seen as one of the best examples of fractal geometry in architecture that have been built long before the development of fractal theory. The self-repeating and self-similar components in their structures reflect a high degree of self-similarity and complex detailing (Joye, 2007). One of the best examples could be the Temples cluster of Khajuraho, which, when seen, seem to be like music in stone (see Figure 2.9).

Works of Antoni Gaudí are perfect examples of an architectural thinking based on fractal geometry that is structurally feasible. The columns of the Sagrada Familia follow the structural rules of nature, as in a tree. Here you witness the construction principles of nature reflected in the structural system of the entire building, and also nature through its ornamental forms (Figure 2.10). Colegio Teresiano by Antoni Gaudí is another example of an ensemble of elements of self-similarity with varying scales. Park Güell, Palau Güell, Casa Milà-La Pedrera, Casa Vicens, Nativity Façade—examples are numerous that reflect fractal geometric formations. Louis

FIGURE 2.9 Khajuraho temples: a reflection of fractal geometry.

FIGURE 2.10 Antoni Gaudí's Sagrada Familia, an example of fractal geometry.

Sullivan, Victor Horta, Otto Wagner in Wien, Charles Rennie Mackintosh, and the Catalan architects Lluís Domènech Montaner and Josep Puig Cadafalch can be seen as architects that have worked with fractal geometric forms. It seems they were the most famous followers of Gaudí. Antoni Gaudí, in fact, was able to obtain such a naturalistic structure by the clever use of Euclidean geometry with the fractal process of self-similarity and self-reproduction. Modern interpretations of fractal geometry in architecture could be the fractal façade designs that could be seen in the Federation Square building in Melbourne and the Grand Egyptian Museum by Heneghan Peng Architects (see Figure 2.11).

FIGURE 2.11 Modern interpretations of fractal geometry.

2.4 IDEATION OF FORM BEYOND GEOMETRY: THE VISUAL GRAMMAR

It is difficult to precisely define aesthetics, but if we see its literal meaning, it has been derived from the Greek word *aesthesis*, meaning "sensory perception," which is used as a philosophical explanation to beauty in arts and nature. Aesthetics in architecture can be seen from two perspectives, firstly, the emotional perspective and, secondly, the intellectual perspective. Works of architects have always been elaborated via these perspectives. Any piece or creation in architecture could be seen as a three-dimensional composition which is conceptualized as a unified whole. An architectural composition, as mentioned, consists of elements from Euclidean geometry, also known as elements of design. Any structure should be conceived in such a way that it is in harmony with its context and makes its presence felt. It is also possible that it creates a new order with its surroundings where it springs out, where a constructive intrusion with the context could be possible. Forms could also be conceptualized as inspired from nature (as in fractal geometry). It can be as simple as possible and should be rational, truthful, and honest.

Form formations, or its ideation, have been described through various aesthetical school of thoughts in architecture: constructivism, expressionism, functionalism, brutalism, modernism, postmodernism, high-tech, deconstructivism, biomimicry, etc. This section shall explore the visual grammar that shapes complicated and complex forms. Simple and basic geometrical forms can also produce more complicated and complex forms by virtue of effects by addition, subtraction, exclusion, division, trimming, and cropping of a given geometric form with other basic geometrical forms.

The intention of any architect is to create a structure that complements its surroundings or conflicts with its surroundings. For this, understanding of the context is particularly important, and one should thoroughly analyze the context in which the ideation of form for a given structure has to be done. In other words, the structure is not an independent identity but is part of the composition that includes its context as well. The aesthetics in architectural composition, apart from elements of design, could also be achieved through applying elements of composition, such as size and scale, proportion, shape and form, space (as in interior spaces), color, texture, light and shade, visual weight or emphasis, pattern, ornamental fenestration, etc. Making these ensembles aesthetically sound, apart from visual grammar, architects use Gestalt principles. Gestalt (derived from the German language) principles are an organized whole that is perceived as more than the sum of its parts or can be seen as a pattern or configuration. The principles of proximity, closure, similarity, continuity, perception, organization, and symmetry have been invariably used in the architectural form formation process.

2.5 IDEATION THROUGH TRANSFORMATION OF SOLIDS

The elements of design as in Euclidean geometry and three-dimensional solids are better idealtion through transformation of solids as it is the fourth generator after point, line, and plane. The transformation of forms in solids can occur through the manipulation of its size and scale (changing dimensions) or by constructive addition or subtraction. During additive form transformation, additional elements are added layer by layer that is a part of the unified whole. In subtractive form transformation, the process is reversed. Constructive addition, or additive transformation, is done through the addition of elements and is dependent upon the elements that are added and lead to the alteration or complete change in the profile of the original solid. The more the variation in the number of added parts, its size, and its placement, the more the change in the profile of the original solid, which sometimes completely loses its original character. The elements added could be through spatial tension through proximity, edge-to-edge, face-to-face, interlocking, centralized, linear, or clustered. Such processes of additive transformation lead to a much complex form.

Similarly *constructive subtraction*, or *subtractive transformation*, involves the removal of elements from the original solid; for example, a pyramid could be converted into a truncated pyramid (when subtracted horizontally), and if it is subtracted with an angular inclination, it can result in a different form. Similar ideation could be done for other solid geometrical forms, like cone, sphere, etc. Also, subtracted elements could be in the form of fenestrations, recessed windows, or entrances. Similarly, process of exclusion, division, trimming, and cropping of a given geometric form can also produce forms that seem to be more complex as compared to its original geometric form. This process of additive and subtractive form formation can range from simple to extraordinarily complex forms. Apart from additive and

FIGURE 2.12 Antoni Gaudí's Sagrada Familia, an example of fractal geometry.

FIGURE 2.13 Subtractive transformation in form ideation.

substrative process, transformation of forms could also be through formal collusion of geometrical solids. Here, the pure geometrical forms get embedded in each other to acquire a different form. The New Design Museum in London by OMA, Allies and Morrison, and John Pawson could be seen as an example of colliding geometric form (Figure 2.12)

Forms do result from the *collision of two or more different geometries* and can be very interesting. Such forms are relatively complex, as compared to the additive and subtractive method. The Orange Cube in Lyon, France, can be seen as examples of subtractive transformation (Figure 2.13). The colliding geometries could vary in size or shape, and the collusion between them could have very varied planer attributes, making it even more complex. This collusion of solid geometries can be symbolic in nature, could be due to a given functional requirement of the form, could make the form asymmetrical or create a symmetrical identity, or could be a response to the contextuality where the built form is placed (Figure 2.14).

FIGURE 2.14 Colliding geometries and form ideation.

2.6 VISUAL CHARACTERISTICS IN AN ARCHITECTURAL COMPOSITION OR ENSEMBLE

A "composition" is an arrangement of elements that are part of the given ensemble having a proper relationship with each other and seem to look as they are constituents of a whole unified by these elements. They, as a result, produce a visual stimulus that evokes emotions. The fundamental for evaluating any composition starts from unity, balance, harmony, contrast, emphasis, movement, and finally, rhythm. This section explores the ideation of forms using elements of design, elements of composition, and principles of visual grammar or design. First, we take up the fundamental characteristics of any composition, then its application in architectural form formation shall be discussed.

Unity, balance, harmony, contrast, emphasis, movement, and rhythm are the basis of any given visual formation or an ensemble and are part of our human psychology that helps us understand our surroundings/context and how to modify it. This could be due to our survival skills, developed during the evolution of humans, and further developed by the emotions in us. Although these pure form ensembles could sometimes bring monotony and could be boring, architects have gone beyond these basic characteristics to innovate on form development that has been able to generate forms that are even stronger emotionally by creating variety within these basics through scale, proportion, color, and texture.

Unity is an essential element of any composition that binds all the constituents of the ensemble together through some or another visual reason. This gives the entire composition a sense of completeness or singleness. The entire composition becomes a unified whole. In architecture, this is important, as since it is a single building or a group of buildings, there should be one visual stimulus, as two visual stimuli can disturb the sense of a unified whole and bring in a sense of duality or multiplicity (in case of more than two stimuli of equal importance). Unity is one of the basic elements necessary for a good composition, but architects have also defied it by adding complexity within the ideation of forms. The Interlace Condominium in Singapore and the Bahai Temple in New Delhi are examples where a single visual stimulus or a unit or an element can bind the whole composition (Figure 2.15). This can also be seen as additive transformation.

FIGURE 2.15 Single visual stimulus binding the whole composition.

FIGURE 2.16 Balancing the elements in design.

Balance is one of the design principles that have varied emotional consequences; it can, at times, be very simplistic, but at times it can generate strong emotions. It has three application strategies, symmetrical, asymmetrical, and radial. Through balance, an ensemble where two visual elements are similar or may be opposing in nature appear to be equal or in harmony either through an axis or proportion. Balance and proportion are complementary to each other and go hand in hand. For example, the Petronas Twin Towers in Kuala Lumpur and the Mika Twin Towers in Iran (Figure 2.16), or it could be the Government Center in Brasilia, Brazil (Figure 2.17), have a symmetrical and asymmetrical balance, respectively. Asymmetrical balance could be achieved using design elements like color, shape, mass, scale and proportion, orientation, placement, materials, light and dark values. Auriville is an example of radial balance (see Figure 2.18), where balance has been achieved through a central focal point and the balancing of elements around it. This is more complex as compared to symmetrical and asymmetrical balance.

FIGURE 2.17 Building with asymmetric balance.

FIGURE 2.18 Radial symmetry: Auroville vision plan.

Contrast is an ensemble of contradictory elements, as these contradictory elements attract attention and portray that a whole is created not just by unity and similar elements but by contradictory elements as well. This creates visual interest in the composition. Contradictions have always impressed humans as they liven up a composition, as both the contradicting elements, in a way, give meaning to each other and break the monotony in a given ensemble (Polatoglu, 2009; Gungor, 2005, 1972). It's the juxtaposition of elements that are strongly different from one another either through shape, color, texture, weight, size, etc. The contrast in each ensemble could be bold or soft, depending on the elements used in the ensemble. The Royal Ontario Museum is a particularly good example of two contrasting styles: an element of its rich past that can be seen along with a very modernist approach. The rich past was built in 1914 in a neo-Romanesque style and was later renovated in an art deco–inspired form. Daniel Libeskind added a multi-million-dollar extension to that art deco–inspired form which was made primarily of glass, aluminum, and steel (Lorenzi and Francaviglia, 2010). Another example is the Sant Francesc Church (Santpedor, Spain), originally built in the early 18th century by Franciscan priests and abandoned by the 19th century. Architect David Closes redesigned it in 2000 by adding an eye-catching entrance to the building that was in sharp contrast with the rest of the built form (see Figure 2.19).

Harmony is one of the ways to create and assess aesthetics, as it is an objective feature that helps in composing elements of design that have participation and harmony among those elements, whether aesthetical or functional. The architecture of any given built form must have harmony between the aesthetics of the spaces and the functional requirements. They must be beautiful and, at the same time, must be useful. So harmony among various design elements present in an ensemble also must complement the functional requirement of the spaces designed. Also, the ensemble needs to be in harmony with its surroundings. Fallingwater by Frank Lloyd Wright is a particularly good example of visual harmony in architecture where it was built to integrate with the natural landscape around (Figure 2.20).

Rhythm is an element taken from music, and when applied to architecture by human cognition of repetition through elements of design, it gives life or movement to the ensemble, as it affects movement and direction. It's a consistent repetition of elements of design that create a sense of order, visual appeal, and movement. A sequential repetition of elements like colors,

FIGURE 2.19 Design with contrast.

FIGURE 2.20 Fallingwater.

forms, shapes, and lines and visualizing variations in the elements make the rhythm more complex. It also creates a sense of balance and harmony with a flow and sometimes fluidity (direction and movement) that has a visual appeal. Rhythm in architecture has the capacity to unify repetitive elements in any given ensemble and to set the dynamic behavior of the given built form, through repetition of shape, color, tone, texture, accents, direction, and dynamic. Rhythm is a morphological organization of elements within the built form that sets the elements into motion through the recurrence of these elements, resulting in an organized movement in space and time. These organization and repetition could be of shape, color, tone, texture, accents, and direction that give the structures an element of motion.

There are many examples of rhythm at all points of time in history. For example, the Indian temples at Rameshwaram and Madurai both have a rhythm in the pillared corridor where repetition through pillars as an element creates a rhythmic order (Figure 2.21). Modern examples of **rhythm i**n architecture could be seen with variation in the sequential repetitions. Heydar Aliyev Centre by Ar. Zaha Hadid consists of swooping curves and flowing spaces that become an intrinsic component of its architectural composition and generate a rhythmic form. The Aqua Tower, on the other hand, generates a rhythmic façade through additive protrusions from the basic, which is a cuboid. Similarly, the Sydney Opera House (Australia) is an example, where repetitive form with a variation is used to give rhythm of repetition in the design (see Figure 2.22).

It is natural among humans that their attention is drawn to many elements in each building at once. To draw attention toward where the architect wants the user to land, emphasis is created through the visual element of domination**. Emphasis** in architecture is when a focal point of any architectural design ensemble is created that draws visual attention. It could also be seen as a domination effect in any given ensemble. This could be through shape, color, material, proportion, and its placement within the ensemble. This principle creates a focus in the ensemble that commands visual

FIGURE 2.21 Design with rhythm.

FIGURE 2.22 Rhythm of repetition in design.

FIGURE 2.23 Dallas's Perot Museum of Nature and Science.

attention but, at the same time, does not dominate the overall form. Emphasis very well goes with other design principles. Dallas's Perot Museum of Nature and Science, by Morphosis, can be seen as a particularly good example of emphasis and also of contrast (see Figure 2.23).

2.7 GESTALT PRINCIPLES AS A VISUAL RULE FOR ARCHITECTURAL COMPOSITION

The visual pleasure that we get from seeing any object is subjective to how we perceive it. In a way, it's all about individualistic perception. The perception theory is given by the Gestalt (means "form and shape" in the German language) perception theory, presented by Wertheimer, Kohler, and Kafka introduced a new definition of aesthetical perception with its various principles (Hancerlioglu, 2008). Gestalt's theory portrays a comprehensive cognitive experience and does not separate or divide the visual perception into small components, as discussed previously, but tries to understand the form and shape that these design elements create (Koffa, 1963; Polatoglu, 2009). Gestalt theory is not a theory but a set of theories on how the human brain perceives visual information and interpretates it. The human brain has the quality of working holistically, parallelly, and analogously, as it has the ability of self-organization. It groups similar elements, recognizes patterns among them, and has a tendency to simplify complex ensembles. The fundamental principles of Gestaltism are as follows.

2.8 PRINCIPLE OF SIMILARITY: HUMAN PERCEPTION OF GROUPING SIMILAR ELEMENTS

The principle of similarity can be explained by the human perception of grouping elements or things of a similar nature. Similarity of elements can be used as a visual tie between elements that might not be right next to each other in a design. Within a composition of elements or in any architectural ensemble, the objects that look alike, our brain groups them (again, this is perceptual, not literal). These elements could be through shape, size, color/texture, or other qualities. The principle of similarity is grounded in the idea that things with similar visual characteristics, such as shape, size, color, texture, value, or orientation, could be perceived as belonging together. It means that if a person perceives the set of elements, he or she tends to group the ones that have one or several features alike as the related whole (Figure 2.24).

2.9 PRINCIPLE OF PROXIMITY

The principle of proximity is the human perception of elements. It is assumed as more related to each other if they are closer.

The principle of proximity emphasizes the physical location of any given component or feature of a given architectural ensemble as it creates a spatial relationship of the given feature with other features. Related components when grouped are viewed as a group, and components that are not similar or are unrelated should be placed in a manner that they don't break the spatial relationship of the similar components. When the components of a given ensemble are close together, they are likely to be perceived as part of the same group or represent the same whole, reflecting similar functionality or traits and, as a result, when together create a more effective visual impact (Figure 2.25). As seen in the figure, the position of doors, although placed

FIGURE 2.24 A house in Japan (4 × 4).

FIGURE 2.25 The Oval, Japan.

at different distances, creates a visual rhythm around the pool by appearing to be in a group. If these features are in a visual order, they tend to become a visual group, attract more attention, and then are understood differently (Hancerlioglu, 2008; Divanlıoglu, 1997). The principle of common fate can also be seen as similar to proximity, as similar forms in a visual order; when they tend to move in one direction, they also are seen as a group.

2.10 PRINCIPLE OF CONTINUITY: CAPACITY TO ALIGN ELEMENTS THROUGH THE PATH OF ANY ENSEMBLE

Humans tend to take the smoothest path, whether it be in any given field. In a visual composition, when the elements of the ensemble are on a given line or curve, they seem to be more related as compared to when they are not, as our brain tends to take the smoothest visual trajectory rather than the disjointed or discontinuous ones (Figure 2.26). This could also be called continuity in visual grammar, wherein elements that stay uninterrupted (maybe through proximity) and continue to be connected on the back of each other, the visual perception of such ensembles has the quality of clarity and distinctiveness, with a more regular and effective form (Divanlıoglu, 1997; Gungor, 1972).

2.11 PRINCIPLE OF CLOSURE: FILL IN THE GAPS AND OPT FOR SIMPLICITY

The principle of closure comes from the ability of our mind to fill the gaps or blanks to perceive an object where the object is not complete in itself. This happens due to the external stimuli that match the object. This brings a sense of closure or completeness, as when we look at a complex architectural ensemble, we try to identify a single recognizable pattern. In other words, when any object, image, or form has missing parts, our mind will fill in those missing parts and make a complete image so a pattern can be recognized.

In the shapes seen in Figure 2.27, if the boundaries of the shapes and forms that we observe are endless and fully covered (where start and endpoints meet) by the form that they represent, the perception is clear and distinct. The figure of the three

FIGURE 2.26 Continuance grouping: continuation of direction.

FIGURE 2.27 Gaps: external stimuli bringing a sense of closure.

FIGURE 2.28 External stimuli fill in the gap to bring a sense of closure.

objects which do have gaps and are not complete in itself, but the external stimuli bring a sense of closure and, by identifying the pattern makes a complete image. A similar principle has been applied in the ideation of forms in architecture (see Figure 2.28).

2.12 PRINCIPLE OF FIGURE AND GROUND

This principle is based on segment visuals and classified into two components: figure and ground. The principle of figure and ground is based on the perception of the

FIGURE 2.29 Figure and Ground.

FIGURE 2.30 Structure of walls and background space creating a relationship.

human mind that identifies an object or figure from its background (Figure 2.29). This means that visual percept of humans perceive objects either by their background or by their foreground. Similarly in architectural ensembles either a given built form stands out or recedes into the background (Figure 2.30). For Each form in such a relationship the ground is part behind the form and does not attract but can establish the superiority of the form and gives a sense of depth or emphasis (Denel, 1981).

2.13 PRINCIPLE OF SYMMETRY: PEOPLE TEND TO SEEK BALANCE

The human brain tends to look for a kind of order in objects when symmetrical elements are perceived as a unified whole. Objects are made of elements of identical or equivalent parts, and the ensemble looks unified due to symmetry and balance. A very common example is our human face, which is a creation that has balance, resulting in symmetry. The Gestalt principle of symmetry perceives that any given symmetrical element is part of a unified group. People tend to look for order in objects through symmetry, which is one of the ways by which the human mind perceives the visual property of elements that are made up of equivalent parts through proportion and balance. A common example would be the human face, which has two equivalent elements, like eyes and ears, and human hands and legs, which create a balance-forming symmetry, but at the same time, the face is seen as a unified whole. The elements in any architectural ensemble could be through form, color, texture, value, direction or movement, etc.

3 Materials and Loads

3.1 CLASSIFICATION OF LOADS

Buildings are exposed to various loads that significantly impact their design, safety, and performance. These loads—such as wind, dead, live, and earthquake—are critical in determining the structural integrity and lifespan of buildings. In addition, buildings are also subjected to accidental loads arising from fire and explosion, causing severe damage. While buildings are primarily designed based on permanent and operational loads, accidental loads are also considered in the analysis to assess safety under untoward incidents (Dadashzadeh et al., 2013; Dellchatsios, 1984).

Loads acting on the buildings are inherently unpredictable, with a high degree of uncertainty. However, load factors recommended by the code help quantify them with reasonable accuracy. But still, accidental loads are estimated only in the probabilistic sense, causing uncertainty in the structural design. Hence, as an architect, one should be aware that the analysis and design process is not deterministic but carried out with some degree of probability/uncertainty. This concern and understanding should ensure a commitment to high-quality construction so that the building or construction project is delivered to the client with the best possible capacity.

Detailed studies based on available data help determine the key parameters that define design load values. According to US and Norwegian regulations, the average recurrence interval for a design event should be 100 years, while British regulations require at least 50 years. The structural design accounts for multiple load combinations, including wind, seismic, dead, live, impact, mass, and damping, ensuring the robustness and resilience of the structure.

Loads acting on a structure can be broadly classified as follows:

- **Permanent loads**, or dead loads, are the ones that act on the structure as a constant static load throughout its lifetime.
- **Operating loads**, or live loads, are the variable loads that depend on operational conditions and can fluctuate over time.
- **Environmental loads** include wind and earthquakes, significantly influencing the structure.
- **Construction and installation loads** are temporary loads encountered during the construction and installation stages of the structure.
- **Accidental loads** refer to the unexpected loads that arise from unforeseen accidents or extreme events, such as fire, explosion, etc.

3.2 GRAVITY LOADS

Gravity loads, also called dead loads, play a crucial role in the design and functional use of buildings. These loads consist of the structure's weight, including any permanently installed equipment. The main challenge with gravity loads is ensuring the

DOI: 10.1201/9781003581314-3

building remains stable under all conditions, which is accomplished through precise design and construction practices. The structure must be engineered to evenly distribute the gravity loads across its foundation, with a clean load transfer path, preventing any one part from bearing excessive weight, which could compromise its integrity. Calculating gravity loads involves determining the weight of each component of the structure, which is as follows:

$$W = m \times g \tag{3.1}$$

Where **W** is the weight (or gravity load in newton), **m** is the mass of the member (in kilogram), and **g** is the acceleration due to gravity (9.81 m/s^2). Once the gravity loads are calculated, they are used to design the structure's foundation. The foundation must be able to support these loads without excessive settlement or movement, as any instability could result in structural failure.

3.3 WIND LOADS ON BUILDINGS

Wind loads can generate substantial lateral forces on buildings. Estimating wind loads on buildings ensures their structural safety, functionality, and cost efficiency. Wind forces, especially in regions prone to high winds, storms, or extreme weather, can exert significant lateral pressure on structures, affecting their stability. Proper wind load estimation helps architects design buildings that can withstand these forces, reducing the risk of collapse or structural damage. It also ensures that non-structural elements, such as facades, remain intact under high wind conditions, preventing potential hazards. Furthermore, accurate wind load calculations balance safety and cost, preventing conservative design and leading to financial and safety implications. By following standardized methods outlined in building codes like IS 875 Part 3, wind loads can be accurately calculated, ensuring compliance with safety standards. It minimizes the risk of structural failures and enhances long-term durability and functionality.

Indian code (Bureau of Indian Standards, 1987) provides guidelines for calculating wind loads on buildings and structures in India. It offers a step-by-step approach to determine the wind pressure and forces acting on a structure. The wind load calculation considers factors like wind speed, terrain, height of the structure, and its shape.

The basic wind speed (V_b) is the speed for a region measured 10 m above ground level in open terrain with few obstructions. The basic wind speed depends on the location and can range from 33 m/s to 55 m/s across India. The design wind speed (V_z) at height z is calculated as:

$$V_z = V_b \cdot k_1 \cdot k_2 \cdot k_3 \cdot k_4 \tag{3.2}$$

where k_1 is the risk coefficient, which accounts for the probability factor; k_2 accounts for the terrain, height, and structure size factor; k_3 accounts for the topography factor; and K_4 is the importance factor to account for cyclonic regions, in particular. The risk coefficient considers the likelihood of the design wind speed being exceeded.

For most structures, it ranges from 0.92 to 1.08 (IS 875), with a value of 1.0 for general structures. The terrain height factor and size factor account for the terrain roughness and its influence on the size of the structure (IS 875). It is classified into four categories:

Category 1: Open terrain with very few obstructions (e.g., coastal areas)
Category 2: Open terrain with a few isolated obstructions
Category 3: Terrain with numerous closely spaced obstructions (e.g., urban areas)
Category 4: Terrain with large, closely spaced obstructions (e.g., city centers)

The topography factor accounts for the effects of hills, valleys, or cliffs. K_3 is taken as 1 for flat terrain. For structures constructed on sloped ground and hilly terrains, it is computed using the code guidelines (IS 875). The importance factor is used when designing structures in cyclonic areas. It ranges from 1.0 to 1.3, depending on the structure's importance. For general structures, k_4 is taken as 1, unless specified otherwise.

3.3.1 Design Wind Pressure (P_z)

The design wind pressure (P_z) at height z is related to the design wind speed using the formula:

$$P_z = 0.6 \cdot V_z^2 \tag{3.3}$$

where P_z is the wind pressure at height z (N/m^2), and V_z is the design wind speed at height z (m/s).

3.3.2 Wind Force on Structures (*F*)

The total wind force on a structure is calculated by multiplying the design wind pressure with the effective area on which the wind acts. The formula is:

$$F = C_f \cdot A \cdot P_z \tag{3.4}$$

where F is the wind force (N), C_f is the force coefficient given in the code (see, for example, IS 875), and A is the effective area (area projected in the direction normal to wind) of the structure (m^2). The force coefficient (C_f) depends on the shape and size of the structure and can be seen in the code (IS 875). For example, flat-roofed buildings have this value in the range 1–1.4, and it depends on the building's aspect ratio. For cylindrical structures, it varies with the wind direction and approach angle of attack.

3.3.2.1 Wind Load on Roofs

For roofs, wind load calculations are slightly different. The wind pressure acting on the roof depends on the slope of the roof. It is calculated using the formula:

$$F_r = C_p \cdot A_r \cdot P_z \tag{3.5}$$

where F_r is the wind force on the roof (N), C_p is the pressure coefficient for roofs (from IS 875), A_r is the projected area of the roof (m^2), and P_z is the wind pressure at height z (N/m²). The pressure coefficient (C_p) varies with roof shape and slope. For a pitched roof, C_p values may range between 0.7 and 1.2, depending on the roof's angle and aspect ratio. The effective roof area (A_r) is the horizontal projection of the roof area that faces the wind. If the structure is slender or flexible, dynamic effects must be considered in the analysis. This includes accounting for wind-induced vibrations, such as vortex shedding, which can cause oscillations and fatigue. Additional considerations for wind-induced sway and deflections are important for tall buildings, where the height-to-width ratio exceeds 5. The dynamic response of such buildings is evaluated as gust factor approach. For structures like towers and chimneys, codes provide specific guidelines to calculate wind loads based on their slenderness and height.

3.3.2.1.1 Example 1

Consider a building with a basic wind speed (V_b) of 47 m/s, located in terrain category 2, and at a height of 20 m. Calculate the wind force acting on the structure, following the steps outlined in IS 875 Part 3.

3.3.2.1.2 Solution

Step 1: Calculate the design wind speed (V_z). The **design wind speed** (V_z) at height z is calculated using the formula:

$$V_z = V_b \cdot k_1 \cdot k_2 \cdot k_3 \cdot k_4$$

Where:

V_b = basic wind speed = 47 m/s
k_1 = risk coefficient = 1.0
k_2 = terrain and height factor = 1.03 (for 20 m height in terrain category 2, from Table 2 of IS 875 Part 3)
k_3 = topography factor = 1.0 (assuming flat terrain)
k_4 = importance factor = 1.0
$V_z = 47 \times 1.0 \times 1.03 \times 1.0 \times 1.0 = 48.41$ m/s

The design wind speed at a height of 20 m is **48.41 m/s**.

Step 2: Calculate the wind pressure (P_z). The wind pressure is calculated using the formula:

$$P_z = 0.6 \cdot V_z^2$$

Where:

P_z = wind pressure (N/m²)
V_z = design wind speed = 48.41 m/s

Substituting the value of V_z:

$$P_z = 0.6 \cdot (48.41)^2 = 0.6 \cdot 2344.56 = 1405.73 \text{ N/m}^2$$

The wind pressure on the structure at 20 m height is **1405.73 N/m²**.

Step 3: Calculate the wind force (*F*). The total wind force acting on the structure is calculated using the formula:

$$F_r = C_f \cdot A_r \cdot P_z$$

Where:

F = wind force (*N*)

C_f = force coefficient = 1.2 (from Table 5 of IS 875 Part 3 for a rectangular building)

A = effective area of the structure = 200 m^2

P_z = wind pressure = 1405.73 N/m^2

Substituting the values:

$$F = 1.2 \times 200 \times 1405.73 = 337375.2 \text{ N}$$

The total wind force acting on the structure is **337,375.2 N**, or approximately **337.38 kN**.

3.4 IMPACT LOADS

Impact loads arise from a sudden collision or contact between two bodies. In the context of buildings, these could be caused by falling objects, vehicle collisions, or any sudden dynamic event that transfers force to the structure. While selecting the construction materials, even ductile materials considered (based on uniaxial tensile tests) can exhibit brittle behavior under specific conditions (Chandrasekaran and Jain, 2016; Chandrasekaran, 2013b). This is especially true when components are exposed to impact or sudden dynamic loads. Therefore, if a part is expected to experience impact loads, assessing the material's impact strength becomes a crucial consideration in the design process.

Impact strengths are measured by Charpy or IZOD tests. Although it is difficult to relate the results of these tests directly to design requirements, in general, materials that have high-impact resistance also have high strength, high flexibility, and high toughness. These impact strengths are more helpful in comparing various types of grades of metals within the same material family and not as beneficial in comparing one metal with another.

Studying impact loads on concrete and steel columns in buildings is crucial for understanding how critical structural elements respond to sudden dynamic forces, such as vehicle collisions (Chandrasekaran, 2019a, 2024). Impact loads can lead to significant internal stresses, deformations, and even progressive collapse, particularly in multi-story structures with different material compositions, such as reinforced concrete columns in underground car parks and steel columns on aboveground floors. By analyzing the behavior of reinforced concrete and steel columns under impact conditions, engineers can better evaluate the strengths and vulnerabilities of these materials (Chandrasekaran and Srivastava, 2017). Concrete columns may respond differently than steel columns, particularly regarding stress distribution and energy absorption (Chang et al., 2006). Using equivalent static and dynamic methods, engineers can simulate impact forces and assess the potential for structural failure, helping to improve building designs for enhanced safety and resilience. These insights are critical in areas such as parking garages or ground floors, where the likelihood of impact is higher. Adhering to design guidelines and standards ensures structures are better equipped to resist accidental effects and reduce the risk of progressive collapse (Chandrasekaran et al., 2021).

3.4.1 Critical Concepts

- **Nature of Load:** Impact loads are dynamic, short-duration forces that can induce local damage (at the point of contact) and global responses (vibrations or displacements throughout the structure).
- **Potential Consequences of Stress Concentration:** The area of contact experiencing a high-stress concentration can lead to localized failures, such as cracking, denting, or penetration. Understanding this potential consequence is vital in structural design.
- **Energy Absorption:** Buildings must be designed to absorb the kinetic energy from the impact without significant damage.

Impact stresses caused by the fall of objects pose substantial difficulties to the integrity and strength of the building components; this is common in industrial structures, where small accidents are quite common. The accidental or unintentional release of things, equipment, or materials from heights, such as cranes, machinery, drill pipes, tools, and equipment, causes impact loads. These loads can apply rapid and concentrated forces on the floor/rook, resulting in localized damage, deformation, and structural failure. They result in excessive slab deformation; sudden collapse can jeopardize building safety. It also imposes risks to personnel safety and leads to potential loss of life. In addition, the downtime, output losses, and cost of repairs/replacements are very high.

3.4.2 Challenges

Structural Design Considerations: Buildings can bear impact loads if designed regarding to material qualities, structural layouts, and energy absorption methods. Traditional materials, such as steel and aluminum, may have limits in terms of energy absorption and impact resistance.

Material Selection: Functionally graded materials (FGMs) present attractive alternatives for improving the impact resistance of industrial constructions. FGMs have distinct properties that vary spatially, enabling material qualities to be adjusted to specific performance requirements. Engineers can improve energy absorption capacities by adding FGMs into topside decks and structural components without increasing thickness or weight.

Modeling and Analysis: Advanced numerical simulation methods, such as finite element analysis (FEA) software like ABAQUS Explicit, simulate and analyze structures' response to impact loads. These simulations allow engineers to forecast structural behavior, deformation patterns, stress distribution, and failure modes, which helps them optimize design parameters and material selection.

3.5 BLAST LOADS

Blast loads refer to the extreme pressure generated by explosions, which propagate from the explosion source in all directions. These loads create high-magnitude, short-duration pressure waves that can cause structural and non-structural damage to buildings.

3.5.1 Critical Concepts

- *Blast pressure* refers to the sudden release of energy, creating a shock wave. It is characterized by a rapid rise in pressure, followed by a gradual decrease, causing the structure to deform and lead to failure.
- *Blast wave phases* possess distinct domains. The *positive phase* is the high-pressure section that hits the structure first, causing an outward push. The *negative phase* is the phase at which a subsequent drop occurs in pressure, causing a suction effect. *Reflected pressure zone* is the section where the blast loads are amplified at surfaces perpendicular to the blast wave. These are known as reflected pressure.

Blast loads can cause explosions for various reasons, including gas leaks, equipment failures, or malicious acts. These explosions unleash a large amount of energy in a short period, resulting in high-pressure shock waves that travel throughout the surrounding environment. It can result in a fire accident. Combustion during fire produces heat, pressure, and potentially explosive gases, which add to blast loads and thermal effects on structural components (Hall, 2014; Hemmatian et al., 2015; Purandare, 2016; Purser, 2002; Thomas et al., 1972).

3.5.2 Effects on Structures

Structural Damage: Blast loads apply abrupt and extreme pressure to structures, causing structural damage, distortion, and failure. Primary structural components, such as columns, beams, and braces, may distort, burst, or collapse under high-loading conditions.

Dynamic Response: Structures respond dynamically to blast loads with vibrations, oscillations, and resonance effects. The structure's dynamic behavior is determined by its intrinsic frequencies, damping properties, and interactions with the surrounding environment.

Fragmentation and Debris: Explosions and blast loads can cause fragmentation and debris, posing extra risks to persons, equipment, and neighboring sites. Fragmentation can spread over long distances, resulting in secondary repercussions, including collateral damage to the offshore facility and surrounding infrastructure.

3.5.3 Design Considerations

Material Selection: Buildings are often built using high-strength materials, like steel or reinforced concrete, to withstand blast pressures and dynamic loading conditions. These materials provide strength, ductility, and fatigue resistance.

Structural Configuration: The design and layout of structures are optimized to reduce the impact of blast loads. The structural components are designed to minimize blast propagation, reduce blast reflection, and effectively dissipate energy.

Blast-Resistant Design: To improve blast load protection, buildings have specialized design features, such as blast walls, blast-resistant windows, and reinforced enclosures. These elements help contain explosions, redirect blast energy, and safeguard vital infrastructure and personnel.

3.6 EARTHQUAKE LOADS

Structures in seismic regions are typically designed for two levels of earthquake intensity, namely, strength-level and ductility-level earthquakes. The structure is designed to respond elastically to the strength-level earthquake, defined as having a "reasonable likelihood of not being exceeded during the platform's life" (mean recurrence interval ~ 200 – 500 years). For the ductility-level earthquake, defined as close to the "maximum credible earthquake" at the site, the structure is designed for inelastic response and adequate reserve strength to avoid collapse.

For strength-level design, seismic loading may be specified either by sets of accelerograms or through design response spectra. Design spectra have advantages over time history solutions (base acceleration input). For this reason, design response spectra are the preferable approach for strength-level designs. Designs for ductility-level earthquakes will usually require inelastic analyses. The seismic input must be specified by sets of three-component accelerograms, natural or artificial, representing extreme ground motions that could shake the platform site. However, the characteristics of such motions may still be prescribed through design spectra, which are usually the result of a site-specific seismotectonic study.

3.6.1 Example Study

This section presents a case study on the Loma Prieta earthquake, a significant seismic event that struck the San Francisco Bay Area in 1989. It was one of the most destructive earthquakes in US history, prompting extensive research into earthquake preparedness and response. The case study will delve into the construction of the response spectrum, a crucial tool in seismic analysis that graphically represents structures' peak responses to various ground motion frequencies. By examining the Loma Prieta earthquake and the application of response spectrum construction, this section aims to provide insights into earthquake engineering and enhance our understanding of structural dynamics during seismic activities.

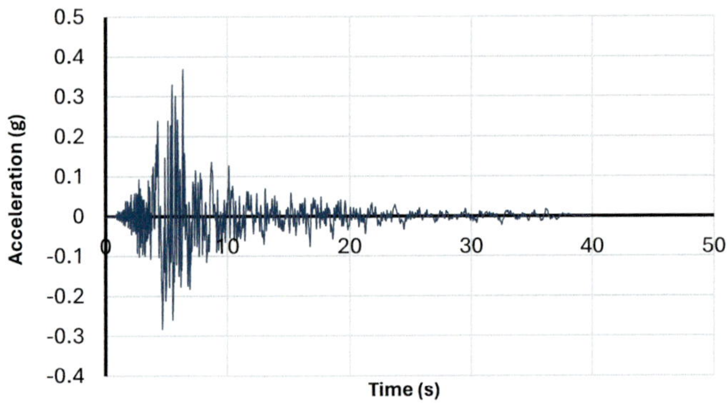

FIGURE 3.1 Accelerogram of Loma Prieta earthquake.

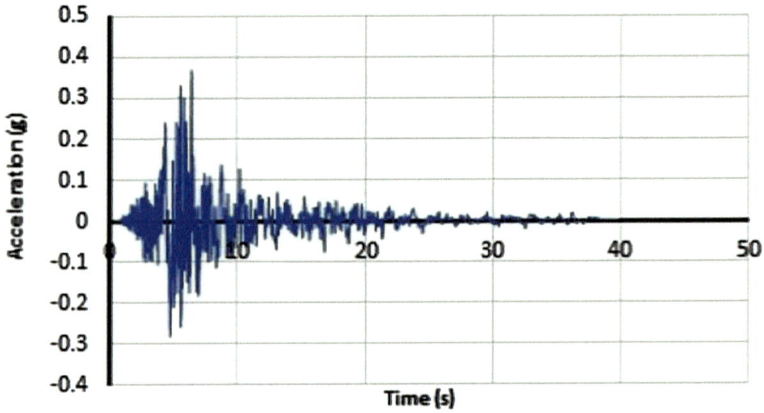

FIGURE 3.2 Power spectral density of Loma Prieta earthquake.

3.6.1.1 Construction of Response Spectrum

1. Define the ground acceleration numerically. Ground motion coordinates are normally defined every 0.02 s.
2. Determine the natural vibration period (T_n) and damping ratio (ζ) of an SDOF system.
3. Compute the deformation response $x(t)$ of this SDF system.
4. Determine x_o, the peak value of x*(t)*.
5. The spectral ordinates are $D = x_o$, $V = \dfrac{2\pi}{T_n} D$, and $A = \left(\dfrac{2\pi}{T_n}\right)^2 D$.

From Figure 3.3, peak deformation x_0 is determined. For example, the value in the figure refers to 32.4 cm. Subsequently, using Figure 3.4, the corresponding

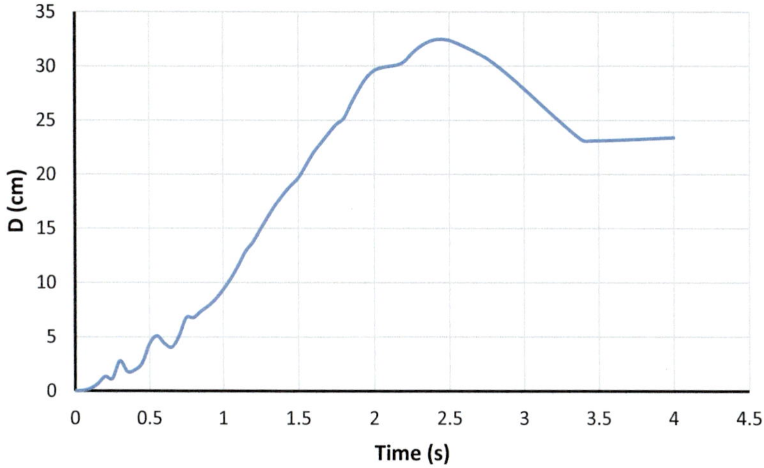

FIGURE 3.3 Deformation response spectrum for the Lomo Prieta ground motion; $\zeta = 5\ \%$.

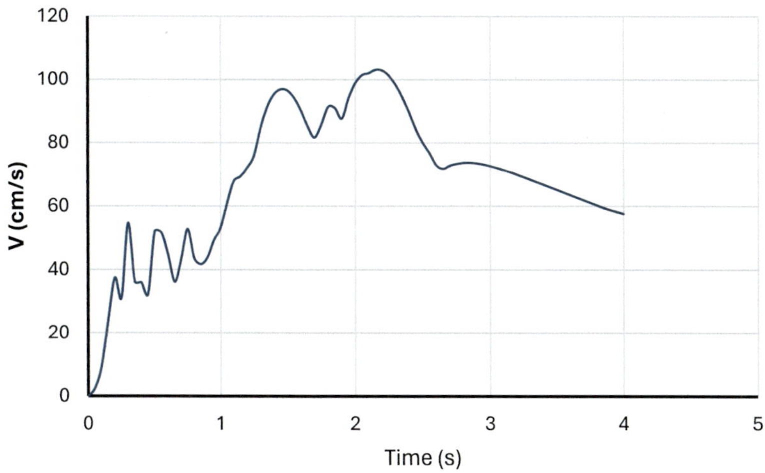

FIGURE 3.4 Velocity response spectrum for the Lomo Prieta ground motion; $\zeta = 5\ \%$.

velocity V is $=$ 84.15 cm/s. Considering natural frequency as 1.5 s, the spectral coordinates are:

$$D = x_0 = 32.4\,cm$$

$$V = \frac{2\pi}{T_n} D = 135.73\,cm/s$$

$$A = \left(\frac{2\pi}{T_n}\right)^2 D = 568.62\,cm/s^2$$

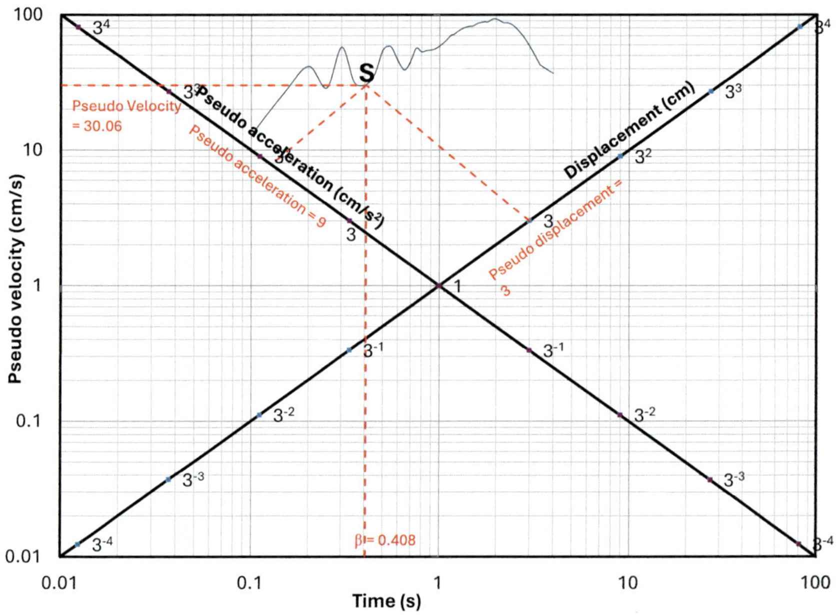

FIGURE 3.5 Combined *D–V–A* response spectrum for the Lomo Prieta ground motion; $\zeta = 5\%$.

The response spectra for a given ground motion, including deformation, pseudo-velocity, and pseudo-acceleration, include identical information. The three spectra provide similar information about structural response. Equations $V = \dfrac{2\pi}{T_n} D$ and $A = \left(\dfrac{2\pi}{T_n}\right)^2 D$ can be used to derive the other two spectra based on the first one (displacement).

One of the advantages of the combined DVA plot is a combined visualization of displacement, velocity, and acceleration response spectrum; each spectrum directly represents a physically significant quantity. The deformation spectrum describes a system's highest deformation. The equation $\left(E = \dfrac{mV^2}{2}\right)$ shows a direct relationship between the pseudo-velocity spectrum and the peak strain energy stored in the system after an earthquake. The pseudo-acceleration spectrum is directly related to the peak value of equivalent static force and base shear ($V_b = f_s = mA$). The second reason is that using all three spectral values together makes it easier to approximate the spectrum shape for design reasons. A combined plot of all three spectral quantities is preciously valuable.

3.6.2 Importance in Building Design

1. **Structural Safety:** Earthquake loads can cause significant structural damage or collapse if not adequately accounted for. Designing a structure to withstand seismic forces ensures the safety of its occupants.

2. **Serviceability:** Even if a structure does not collapse, earthquake loads can cause excessive deformation, rendering the building unsafe or unusable. The seismic design ensures that the structure remains functional after an earthquake.
3. **Economic Loss:** Poorly designed buildings are more likely to suffer extensive damage during an earthquake, leading to costly repairs or replacements. Earthquake-resistant design minimizes the financial impact of such events.
4. **Code Compliance:** In seismically active regions, building codes mandate that structures be designed to resist earthquake loads. In India, **IS 1893 Part 1 (2016)** provides guidelines for calculating earthquake loads and designing buildings to resist them.

3.7 EARTHQUAKE LOAD CALCULATION

Indian Standard **IS 1893 Part 1 (2016)** provides the procedure for calculating earthquake forces on buildings. It includes methods such as the equivalent static method and the response spectrum method to determine seismic forces.

Its key parameters are thus:

1. **Zone Factor (Z):** India is divided into seismic zones with different seismic hazard levels. The zone factor represents the expected intensity of ground shaking. It ranges from **0.10 (low seismicity)** to **0.36 (high seismicity)**, per Table 2 of IS 1893 (2016).
2. **Importance Factor (I):** This reflects the importance of the structure. Important buildings like hospitals and schools are of higher importance. The value of **I** typically ranges from **1.0 to 1.5**.
3. **Response Reduction Factor (R):** This factor accounts for the structure's ductility and energy absorption capacity. Ductile materials and systems can dissipate energy more effectively. The value of **R** ranges from **3 to 5**, depending on the type of structure.
4. **Soil Type:** The soil type beneath the building also influences the seismic response. The soil types are classified as type 1 (rock or hard soil), type 2 (medium soil), and type 4 (soft soil).
5. **Fundamental Natural Period (T):** The natural period is when the building completes one vibration cycle. It is calculated based on the height and type of structure. For moment-resisting frames, the empirical formula is:

   $\mathrm{T} = 0.075 \times h^{0.75}$ (for RC moment-resisting frames)

   $\mathrm{T} = 0.085 \times h^{0.75}$ (for steel moment-resisting frames)

   where h is the height of the building in meters.

6. **Design Horizontal Seismic Coefficient (Ah):**

$$A_h = \frac{Z}{2} \cdot \frac{I}{R} \cdot \frac{S_a}{g} \tag{3.6}$$

   where Z is the zone factor, I is the importance factor, S_a is the average response acceleration coefficient, and R is the response reduction factor.

7. **Base Shear (V_B):** The total design lateral force, or base shear (V_B), is given by:

$$V_B = A_h \cdot W \quad (3.7)$$

where **W** is the seismic weight of the building (total dead load and appropriate portions of live load).

3.7.1 Importance of Base Shear in Building Design

Base shear is the total horizontal force exerted at the base of a structure due to seismic activity. It represents the overall seismic load that a building must resist during an earthquake. Calculating the base shear is essential in the structural design of buildings, particularly in earthquake-prone areas, to ensure the structure's stability and safety. Here is why the base shear calculation is crucial.

3.7.2 Distribution of Seismic Forces

The base shear represents the total seismic force acting on the structure, which is distributed along the height of the building. Engineers can determine how these forces are distributed to different building parts, such as columns, beams, and shear walls, by calculating the base shear. The forces are typically most significant at the base and decrease toward the top. This helps design each structural element to withstand the forces they encounter during an earthquake.

3.7.3 Designing for Lateral Forces

Earthquakes primarily exert lateral forces (side-to-side motion) on structures. These forces can cause buildings to sway, tilt, or collapse. Base shear is the starting point for understanding the magnitude of these lateral forces. By calculating the base shear, structural engineers can:

- Design bracing systems or shear walls to resist lateral loads
- Ensure that foundation elements can support the lateral forces without excessive movement
- Prevent structural elements like columns and beams from failing due to excessive bending, twisting, or shearing

3.7.4 Reinforcement Requirements

The base shear helps determine how much reinforcement is needed in the structure. For example, reinforced concrete buildings require proper steel reinforcement in beams, columns, and slabs to resist the lateral forces caused by earthquakes. By calculating the base shear, engineers can:

- Specify the correct amount and placement of reinforcement
- Ensure the structure has sufficient ductility, allowing it to absorb and dissipate seismic energy without collapsing

- Design for seismic joints and flexible connections between building parts to accommodate movement during an earthquake

3.7.5 Ensuring Stability and Preventing Collapse

Buildings designed without considering base shear and seismic forces are more likely to collapse during an earthquake. By calculating the base shear, engineers can:

- Ensure the structural stability of the building under seismic conditions
- Prevent progressive collapse, where the failure of one structural element leads to a chain reaction of failures
- Provide sufficient rigidity and stiffness to resist deformation under lateral loads

3.7.6 Compliance with Seismic Building Codes

In earthquake-prone regions, building codes, such as IS 1893 Part 1 (2016) in India, require buildings to be designed to resist seismic forces. These codes specify the method for calculating base shear and other seismic design parameters. Compliance with these codes ensures that:

- The building is designed to meet minimum safety standards for earthquake resistance.
- The building will be able to withstand seismic events without significant damage.
- Legal and safety requirements are fulfilled, reducing the risk of fines or legal action due to non-compliance.

3.7.7 Foundation Design

The base shear directly impacts the design of the building's foundation. An adequately designed foundation must resist both the vertical forces (due to the building's weight) and the lateral forces generated by seismic activity. By calculating the base shear, engineers can:

- Ensure the foundation is strong enough to resist lateral forces without sliding or overturning.
- Design pile foundations, raft foundations, or mat foundations that provide stability even in cases of firm ground shaking.
- Avoid excessive settlement or failure of the foundation during an earthquake.

For example, consider a five-story reinforced concrete building located in seismic zone III. The height of the building is 15 m, and its seismic weight is 2000 kN. The structure is classified as an ordinary moment-resisting frame (OMRF), and the soil is classified as type II (medium soil). Calculate the base shear for the building.

3.7.7.1 Solution

1. **Zone Factor (*Z*):** For **seismic zone III**, the zone factor (**Z**) from Table 2 of IS 1893 (2016) is **0.16**.
2. **Importance Factor (*I*):** For a general residential or commercial building, **I = 1.0**.
3. **Response Reduction Factor (*R*):** For an ordinary moment-resisting frame (OMRF), **R = 3.0** (Table 7 of IS 1893 [2016]).
4. **Natural Period (*T*):** Using the empirical formula for RC moment-resisting frames:

$$T = 0.075 \times h^{0.75}$$

Given **h = 15 m**:

$$T = 0.075 \times 15^{0.75} = 0.552 \text{ seconds}$$

5. **Response Acceleration Coefficient ($\frac{S_a}{g}$):** From Figure 2 of IS 1893 (2016),

for $T = 0.0552$ seconds and type II soil, $\frac{S_a}{g} = 2.5$.

6. **Design Horizontal Seismic Coefficient (*Ah*):** Using the formula:

$$A_h = \frac{Z}{2} \cdot \frac{I}{R} \cdot \frac{S_a}{g}$$

Substituting the known values:

$$A_h = \frac{0.16}{2} \cdot \frac{1}{3} .2.5 = 0.0667$$

7. **Base Shear (V_B):** The total design base shear is:

$$V_B = A_h \cdot W$$
$$V_B = 0.0667 \cdot 2000 = 133 \cdot kN$$

This base shear is now distributed along the height of the building using a suitable method (e.g., the linear distribution method or the parabolic distribution method), and the design of structural components—such as shear walls, columns, beams, and foundations—is adjusted accordingly. This ensures that:

- Each component can carry the expected seismic loads.
- The building is structurally sound and can withstand seismic forces without failure.

3.8 FIRE LOAD

Fire load measures the total amount of combustible material within a building or structure that can contribute to a fire (ASCE, 2005; Assael and Kakosimos, 2010;

Babrauskas, 1981; Babrauskas and Grayson, 1992). It is typically expressed as the energy (in megajoules, or MJ) released during combustion per unit area (MJ/m^2). This value accounts for the fire load energy density (FLED), including the building materials and the contents within a space, such as furniture, equipment, and other combustibles (Wickstrom, 1986, 2016). Fire load plays a critical role in fire-resistant building design, as it helps determine a fire's potential severity and duration in a given area (Behnam, 2019; Bergman et al., 2011; BIS, 1979; Brode, 1959; Arturson, 1987). Understanding the fire load allows architects and engineers to plan for fire protection measures, such as fire-rated walls, sprinkler systems, and evacuation strategies, while complying with safety standards and regulations (Buchanan and Abu, 2017; Budnick et al., 1997; Buchanan, 2001; NBC, 2016; NFPA, 2020, 1997).

Steel is a widely preferred construction material for industrial structures due to its strength and versatility (BS EN 10025–2, 2019). While design methods for conventional loads, such as dead loads, live loads, and environmental forces, like wind, waves, currents, and earthquakes, are well-documented, special attention is required for designing against extraordinary loads, like blasts, fire, and impact. Industrial structures are particularly vulnerable to accidental explosions, and although such incidents are rare, they carry high risks, with potentially devastating consequences. These accidents can lead to severe financial losses, loss of life, and significant environmental and public safety damage.

Fire occurs due to a rapid, exothermic oxidation reaction fueled by solid, liquid, or gaseous materials (Drysdale, 1998). The onset of fire releases energy, intensifying over time until it reaches its peak. Explosions, which can result from fire, involve the rapid expansion of gases, often triggered by pressure or shock waves. These waves move at high speeds, leading to adiabatic expansion (Chandrasekaran and Srivastava, 2022). Explosions can be mechanical or chemical, representing a significant hazard, particularly in offshore facilities where hydrocarbons are involved. It is estimated that around 70% of fire accidents result in explosions, with severe consequences (Jin and Jang, 2015).

A primary concern for designers is ensuring fire resistance. The risks are significantly heightened when dealing with liquefied natural gas (LNG), as its physical and chemical properties differ significantly from those of other hydrocarbons (Quintiere, 2006). Common oversights contributing to fire-related incidents include improper safety measures, inadequate fire protection systems, and a lack of risk assessments (Manco et al., 2014). These factors emphasize the importance of stringent safety protocols and robust design strategies to mitigate the risks of fire and explosions in offshore and industrial structures.

- All heat transfer systems should be carefully inspected to ensure no leaks or signs of smoke.
- If smoke is detected, avoid disturbing smoldering insulation; doing so may introduce excess air and cause auto-ignition.
- Prevent fluids from contacting heat sources, which could lead to ignition.
- If fluid leaks into the system and becomes trapped, it can oxidize and generate heat through an exothermic reaction.

Following these precautions can help prevent fire accidents. However, it remains essential to implement fire-resistant designs for structures at risk of fire incidents (Chandrasekaran and Srivastava, 2017).

3.8.1 Importance of Fire Load in Building Design

1. **Fire Resistance Design:** The fire load helps assess the fire resistance level required for building materials and structural components. Higher fire loads mean materials with higher fire resistance ratings must be used.
2. **Fire Safety Measures:** Knowing the fire load helps in designing appropriate fire suppression systems (e.g., sprinklers, extinguishers), as well as fire detection systems, like smoke alarms.
3. **Structural Integrity:** In the event of a fire, the structure should withstand the heat long enough for occupants to evacuate and for fire services to control the fire. A high fire load can lead to faster temperatures, putting more stress on structural elements.
4. **Evacuation Planning:** Fire load calculations also impact the planning of evacuation routes and fire doors, ensuring that escape paths remain safe for as long as possible during a fire.

3.8.1.1 Fire Load Energy Density (FLED)

The fire load energy density (FLED) can be calculated using the following formula:

$$\text{Fire Load Energy Density (MJ/m}^2\text{)} = \frac{\sum\left(\text{Mass of combustibles} \times \text{Calorific value}\right)}{\text{Floor area}}$$

Where:

The mass of combustibles is measured in kilograms (kg).

Calorific value is the energy released when the combustible materials burn, expressed in MJ/kg.

Floor area is the total area of the space where the combustibles are located, in square meters (m^2).

The following methodology is used to estimate the fire load energy density.

3.8.1.1.1 Direct Weighing

Weighing is the most direct way to obtain the mass of items. This method needs surveyors to physically enter the building to weigh the mass of explosive objects. It is easy to weigh small items, such as books and chairs. Large items, like heavy furniture, are difficult to weigh. Besides, this technique applies to movable fire load, which is impractical for fixed combustibles.

3.8.1.1.2 Inventory Survey

In the inventory survey, there are two ways to obtain the mass of an item. One way is to measure the object's dimensions, and then the mass is determined by multiplying the measured volume with the corresponding density. The other way is to weigh the commonly encountered items in advance or get their weight from the manufacturers and then create an inventory list. The problem is that sometimes, it is difficult to determine the volume of irregularly shaped items and the density of various materials.

3.8.1.1.3 The Questionnaire Survey

A printed paper questionnaire or an electronic one is designed first consisting of questions to be answered by occupants.

3.8.1.1.4 Website Review Survey

Fire load surveys can be made through real estate website information. This technique estimates the fire load according to the photographs uploaded to the website.

3.8.1.2 Example 1

Consider a 50 m^2 office room containing 200 kg of wooden furniture. The calorific value of wood is 18 MJ/kg. Calculate the fire load of the room.

3.8.1.2.1 Solution

Given:

- Mass of combustibles (wood) = 200 kg
- Calorific value of wood = 18 MJ/kg
- Floor area of the room = 50 m^2

Fire load is given by:

$$\text{Fire load} = \frac{200 \times 18}{50} = 72\,\text{MJ/m}^2$$

The fire load of the office room is $72\,\text{MJ/m}^2$.

3.8.1.3 Example 2

A storage room has an area of 100 m^2 and contains 500 kg of paper and 100 kg of plastic. The calorific value of paper is 16 MJ/kg, and the calorific value of plastic is 35 MJ/kg. Calculate the fire load of the room.

3.8.1.3.1 Solution

Given:

- **Mass of paper** = 500 kg
- **Calorific value of paper** = 16 MJ/kg
- **Mass of plastic** = 100 kg
- **Calorific value of plastic** = 35 MJ/kg
- **Floor area of the room** = 100 m^2

Step 1: Calculate the total energy contribution from the paper and plastic.

$$\text{Energy from paper} = 500 \times 16 = 8000\,\text{MJ}$$

$$\text{Energy from plastic} = 100 \times 35 = 3500\ \text{MJ}$$

$$\text{Total energy} = 8000 + 3500 = 11{,}500\ \text{MJ}$$

Step 2: Calculate the fire load.

$$\text{Fire load} = \frac{11500}{100} = 115\ \text{MJ/m}^2$$

The fire load of the storage room is 115 MJ/m^2.

3.9 CLASSIFICATION OF FIRE

Fire is initiated when a leak or spill of a flammable substance occurs in the presence of an ignition source. Fires can be categorized into four main types: **pool fire**, **jet fire**, **fireball**, and **flash fire** (Chandrasekaran, 2016a). Additionally, fires can be further sub-classified into specific types, such as **flares**, **fire on the sea surface**, and **running liquid fire**. These sub-classifications can be incorporated into the main categories: (1) Flares can be considered jet fires for modeling purposes, and (2) fire on the water surface and running liquid fires can be considered pool fires.

3.9.1 Pool Fire

A **pool fire** is a turbulent diffusion fire that burns above a pool of vaporizing hydrocarbons with low momentum. When liquid fuel is released, it forms a pool on the surface which evaporates and ignites, resulting in a pool fire. The likelihood of a pool fire on an offshore platform is exceptionally high due to the constant handling of hydrocarbons. Pool fires can also occur accidentally during overfilling storage tanks or from ruptured pipelines and cracks in storage tanks caused by metal corrosion. The pool's diameter is determined by the diameter of the bund, which is constructed to contain the spread of the pool fire and is expressed through the following relationship:

$$D_p = \sqrt{\frac{4A}{\pi}} \tag{3.8}$$

where A is the area of the bund in m^2, and D_p is the pool's diameter. The following relationship gives the pool fire length:

$$L = 42\, D_p \left[\frac{\text{Burning Rate}}{\rho_{air}\sqrt{9.81 D_p}} \right]^{0.61} \tag{3.9}$$

3.9.2 Jet Fire

A **jet fire** is characterized by the turbulent diffusion of flames caused by fuel's continuous release and combustion, which has significant momentum to propagate in the downwind direction. Jet fires can severely affect offshore installations, even far from the fire's source. As the jet fire propagates, it releases gases, which can occur in either the horizontal or vertical direction. Of the two, horizontal jet fires are more dangerous as they can cause extensive damage on the downwind side. The potential consequences of a jet fire include structural failure, failure of storage vessels, and failure of pipework. The heat flux released during a jet fire ranges from **200 to 400 kW/m²**, depending on the type of fuel involved. Pressurized gas pipelines are a common source of jet fires, and in the event of a leak, the initial gas release rate is given by the following equation:

$$Q_0 = C_D A p_0 \sqrt{\frac{MV}{RT_0}\left(\frac{2}{r+1}\right)^{(r+1)(r-1)}} \quad \text{if } p_0 > p_a\left[\frac{2}{r+1}\right]^{\left[\frac{r-1}{r}\right]} \tag{3.10}$$

where C_D is the discharge coefficient, A is the area in m², P_0 is the operating pressure of the gas, M is the molecular weight of gas in g/mol, V is the rate of specific heat, R is the universal gas constant $(= 8314 \text{ J/kg mol k})$, T_0 is the operational temperature in Kelvin, and p_a is the absolute pressure.

The Chamberlain equation gives the flame length of a jet fire:

$$m = 11.14(Q_0)^{0.447} \tag{3.11}$$

where Q_0 is the initial release rate in kg/s.

Jet fire length and the corresponding time frame are estimated based on the following relationship:

$$Q_t = Q_0 e^{\left[\frac{Q_0}{M_G}\right]t} \tag{3.12}$$

$$M_G = \frac{PM}{0.08314}\pi r^2 L \tag{3.13}$$

where M_G is the mass of gas in kilograms, P is the operating pressure of gas in pa, M is the molecular weight of gas in gm/mol, r is the diameter of the pipe in meters, L is the length of the pipe in meters, and t is the time of release in seconds.

3.9.3 Fireball

A **fireball** is a rapid, turbulent fuel combustion, typically resulting in a rising and expanding ball of radiant fire. When a fireball impacts a vessel or tank containing pressurized liquefied gas, the internal pressure increases, potentially leading to the catastrophic failure of the vessel or tank. This can result in the loss of the entire inventory stored within. In boiling liquid expanding vapor explosion (BLEVE) cases, the released material is flammable and may ignite, leading to a blast and significant thermal radiation hazards. The heat pulse generated during a BLEVE event typically lasts 10–20 s, causing severe damage. The maximum emissive power resulting from a BLEVE ranges from 270 to 333 kW/m^2 in up- and downwind directions and from 278 to 413 kW/m^2 in the crosswind direction (Hemmatian et al., 2017).

3.9.4 Flash Fire

A **flash fire** is a transient fire that occurs when a gas or vapor cloud ignites. It results from a significant delay between flammable materials' release and subsequent ignition (Feasey and Buchanan, 2002). Initially, a vapor cloud forms over a large area, expanding radially before igniting, leading to an explosion. Flash fires are highly destructive and can cause widespread damage. A distinctive feature of a flash fire is the wall of flame it produces. Like fireballs, a flash fire can ignite and persist as a continuous flame. It can also occur due to delayed ignition and lasts for an extended period. The immediate effect of a flash fire is intense thermal radiation, often triggering "knock-on" events, such as pool fires, jet fires, and BLEVEs. The severity of a flash fire is notably high, making it a significant hazard.

Fire can be classified based on the type of fuel burning; it is also sometimes referred to as fuel classification. Class A includes wood, paper, cloth, trash, and plastics, but not metals. Class B includes flammable liquid, gases, oil, and grease. Class C includes fully charged and energized electrical equipment, or those connected to a live power line. Class D includes metals, potassium, sodium, aluminum, and magnesium. Fire classification helps choose an appropriate extinguisher type. The wrong choice shall worsen the fire scenario. Fire is mainly classified as pool fire, jet fire, fireball, and flash fire, while the sub-classifications include flare, fire on liquid surface, and running liquid fire (Chandrasekaran, 2016a, 2016b). However, these sub-classifications can be treated similarly to the main modelling and numerical simulation (Chandrasekaran and Kiran, 2014a, 2014b). For example, flare is idealized as a jet fire, and fire on liquid surface as a pool fire (Chandrasekaran and Kiran, 2015).

3.10 EXPLOSION

Explosion behavior is influenced by many parameters, namely, ambient temperature and pressure, explosive material composition and its physical properties, nature of ignition source, duration of ignition, amount and turbulence of the explosive material, time before ignition, and the release rate of the explosive material (Crowl and Louvar, 2002).

To understand the consequences of explosion impacts, one needs to understand the dynamics of pressure waves. They propagate in air, termed as blast waves, which is followed by a strong wind. A shock front is expected from highly explosive materials, such as TNT, which generally occurs as a result of sudden rupture of a pressure vessel. In such cases, the maximum pressure above the ambient pressure is generated and termed as peak overpressure. It is very important to understand how this pressure variation is quantified as the blast wave passes. If it is measured normal to the blast wave propagation direction, then it is termed as side-on overpressure, also known as free-field overpressure. If it is measured facing the incoming shock waves, then it is termed as reflected overpressure. The latter is generally twice in magnitude of the former and can reach as high as even eight times. Explosions result in a blast or a pressure wave moving out from the explosion center at the speed of sound. Shock wave or overpressure is the basic cause for damage. Missiles or projectiles are other important source of damage. *Damage* is the function of the rate of pressure increase and duration of blast wave. Blast damage is estimated based on the peak side-on overpressure.

Figure 3.6 shows the variation of peak overpressure with time. The area under this curve is the measure of severity of explosion. As seen in the figure, an explosion generates a rapid rise in pressure, while the propagating wave causes damage to the objects located on its path. It is then followed by negative pressure wave, which causes further damage before pressure returns to atmospheric. Thus, damage depends on the maximum pressure reached, the velocity of propagation, and environmental characteristics.

Explosive force is released in a direction normal to the surface of the explosive, but if released in mid-air, then the direction of the blast will be 360°. The speed of an explosive reaction is very rapid, resulting in the thermal expansion of gases, causing high pressure (Dubnikova et al., 2005). The generation of heat in large quantities accompanies most explosive chemical reactions. The rapid liberation of heat causes instantaneous expansion of gases, generating high pressure. It is important to note that the liberation of heat at a slower pace will not cause an explosion. A substance that burns less rapidly (slow combustion) evolves more heat in comparison to those that detonate rapidly (fast combustion). In the former, slow combustion converts more of the internal energy of the burning substance into heat released to the

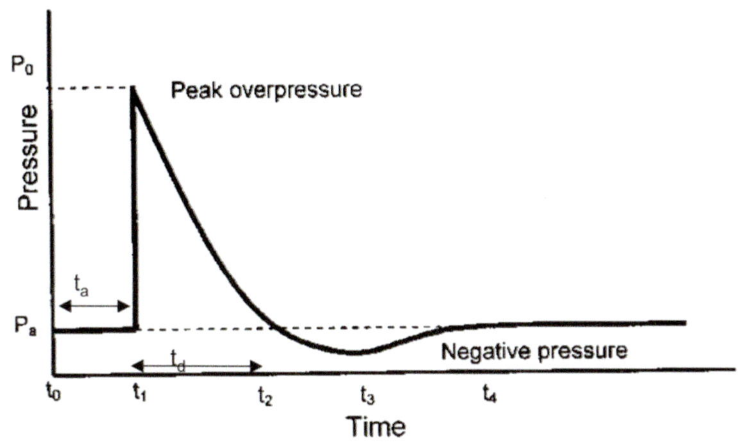

FIGURE 3.6 Variation of overpressure with time.

Source: Courtesy of Chandrasekaran (2016a, 2016b).

TABLE 3.2
Human Effects Caused by Explosion

Effect of Explosion	Damage
Probability of eardrum injury	Eardrum rupture is a direct effect of overpressure during an explosion.
Probability of death due to lung damage	Explosions can cause a sudden pressure difference between the inside and the outside of the lungs, due to which the thorax is pressed inward, causing lung damage and possible death.
Probability of death from head impact	A shock wave can push the head of a person backward, resulting in skull rupture or fracture, or even the collision of the head with another stationary or non-stationary object.
Probability of death due to whole-body displacement	A shock wave can throw the whole body backward, causing death because of its impact with other objects.

surroundings. In the latter, fast combustion converts more internal energy into work on the surroundings. A chemical explosive, under heat or shock, decomposes very rapidly and yields gas and heat. An explosive reaction must be capable of being initiated by the application of shock, heat, or a catalyst to a small portion of the mass of the explosive material. Major human effects caused by fire and explosion are summarized in Table 3.2 (Assael and Kakosimos, 2010).

Studies also showed that the overpressure that resulted from the explosion is not significant in open areas. Further, a complex geometric layout of a process plant or an offshore drilling unit can initiate high-explosion overpressure (Dadashzadeh et al., 2013).

Table 3.3 summarizes the main causes and frequent locations of accidents in such plants and various contributing factors (Lees, 1996). It is evident that about 18% of fires are due to release and overflow of flammable gases and/or liquids. Fire contributes about 20% to the total loss, while the rest of the major contributions come from the explosions (Faisal and Abbassi, 1999; ISO, 2012; James, 2016). Fire and explosion accidents in the maritime industry arise from four major groups of activities, namely, human error, mechanical failure of plants and equipment, thermal reaction, and electric faults (Til et al., 2018; Kawagoe, 1958); their percentage contributions are reported as 48, 22, 14, and 7, respectively. The major factors that contribute to fire and explosions due to human error are poor maintenance scheduling, inadequate training and education, and lack of situation awareness. Other factors that contribute to mechanical faults are error in maintenance works, metal fatigue, excess heat generation, and faulty operational conditions. Short circuit, poor connection and maintenance, and use of defective devices are major reasons for electric faults causing fire and explosions.

3.11 DAMAGE CONSEQUENCES OF FIRE AND EXPLOSION

One of the most common methods used to quantify the consequences that arise from explosions is in terms of trinitrotoulene (TNT) (Chandrasekaran, 2016a, 2016b; Khan and Srivastava, 2018). TNT is an explosive with a few special characteristics. It can rapidly change its state from solid to hot expanding gas state (Aven and Vinnem, 2007). Elements of carbon, oxygen, and nitrogen, present in TNT, produce highly

TABLE 3.3
Causes and Factors of Explosion in Process Industries

Description	Proportion
Main cause	
Chemical reaction uncontrolled	20.0
Chemical reaction accidental	15.0
Combustion explosion in equipment	13.3
Unconfined vapor cloud	10.0
Overpressure	8.3
Decomposition	5.0
Combustion sparks	5.0
Pressure vessel failure	3.3
Improper operation	3.3
Others	16.8
Frequent location of occurrence	46.7
Enclosed process or manufacturing buildings	31.7
Outdoor structures	6.7
Yard	3.3
Tank farm	3.3
Boiler house	8.3
Others	26.7
Contributing factors	18.3
Rupture of equipment	18.3
Human element	11.7
Improper procedures	11.7
Faulty design	11.7
Vapor-laden atmosphere	8.3
Congestion	6.7
Flammable liquids	5.0
Long replacement time	5.0
Inadequate combustion controls	
Inadequate explosion relief	

stable substances with a strong bonding between them while it burns. However, as TNT explosions are chemically unstable, significant energy is not required to break this bond. In an explosion damage scenario, it is a common practice to estimate the scaled distance of the spread of explosion energy. To compute this, the equivalent mass of TNT needs to be estimated. Following are the steps involved in estimating the scaled distance.

Step 1: Estimate the flammable inventory present in the scenario. This is done by calculating the total mass (m) of the flammable fuel participating in the explosion.

Step 2: Determine the energy of explosion ΔH_c.

Step 3: Fix the efficiency of explosion process (η), which depends on the site-dependent parameters that initiate the explosion process. In the absence of any detailed studies, this factor can be assumed in the range of 1 to 15%.

Step 4: Compute the equivalent mass of TNT using the following relationship:

$$m_{TNT} = \frac{\eta\, m\, \Delta H_c}{E_{TNT}} \tag{3.14}$$

where *ETNT* is the energy of explosion of TNT $(= 4686\ \text{kJ/kg})$.

Step 5: Determine the scaled distance (*Ze*) using the following relationship:

$$Z_e = \frac{r}{\sqrt[3]{m_{TNT}}} \tag{3.15}$$

where *Ze* is expressed in m/kg1/3, where *r* is the distance of the point of concern from the explosion source (in meters).

Step 6: Based on the scaled distance estimated in the previous step, one can estimate the resulting overpressure (P_0), either graphically (please see Figure 1.4) or using the following relationship:

$$P_0 = P_s \times P_a \tag{3.16}$$

where P_s is the scaled overpressure, which can be calculated using Figure 3.7 for the corresponding scaled distance. P_a is the atmospheric pressure, whose equivalent values for a variety of explosion activities are given in Table 3.4 (Chandrasekaran, 2016a, 2016b).

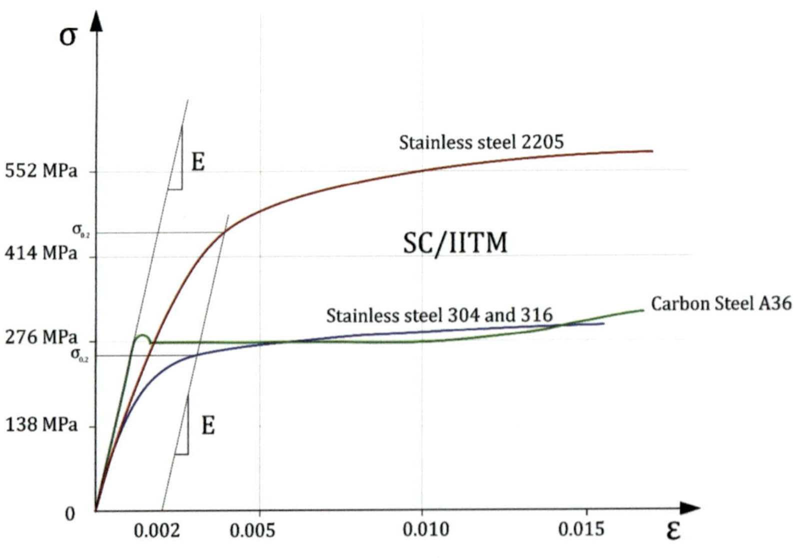

FIGURE 3.7 Stress–strain curve for carbon and stainless steel.

TABLE 3.4
Equivalent Atmospheric Pressure for Explosion Activities

Pressure P_a (kPa)	Quantifiable Damage Arising from Explosion
0.28	Loud noise, like glass breaking (about 143 dB)
0.69	Breakage of small windows
3.4–3.6	Windows shattering; occasional damage to window frames
4.8	Minor damage to structure
6.9–13.8	Significant damage to wooden furniture and asbestos sheets
15.8	Lower limit of significant damage
17.2	Complete destruction of brick masonry buildings
27.6	Rupture of claddings in industrial structures, oil tank ruptures; 50% probability of human fatality
34.5–48.2	Partial destruction of framed structures
62	Damage of loaded wagons of cargo trains
68.9	Total destruction of RC-framed structures; damage of heavy machinery
75	Probability of 90% human fatality; complete damage of steel structures

Alternatively, overpressure can also be computed using the following relationship:

$$P_0 = P_a \left\{ \frac{1616\left[1+\left(\frac{Z_e}{4.5}\right)^2\right]}{\sqrt{\left[1+\left(\frac{Z_e}{0.048}\right)^2\right]\times\left[1+\left(\frac{Z_e}{0.32}\right)^2\right]\times\left[1+\left(\frac{Z_e}{1.35}\right)^2\right]}} \right\} \tag{3.17}$$

3.12 MATERIALS

Material properties are crucial to enable an appropriate design technique. Although many modern structural systems are designed to develop their capacity from geometric form, material strength is inherent to achieving the desired capacity. Though steel is one of the most favorable materials for heavy industrial structures, alternate materials also show the equivalent performance of steel at elevated temperatures (Chandrasekaran et al., 2021). Form-dominant structures, like large domes, shell roofs, folded plates, cable-stayed bridges, floating offshore platforms, semi-submersibles, and nuclear power plants, are unique and expensive in design, installation, and functions. In addition to the complexities that arise from the innovative geometry, the combination of several environmental loads makes the design more complex.

The legislation calls for periodic certification of their functional safety, which, by and large, depends on the fact of material strength. While design procedures ensure no immediate collapse and catastrophic failure of structures even under critical loads, material degradation under the environment's influence poses a consistent challenge in assessing their reliability. All these factors result in a heavy penalty on the existing weight, which otherwise governs the design of form-dominant structures. Furthermore, using the input data of materials at elevated temperatures and fire is also one of the critical design considerations to achieve a high level of reliability.

Different types of materials used in building construction serve a variety of purposes, namely, installation, repair, rehabilitation, corrosion protection, etc. The choice of material from a wide range, like composites, concrete, mild steel, copper, aluminum, fiber-reinforced plastics, titanium, and buoyancy materials, depends on the purpose of construction; all the aforementioned cannot be used for the same purpose (Chandrasekaran and Jain, 2016). Therefore, the selection of materials for building construction is a complex task, as no single material characteristic shall help rank the materials. Detailed knowledge about their structural characteristics and relevant code compliance is necessary to arrive at a practical design.

The behavior of materials under high temperature, fire, explosion, and impact is another critical design consideration, as structures are designed for a high level of reliability (Chandrasekaran and Srivastava, 2017; Chandrasekaran, 2017, 2015b, 2016a, 2016b; Wielgosz et al., 2014). Materials used in form-dominant structural systems should possess unique characteristics, besides being compatible with extreme weather conditions. In addition, materials used in marine construction need to possess unique characteristics besides being compatible with extreme weather conditions. Therefore, design procedures are also dominated by material characteristics in addition to the critical combination of various environmental loads (Chandrasekaran and Bhattacharya, 2012; Chandrasekaran, 2014).

A strong bond exists between engineering materials and the type of structures constructed using them. Referring to the guidelines suggested by various international and Indian codes of practice for structural design, material selection is only desirable but not mandatory. Under the given choice of a wide variety of materials available for use, a few parameters govern their choice: type of applications, properties under regular cyclic loads, high and very low temperature, recycling characteristics, non-toxic nature and bio-friendly characteristics, sustainability for the service life, and code compliance. In particular, under elevated temperatures, these complexities increase manifold. It is important to note that conventional design procedures are governed by the codes, but the use of advanced materials is always welcome. Therefore, the choice of appropriate material for construction becomes a critical engineering decision. The use of advanced materials is always welcome, even if their strength-related characteristics are referent values in a particular model.

As it is well understood that, under fatigue loads, the magnitude of the stress only follows the severity of failure induced by the number of stress cycle reversals, even low-magnitude stress under a large number of cycles can cause failure that can be lesser than the yield strength. The degradation of material properties in the marine environment is a major concern. Some of them are, namely, chemical and strength degradation, loss of fatigue strength, poor performance due to high stress concentrations, corrosion, and bio-fouling effects. Therefore, materials should possess survivability under accidental loads and the capacity to withstand large hydrostatic pressure generated under hurricanes. An intelligent choice of material under such special loads as fire should focus on the fact that the material should not initiate a catastrophic failure. Even if the failure occurs at a lesser probability, it should not be sudden or immediate.

3.13 STRUCTURAL STEEL

The properties of structural steel result from its chemical composition and method of manufacturing. The significant properties of steel from a designer's perspective are strength, toughness, ductility, durability, and weldability. These mechanical properties vary based on their chemical composition, heat treatment, and manufacturing processes (Chen et al., 2017). The addition of small quantities of other elements to its major constituent, iron, influences the mechanical properties of steel. While the strength can be increased by the addition of alloys, such as manganese, niobium, and vanadium, ductility, toughness, and weldability will be compromised. Minimizing sulfur enhances ductility, while the addition of nickel improves toughness. It is therefore important to arrive at a carefully balanced chemical composition to ensure that desired structural characteristics are achieved. Similar to that, the alloying elements, heat treatment, and manufacturing processes are of critical importance to the performance of steel. Rolled or formed steel shows higher strength, but increased thickness reduces its yield strength.

The effect of heat treatment also governs its strength and toughness. The common manufacturing processes are, namely, as-rolled, normalized, normalized-rolled, thermo-mechanically rolled (TMR), and quenched and tempered steel (Q&T). Steel cools as it is rolled, with a typical rolling finish temperature of about 750°C. If it is allowed to cool naturally, then it is called *as-rolled steel. Normalized steel* is manufactured by heating as-rolled steel up to about 900°C, held at that temperature for a specific time before being allowed to cool naturally. Normalizing refines the grain size and enhances toughness. *Normalized-rolled steel* is manufactured by rolling steel at a temperature above 900°C, which eliminates the extra process of reheating the material. Normalized and normalized-rolled steels have an "N" designation.

Thermo-mechanically rolled steel is manufactured at a lower rolling finish temperature of about 700°C. A greater force is required to roll the steel at lower temperatures, but the properties remain unaltered (BS EN 1993, 2005). This classification of steel has an "M" designation. *Q&T steel* is manufactured at about 900°C, at which it is rapidly cooled or "quenched" to produce steel with high strength and hardness but low toughness. This classification has a "Q" designation. Quenching involves cooling a product rapidly by immersion directly into water or oil. Tempering is the second stage of heat treatment to temperatures below the austenitizing range, which softens the steel and also makes them tougher and more ductile. The use of high-tensile steel can reduce the volume of steel, but the steel needs to be tough at operating temperatures. Therefore, higher-strength steels require improved toughness and ductility, which can be achieved only with low-carbon clean steels under the thermo-mechanical rolling process (TMR). Table 3.5 shows the minimum yield and tensile strength of commonly used hot-rolled steel grades in structural design (BS EN 10025–2, 2019). An *S* designation, as seen in the table, indicates a carbon steel. The minimum yield strength and the tensile strength are taken as

TABLE 3.5
Minimum Yield and Tensile Strength of Hot-Rolled Steel

	Yield Strength (N/mm²) For Nominal Thickness (in mm)				Tensile Strength (N/mm²) For Nominal Thickness (in mm)	
Grade	**t ≤ 16**	**16 ≤ t ≤ 40**	**40 ≤ t ≤ 63**	**63 ≤ t ≤ 80**	**3 ≤ t ≤ 100**	**100 ≤ t ≤ 150**
S275	275	265	255	245	410	400
S355	355	345	335	325	470	450

characteristic yield and characteristic ultimate strength for design purposes (BS EN 1993–1–1, 2005).

Stainless steel used for structural applications is classified into four groups, namely, austenitic, ferritic, duplex, and martensitic. The former is the most commonly used stainless steel for construction. For example, alloy 310 is an austenitic stainless steel used for high-temperature and corrosion-resistant applications. The strength of stainless steel ranges from 170 to 450 N/mm^2. The austenitic type has yield strength lesser than that of carbon steel, while the duplex type has more. However, the strength ratio, which is the ratio of ultimate to yield strength, is higher in carbon steel compared to stainless steel. Figure 3.7 shows the stress–strain curve for carbon steel and stainless steel. The variations in the ductility and strength can be seen in the figure. *Ductility* indicates the strain or elongation capability under tensile loading. Ductility is desirable for various reasons, namely, redistribution of stress at the ultimate limit state, reduced risk of fatigue crack propagation, and in the fabrication processes of welding, bending, and straightening.

3.14 REINFORCING STEEL

3.14.1 Mechanical Properties

Eurocode specifies a robust temperature-dependent constitutive behavior for reinforcing steel. The relations are not reproduced directly. It is parametrized by six parameters, as shown in Figure 3.8. A MATLAB program to generate the stress–strain relation is provided in rebar_constitutive.m. These relations at different temperatures are plotted in Figure 3.9.

MATLAB PROGRAM TO PLOT THE CONSTITUTIVE BEHAVIOR OF REBAR

rebar_constitutive.m

```
function [stress, strain] = rebar_constitutive(T)
% Uses Eurocode 3 definition
% T: temperature for which constitutive relation is
needed (Celsius)
```

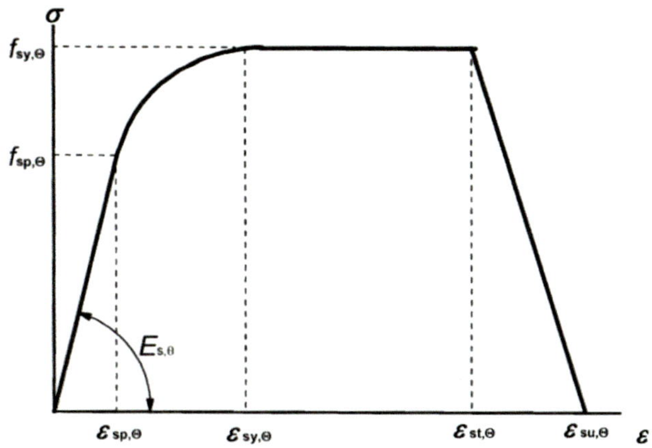

FIGURE 3.8 Constitutive behavior of reinforcing steel.

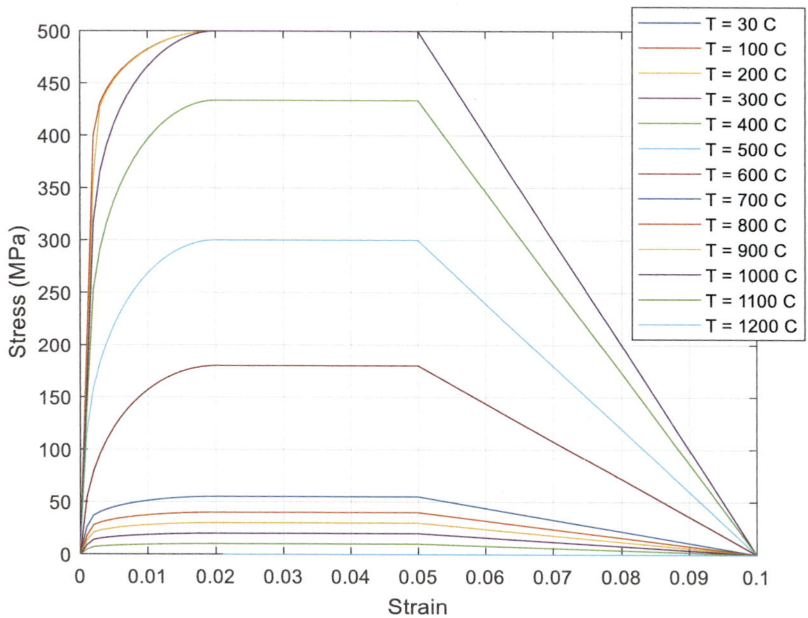

FIGURE 3.9 Constitutive behavior of reinforcing steel at different temperatures

```
% Cold rebar constitutive parameters
fp_20 = 400; %proportional limit, MPa
fy_20 = 400; %yield strength, MPa
E_20 = 200e3; %Young's modulus, MPa
% Calculate reduction factors for current temperature
```

```
[k_fy, k_fp] = rebar_strength_reduction(T);
kE = steel_modulus_reduction(T);
% Parameters of the constitutive relation curve (from
the code)
fsp_T = k_fp * fp_20;
fsy_T = k_fy * fy_20;
E_s_T = kE * E_20;
e_sp_T = fsp_T / E_s_T;
e_sy_T = 0.02;
e_st_T = 0.05;
e_su_T = 0.10;
% Don't change beyond this line
c = (fsy_T - fsp_T)^2 / ((e_sy_T - e_sp_T) * E_s_T - 2 *
(fsy_T - fsp_T));
b = sqrt(c * (e_sy_T - e_sp_T) * E_s_T + c^2);
a = sqrt((e_sy_T - e_sp_T) * (e_sy_T - e_sp_T +
c/E_s_T));
strain = 0:1e-3:e_su_T;
n = length(strain);
stress = zeros(n, 1);
for i = 1:n
  e = strain(i);
  if e < e_sp_T
    stress(i) = e * E_s_T;
  elseif e <= e_sy_T
    stress(i) = fsp_T - c + (b/a) * (a^2 - (e_sy_T - e)^2)
    ^0.5;
  elseif e <= e_st_T
    stress(i) = fsy_T;
  elseif e < e_su_T
    stress(i) = fsy_T * (1 - (e - e_st_T)/(e_su_T -
    e_st_T));
  else
    stress(i) = 0;
  end
end
```

Eurocode provides different values of reduction factor for class N when no experimental data is available, and class X when experimental data is available. Figure 3.10 shows the strength reduction factors for reinforcing and prestressing steel at high temperature. A MATLAB program used to plot the figure is also given in the following.

rebar_strength_reduction.m

```
function [ks, kp] = rebar_strength_reduction(T)
% Computes temperature-dependent reduction in strength
of rebar steel.
```

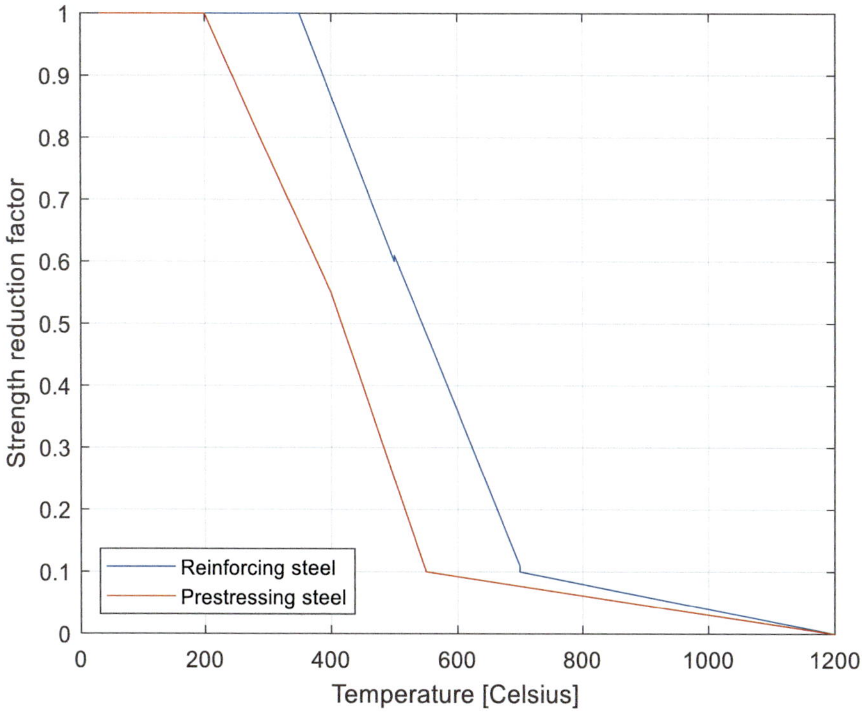

FIGURE 3.10 Strength reduction factor of reinforcing and prestressing steel.

```
% Uses Eurocode 3 definition
% T: vector of temperatures where conductivity is needed
(Celsius)
% ks: vector of reduction factors for reinforcing steel
strength (dimensionless)
% kp: vector of reduction factors for prestressing steel
strength (dimensionless)
n = length(T);
ks = zeros(n, 1); kp = ks;
% reinforcing steel
for i = 1:n
  if T(i) <= 350
    ks(i) = 1;
  elseif T(i) <= 500
    ks(i) = 1 - 0.4 * (T(i) - 350)/150;
  elseif T(i) <= 700
    ks(i) = 0.61 - 0.5 * (T(i) - 500)/200;
  else
    ks(i) = 0.10.1 * (T(i) - 700)/500;
  end
end
```

```
% prestressing steel
for i = 1:n
  if T(i) <= 200
    kp(i) = 1;
  elseif T(i) <= 400
    kp(i) = 1 - 0.45 * (T(i) - 200)/200;
  elseif T(i) <= 550
    kp(i) = 0.55 - 0.45 * (T(i) - 400)/150;
  else
    kp(i) = 0.1 - 0.1 * (T(i) - 550)/650;
  end
end
```

3.15 CONCRETE

3.15.1 Mechanical Properties

Concrete behaves differently under compression and tension. The parametrized constitutive behavior of concrete under compression is shown in Figure 3.11. It has three parameters, as mentioned in what follows. The post-peak portion can be idealized as a linear or non-linear curve, depending on the available data. Eurocode's constitutive relation is relatively straightforward and practical when compared with

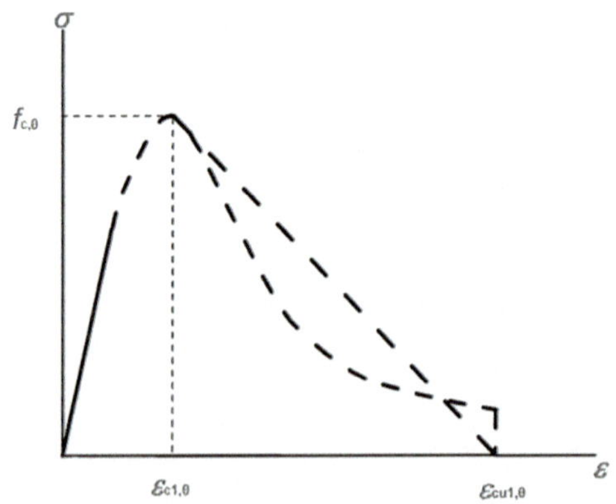

Range	Stress $\sigma(\theta)$
$\varepsilon \le \varepsilon_{c1,\theta}$	$\dfrac{3\varepsilon f_{c,\theta}}{\varepsilon_{c1,\theta}\left(2+\left(\dfrac{\varepsilon}{\varepsilon_{c1,\theta}}\right)^{3}\right)}$
$\varepsilon_{c1(\theta)} < \varepsilon \le \varepsilon_{cu1,\theta}$	For numerical purposes a descending branch should be adopted. Linear or non-linear models are permitted.

FIGURE 3.11 Constitutive behavior of concrete under compression.

the relations given by other researchers (Chang et al., 2006; Kodur, 2014). Eurocode enables both linear and non-linear post-peak behavior, while ASCE suggests non-linear behavior for the post-peak portion (Kodur, 2014). As per Eurocode, concrete tensile strength reduces to zero at 600°C, although the experiments conducted by Chang et al. (2006) suggest that the tensile strength of concrete is reduced to 35–40% at this temperature.

As concrete is generally weak in tension, tensile strength is neglected for design-related calculations. Hence, such an assumption does not have any impact on the design calculations through code-based approaches. Furthermore, detailed constitutive behavior is generally not considered in usual design considerations; instead, only tensile strength is utilized where required.

The compressive and tensile strengths of concrete decrease with the rise in temperature. While Eurocode suggests different compressive strength reduction factors for siliceous (granite, sandstone) and carbonate aggregates (dolomite, limestone), ASTM prescribes a single reduction factor. For tensile strength, both the codes provide the use of a common reduction factor regardless of the kind of aggregate used. MATLAB codes for computation of reduction factors for compressive and tensile strength of concrete are shown in **concrete_compression_reduction.m** and **concrete_tensile_reduction.m**, respectively. A graphical representation of these reduction factors is shown in Figure 3.12.

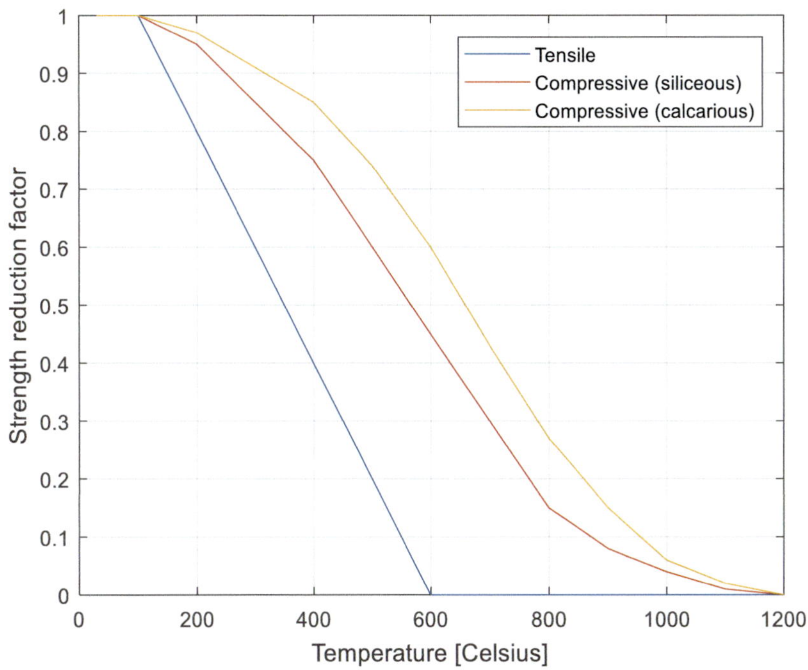

FIGURE 3.12 Reduction factors for compressive and tensile strength of concrete at high temperature.

```
concrete_compression_reduction.m
function [kcs, kcc] = concrete_compression_reduction(T)
% Uses Eurocode 3 definition
% T: vector of temperatures where conductivity is needed
(Celsius)
% kcs: vector of compression strength reduction factor
(dimensionless)
% (siliceous aggregate)
% kcc: vector of compression strength reduction factor
(dimensionless)
% (calcarious aggregate)
n = length(T);
% Eurocode 3 Table:
Tpoints = [20; [100:100:1200]'];
kcspoints = [1; 1;.95;.85;.75;.6;.45;.3;.15;.08;.04;.01;
0];
kccpoints = [1; 1;.97;.91;.85;.74;.6;.43;.27;.15;.06;
.02; 0];
kcs = interp1(Tpoints, kcspoints, T);
kcc = interp1(Tpoints, kccpoints, T);
concrete_tensile_reduction.m
function kt = concrete_tensile_reduction(T)
% Uses Eurocode 3 definition
% T: vector of temperatures where conductivity is needed
(Celsius)
% kt: vector of tensile strength reduction factor
(dimensionless)
n = length(T);
kt = zeros(n, 1);
for i = 1:n
  if T(i) <= 100
    kt(i) = 1;
  elseif T(i) <= 600
    kt(i) = 1 - (T(i) - 100)/500;
  end
end
```

It can be noted that concrete loses about half of its compressive strength at 600°C. The Eurocode model is generally conservative when compared to the experimental values. Further, the reduction of compressive strength given by Eurocode is not linear with respect to rise in temperature, unlike the ASTM model. The compressive strength of concrete reduces from 100°C as per Eurocode, while ASCE suggests no reduction in the compressive strength up to 400°C (Kodur, 2014). ASCE also suggests that the compressive strength of concrete reduces to zero at about 900°C. Thus, the temperature–rate of strength reduction as per the ASCE model is more rapid as compared to that of the Eurocode model.

3.15.2 Thermal Properties

The specific heat of concrete depends on its moisture content within the temperature range when water exists in vapour phase (Kodur, 2014; Lie, 1995; Lin, 2005). Evaporation begins at about 100°C, resulting in a sharp rise in specific heat. It gradually reduces as the temperature reaches 200°C (Pan et al., 2017). The specific heat model as per Eurocode can be calculated for different moisture content and temperature conditions for concrete, as given in the MATLAB program **concrete_specific_heat.m**. A graphical representation of the specific heat of concrete for different moisture contents is shown in Figure 3.13.

concrete_specific_heat.m

```
function c = concrete_specific_heat(T, u)
% Uses Eurocode 3 definitionconcr
% T: vector of temperatures where conductivity is needed
(Celsius)
% u: moisture content of concrete
% c: vector of specific heat (J/kgK)
n = length(T);
c = zeros(n, 1);
% Table for cpeak interpolation
cp_peak_moisture_points = [0; 1.5; 3];
```

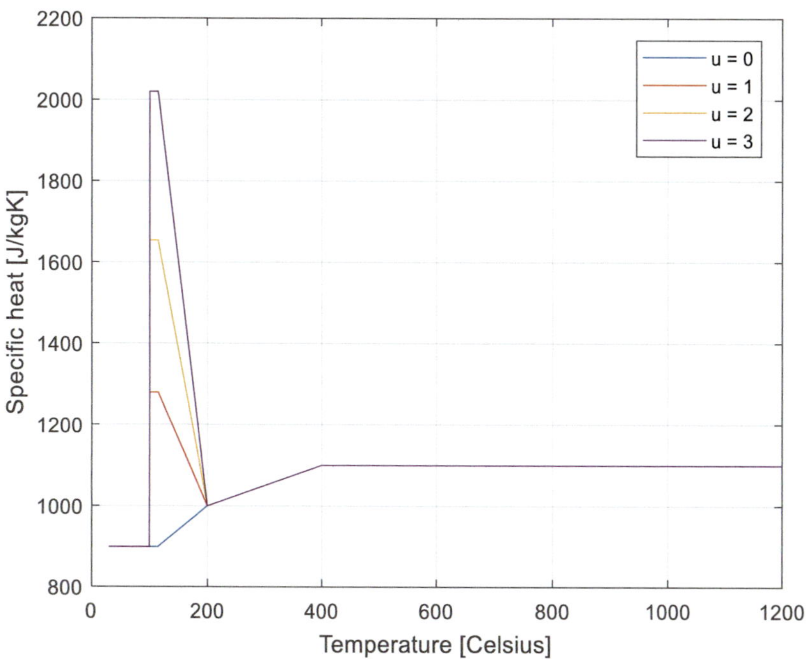

FIGURE 3.13 Specific heat of concrete.

```
cp_peak_values = [900; 1470; 2020];
cpeak = interp1(cp_peak_moisture_points, cp_peak_values, u);
for i = 1:n
  if T(i) <= 100
    c(i) = 900;
  elseif T(i) <= 200
    % values depend on moisture content in this
    temperature range
    if T(i) <= 115
      c(i) = cpeak;
    else
      c(i) = interp1([115; 200], [cpeak; 1000], T(i));
    end
  elseif T(i) <= 400
    c(i) = 1000 + (T(i) - 200)/2;
  else
    c(i) = 1100;
  end
end
```

While ASCE provides different models for assessing the thermal capacity (i.e., ρc) of concrete, Eurocode provides different models for assessing specific heat (C) and density (ρ) separately. The model for reduction in density as per Eurocode is given in the MATLAB program **concrete_density_reduction.m** and shown in Figure 3.14.

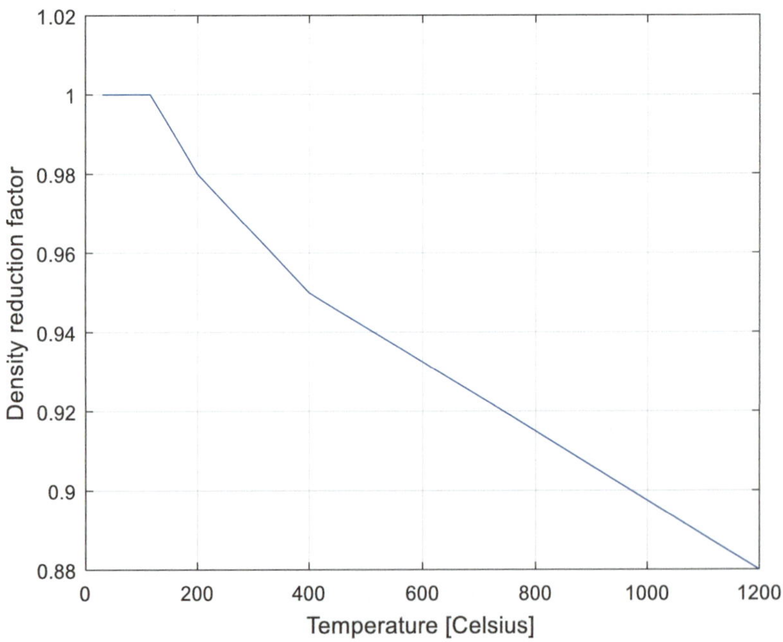

FIGURE 3.14 Density reduction of concrete.

concrete_density_reduction.m

```
function p = concrete_density_reduction(T)
% Uses Eurocode 3 definition
% T: vector of temperatures where conductivity is needed
(Celsius)
% p: vector of density reduction factor (dimensionless)
n = length(T);
p = zeros(n, 1);
for i = 1:n
  if T(i) <= 115
    p(i) = 1;
  elseif T(i) <= 200
    p(i) = 1 - 0.02 * (T(i) - 115)/85;
  elseif T(i) <= 400
    p(i) = 0.98 - 0.03 * (T(i) - 200)/200;
  else
    p(i) = 0.95 - 0.07 * (T(i) - 400)/800;
  end
end
```

The thermal conductivity of concrete decreases gradually with the rise in temperature. Eurocode suggests relations for the upper and the lower limits of the thermal conductivity of concrete; the model is given in **concrete_thermal_conductivity.m** and shown in Figure 3.15. Factors affecting the thermal conductivity of concrete

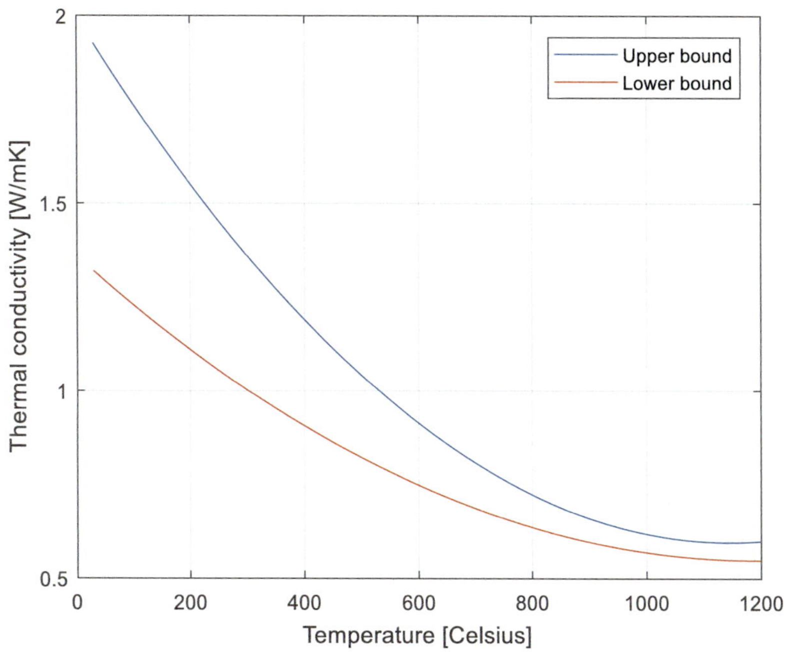

FIGURE 3.15 Thermal conductivity of concrete.

(in order of effectiveness) are, namely, moisture content, recycled coarse aggregate replacement ratio, temperature, volume fractions of aggregates, types of aggregate, and types of admixtures (Pan et al., 2017).

concrete_thermal_conductivity.m

```
function [kUpper, kLower] = concrete_thermal_
conductivity(T)
% Uses Eurocode 3 definition
% T: vector of temperatures where conductivity is needed
(Celsius)
% kUpper: vector of upper limit of thermal
conductivities (W/mK)
% kLower: vector of lower limit of thermal
conductivities (W/mK)
kUpper = 2 - 0.2451 * (T/100) + 0.0107 * (T/100).^2;
kLower = 1.36 - 0.136 * (T/100) + 0.0057 * (T/100).^2;
```

The thermal strain of concrete also depends on the type of aggregates used. The Eurocode model is given in **concrete_thermal_strain.m**.

concrete_thermal_strain.m

```
function [es, ec] = concrete_thermal_strain(T)
% Uses Eurocode 3 definition
% T: vector of temperatures where conductivity is needed
(Celsius)
% es: vector of thermal strain for concrete with
siliceous aggregate (dimensionless)
% ec: vector of thermal strain for concrete with
calcareous aggregate (dimensionless)
n = length(T);
es = zeros(n, 1); ec = es;
for i = 1:n
  if T(i) <= 700
    es(i) = -1.8e-4 + 9e-6 * T(i) + 2.3e-11 * T(i)^3;
  else
    es(i) = 14e-3;
  end
end
for i = 1:n
  if T(i) <= 805
    ec(i) = -1.2e-4 + 6e-6 * T(i) + 1.4e-11 * T(i)^3;
  else
    ec(i) = 12e-3;
  end
end
```

3.16 FUNCTIONALLY GRADED MATERIALS (FGM)

FGM possess specific benefits:

1. *Improved energy absorption.* FGMs have better energy absorption properties than traditional materials, such as steel and aluminum. The spatial variety of material characteristics of FGMs enables efficient impact energy dissipation, which reduces structural damage and deformation (Yeo et al., 1998; Chandrasekaran et al., 2022, 2020; Chandrasekaran, 2022).
2. *Reduced plastic strain and deflection.* FGMs can reduce plastic strain and distortion in offshore structural components exposed to impact stresses. FGMs' customized material qualities aid in reducing localized yielding and permanent deformation while preserving structural integrity and performance.
3. *Enhanced mechanical properties.* FGMs have outstanding mechanical qualities, including high strength, stiffness, and toughness, which help them resist impact loads. Engineers can use the unique features of FGMs to design offshore structures with improved impact resistance and structural performance.

4 Analysis of Structures

4.1 INTRODUCTION

Structural analysis is a mathematical method to determine the behavior of structural members under various loads. For architects, understanding the fundamental principles of structural analysis is essential, as it allows them to design buildings that are not only aesthetically pleasing but also structurally sound. Structural analysis enables an architect to predict how a building will respond under the action of various loads and their combinations, while design helps them arrive at the cross-sectional dimensions of various structural components, such as beams, slabs, columns, and foundations. Though detailed analysis and design do not fall within the scope of an architect, understanding them is imperative to gain confidence in the overall control of a construction project.

Input to the classical structural analysis is, namely, geometric composition of elements, support conditions, and material properties. For example, if one has to analyze a beam, then the corresponding input to analysis shall be the span, support conditions, and loads. Various loads that are common to act on building and structural systems are gravity (dead loads), occupants (live loads), wind, and even seismic forces. The study of structural behavior directly informs key design decisions, ensuring the safety, stability, and durability of the structure over its lifespan.

A successful architectural design integrates structural considerations from the beginning, rather than treating structure as an afterthought. This integration allows for innovative designs that push the boundaries of traditional construction methods, such as long-span structures, floating cantilevers, or slender high-rises. As an architect, one must be familiar with the principles of equilibrium, stability, and material behavior, which will empower one to work closely with structural engineers and design efficiently without compromising artistic vision. For example, the Sydney Opera House and the Eiffel Tower (Figure 4.1) are quintessential examples of balancing form and structure in architecture. The Sydney Opera House merges an aesthetic appeal with engineering functionality through its distinctive sail-like shells. These shells not only create a visually striking appearance but also contribute to structural stability by distributing weight efficiently. The combination of the chosen geometric form and function demonstrates not only the innovative concept but also meeting the structural demands.

Similarly, the Eiffel Tower showcases the use of steel in an attractive geometric form and gives the full advantage of its strength. The lattice iron framework provides strength and resilience and ensures stability while maintaining an elegant, airy silhouette. This design exemplifies the principles of structural analysis, where the arrangement of elements enhances both the durability and aesthetic quality of the structure.

These architectural landmarks illustrate the fundamental concepts of structural design, where the harmonious blend of artistic vision and engineering principles

 DOI: 10.1201/9781003581314-4

FIGURE 4.1 Aesthetic forms of architectural design.

results in buildings that are both beautiful and structurally sound. By understanding the balance between form and structure, architects can create designs that are not only visually appealing but also capable of withstanding various forces and stresses. It is important to note that the diagram of an architect's workflow shows the relationship between architectural design, structural engineering, and construction.

Every structure must meet two essential criteria: safety and serviceability. *Structural safety* ensures that a building can carry the expected loads without risk of collapse or excessive deformation. This is particularly critical in situations where loads can change rapidly, such as during an earthquake or a hurricane. *Serviceability*, on the other hand, addresses how well the structure performs under normal conditions. Even if a structure is technically safe, excessive deflections, vibrations, or cracking could make it uncomfortable or even unusable for occupants.

Architects must carefully consider both of these factors during the design process. A building may be designed to withstand extreme loads, like wind or seismic forces, but if its floors vibrate excessively during everyday use, it won't serve its intended function. The balance between structural safety and comfort for users is a key aspect of architectural design.

The first and fundamental step in structural analysis is the formulation of the problem using an appropriate mathematical model. Commonly used methods in classical structural analysis are, namely, (1) statically determinate structural model and (2) statically indeterminate structural model. Statically determinate models are relatively easier to solve for their unknowns in comparison to the latter; a free-body diagram is one of the effective tools to address this problem. It is also not redundant to state that computer methods are more necessary to solve indeterminate structures with a higher order of indeterminacy. In parallel, it would also be interesting and vital to formulate a general procedure to solve statically indeterminate systems, making the approach non-problem-dependent but generic. This objective will be fulfilled if the problem is formulated and solved only using the standard equation of statics. To achieve this, a grouping of formations is necessary. Grouping is done by two methods, namely, (1) flexibility method and (ii) stiffness method.

Both methods are useful for simplifying the model to be solved using the equation of statics. Furthermore, both methods are equally efficient and useful, while no supremacy prevails of one over the other. Any one of the methods can be used for grouping as per the user's preference. However, it is important to note that the chosen method should be easily programmable using computer coding. In this context, both methods differ significantly in identifying the unknowns while formulating the problem: In the flexibility method, unknowns are actions such as shear force, axial force, and bending moment, while unknowns in the stiffness method are displacements (both translational and rotational). Basic assumptions that apply to both methods are as follows:

- A linear relationship exists between applied load and the resulting displacement of the structure. This makes the principle of superposition valid through the formulation.
- The material of the structure must obey Hooke's law, which emphasizes the fact that material should not be stressed beyond its elastic limit.
- Equations of static equilibrium shall be developed using the geometry of the un-deflected model. Change in geometry caused by the imposed loads is negligible in comparison to the original geometry.

4.2 TERMINOLOGY

Before delving into specific types of structures, loads, or analysis methods, it's important to get familiar with the basic terms used in structural analysis.

Load: A force acting on a structure, such as a person standing on a floor (live load), the weight of the building materials themselves (dead load), or wind pushing against a wall.

Reaction: The forces that develop at the supports of a structure in response to loads. For example, when you place a beam on two supports, the supports provide reactions that prevent the beam from moving or falling.

Internal Forces: Forces that develop within structural elements to resist loads. These include:

Axial Force (Tension/Compression): A force that stretches or compresses a member along its length.

Shear Force: A force that causes one section of a material to slide past an adjacent section.

Shear Force Diagram (SFD): A shear force diagram is a graphical representation of the shear force that exists at various points along a beam. It is a plot of shear force (V) against the distance along the beam (x).

Bending Moment: A force that causes a structural element to bend.

Bending Moment Diagram (BMD): A bending moment diagram is a graphical representation of the bending moment that exists at various points along a beam. It is a plot of the bending moment (M) against the distance along the beam (x). An example of BMD and SFD is illustrated in Figure 4.2.

Stress: *Stress* refers to the internal force per unit area within a material. When a load is applied, the material develops stress to resist deformation.

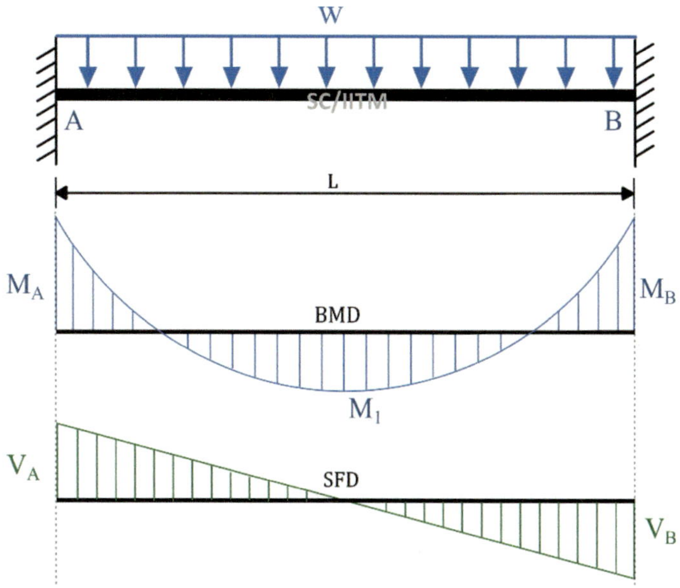

FIGURE 4.2 Fixed beam with external load: BMD and SFD.

Strain: Strain measures the deformation of a material in response to stress. It is the ratio of change in length to the original length.

Equilibrium: A state where all the forces and moments acting on a structure are balanced. For a structure to be stable, it must be in equilibrium—meaning, the sum of forces and moments equals zero.

Stability: Refers to the ability of a structure to remain in equilibrium under a given set of loads. A structure that is unstable may collapse or undergo large deformations when loaded.

Determinacy: This describes whether a structure can be analyzed using the equations of static equilibrium alone. A statically determinate structure can be solved with basic equilibrium equations, while a statically indeterminate structure requires more complex analysis techniques.

4.3 EQUILIBRIUM AND STABILITY

At the heart of structural analysis is the concept of equilibrium. A structure is in equilibrium when the sum of all forces and moments acting on it is zero; it ensures that the structure remains in static equilibrium. This is essential to ensure a structure's stability under various loads. There are two key types of equilibrium:

- **Translational Equilibrium:** The sum of forces in any direction (horizontal and vertical) must equal zero.
- **Rotational Equilibrium:** The sum of all moments about any point must equal zero.

For a structure to be stable, it must resist not only external loads but also the risk of tipping over or collapsing under its weight. Stability is a major concern, especially for tall buildings, cantilevered structures, or slender elements. In structural analysis, one evaluates stability by ensuring that the structure has enough constraints (supports and connections) to stay (or continue to stay) in equilibrium. A structure that lacks proper constraints is unstable, and this can lead to structural failure, though there may not be any functional failure. However, too many constraints can lead to a statically indeterminate structure, which requires more advanced methods of analysis to solve.

4.4 FORM, FUNCTION, AND STRUCTURE: RELATIONSHIP

In architecture, the famous dictum "form follows function" by Louis Sullivan is often used to convey the idea that the design of a building should reflect its purpose. However, for architects, it is just as important to understand that form follows structure. The structural system chosen for a building influences its form, and conversely, the desired form may dictate the type of structure needed to support it.

For example, an architect aiming to design a long-span auditorium may need to choose a truss or arch structure, as these systems are best suited for large, open spaces with minimal internal supports. Similarly, a slender high-rise will require a different structural system, such as a core-supported or braced-frame structure, to resist wind and earthquake forces. While some architects may focus more on aesthetics and leave structural considerations to engineers, the most innovative and successful architectural designs are those where form, function, and structure are considered as one. This results in buildings that are not only visually striking but also functionally efficient and structurally sound. Figure 4.3 shows the **Fallingwater**, a masterpiece designed by Frank Lloyd Wright. The cantilevered design, coupled with the use of local stone and timber, creates a sense of organic unity between the building and its site. Aesthetically, Fallingwater exemplifies Wright's prairie style, characterized by horizontal lines, open floor plans, and a strong connection to the earth. The interplay of light and shadow, achieved through carefully placed windows and overhangs, further enhances the building's visual appeal.

Structural analysis is the mathematical process of determining the effects of loads on structures and their components. Structures can range from simple beams and columns to complex frameworks, like bridges and skyscrapers. The primary goal is to ensure that structures can withstand the applied loads without failure during their intended lifespan. **The Bandra–Worli Sea Link (BWSL)**, shown in Figure 4.4, a majestic cable-stayed bridge spanning the Arabian Sea in Mumbai, India, is a testament to modern engineering marvels. Its innovative design, characterized by twin towers and a slender central span, showcases the importance of structural analysis in large-scale infrastructure projects. The bridge's construction required meticulous calculations to ensure its stability and resilience against various environmental factors, including strong winds and seismic activity. Aesthetically, the BWSL complements the Mumbai skyline, offering stunning panoramic views of the city and the sea. Its elegant design and efficient traffic flow have made it a vital link in Mumbai's transportation network, highlighting the significance of structural analysis in creating a functional and visually appealing infrastructure.

FIGURE. 4.3 Fallingwater.

FIGURE 4.4 Bandra–Worli Sea Link, Mumbai.

4.5 STATIC DEGREES OF FREEDOM

While structural systems can be derived from any geometric form, the basic structural system is referred to as a statically determinate model. Structural members such as beams can be supported at one or more sections. In structural analysis, understanding static DOF is essential for determining the stability and equilibrium of a structure. A beam with only one end supported is referred to as a cantilever (examples are sunshades and chajas). A beam with two supports is referred to as a simple supported beam, and more than two supports are termed as a continuous beam. *Static degree of freedom* (DOF) is the number of independent reactions offered by the supports. There are three types of supports, namely, roller support, hinged support, and fixed support (Figure 4.5).

As seen in the figure, while roller support offers a reaction only in one direction (vertical), hinged support offers a reaction in two directions (both horizontal and vertical). A fixed support offers reactions in two directions, the same as a hinged support; in addition, it also offers a moment constraint. Hence, if one has to classify the preceding mathematically, the following conditions hold good:

No. of reactions (unknowns) of a roller support = 1 (V_y)
No. of reactions (unknowns) of a hinged support = 2 (H_x, V_y)
No. of reactions (unknowns) of a fixed support = 3 (H_x, V_y, M_z)

As it is understood that a structure can possess more than one support, the total number of static degrees of freedom is the sum of the unknowns offered by each support in a structure. For a 2D analysis, the number of equations of static equilibrium are as follows:

- The algebraic sum of all forces acting along the X-axis is set to ZERO: $\sum H_X = 0$.
- The algebraic sum of all forces acting along the Y-axis is set to ZERO: $\sum V_Y = 0$.
- The algebraic sum of moments of forces about any section about the Z-axis is set to ZERO: $\sum M_Z = 0$.

The *static degree of indeterminacy* is the difference between the static degrees of freedom and the number of static equations of equilibrium. For example, let us consider a simply supported beam shown in Figure 4.6. The total number of reactions

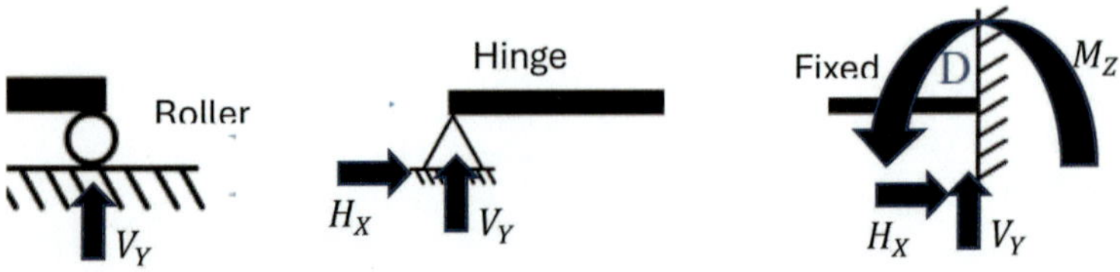

FIGURE 4.5 Types of supports and reactions.

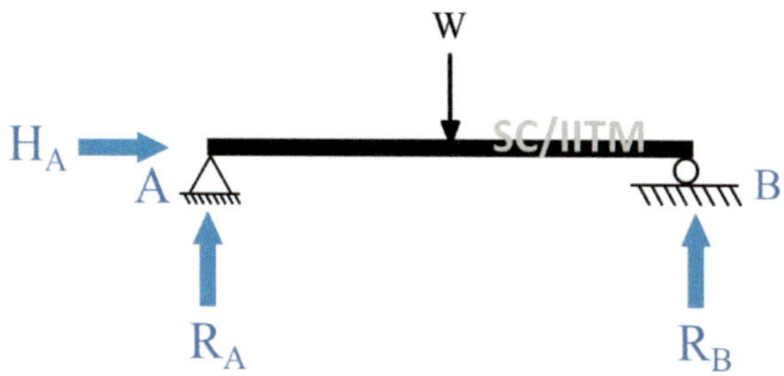

FIGURE 4.6 Simply supported beam.

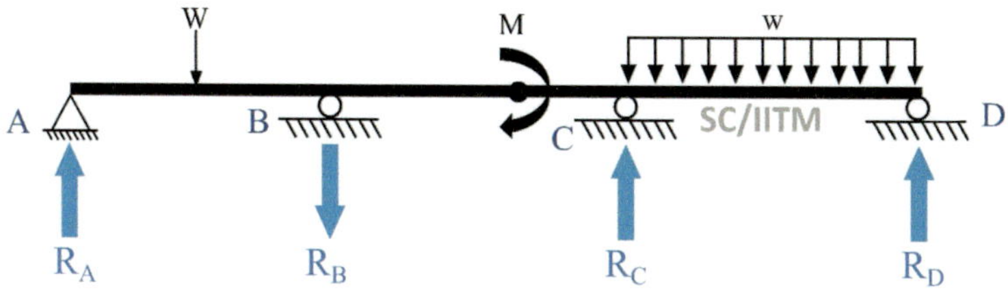

FIGURE 4.7 Continuous beam.

offered by two supports (*A*, *B*) is three, as one end is hinged and the other is on a roller. As the total number of static equations of equilibrium is also three, the static degree of indeterminacy is ZERO. Thus, a simply supported beam is said to be statically determinate. In case the number of reactions is more than the number of equations of static equilibrium, then the member is said to be statically indeterminate. The degree of static indeterminacy can vary from one to infinity.

Let us consider an example of a continuous beam, as shown in Figure 4.7. The total number of supports is four (A, B, C, D). While the support is marked as a hinged support, the rest are on rollers. As explained earlier, roller support has one reaction, marked as R_B, R_C, R_D. The hinged support has two reactions, but only one is marked in the figure (R_A). It is important to note the assumption made in this case: *Axial deformations are neglected.* Hence, no horizontal reaction at support is marked. It is also wise to correlate the valid number of equations of static equilibrium. Instead of three, it is only two, as $\sum H_X = 0$ is not applicable in this case. Hence, the static degree of freedom of the continuous beam is four (as there are only four reactions offered by the supports). The static degree of indeterminacy is two (4 – 2). *The beam is statically indeterminate to the order of TWO.*

Let us now look at one more example, shown in Figure 4.8, which is also a continuous beam. The total number of supports is five (one at each support *A*, *B*, *C*, and two at *D*). Please note the assumption made in the analysis, as *axial deformation was neglected*, and hence, support D has only two reactions (R_D, M_D). Because the equations of static equilibrium are only two, since $\sum H_X = 0$ is not applicable in this case, the static degree of freedom of the continuous beam is five. The static degree of indeterminacy is three (5–2). *The beam is statically indeterminate to the order of THREE.*

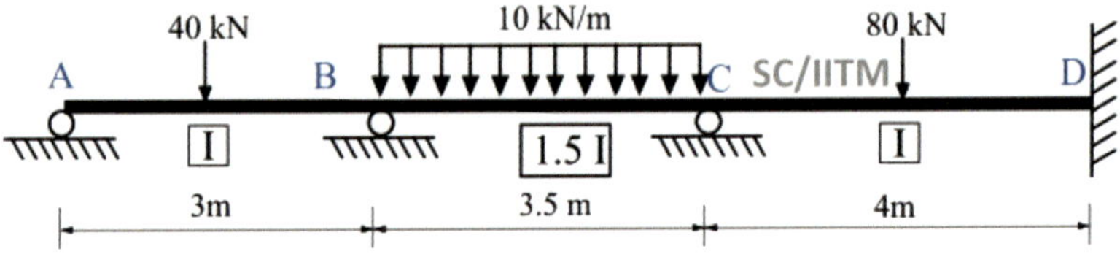

FIGURE 4.8 Continuous beam with a fixed support.

4.6 KINEMATIC DEGREES OF FREEDOM

Kinematic degrees of freedom are related to the displacement capabilities of a structure without considering the forces. Kinematic analysis helps in understanding the possible displacements of a structure under the applied loads. It is important to note that displacements are also measured only at the supports, as reactions. There are two types of displacements possible in a joint/support: They are rotational and translational. Further, translational displacements can be along the X-axis and Y-axis, while rotational displacements are about the Z-axis only.

Similar to the static degree of freedom, it is crucial to estimate the kinematic degrees of freedom of a structural member. It helps assess the ability of a structure to identify the potential mechanisms or modes of failure. For instance, a planar truss can be analyzed for its kinematic degrees of freedom to ensure that it remains stable and does not collapse under load. The kinematic analysis involves checking the geometric configuration and connections of the structure to ensure that it is rigid and stable. Similar to the static degrees of freedom at each support, one can also associate possible kinematic degrees of freedom with each support. Figure 4.9 shows the types of support and the displacements associated with each of them. While identifying the displacements, please note that rotations are about the Z-axis, while translations can be along the X-axis and the Y-axis, in a 2D analysis.

Table 4.1 helps understand the kinematic degrees of freedom for each support. In the figure, unrestrained (free-to-move) degrees are marked in green, while restrained degrees are marked in red, for clarity. All sign conventions marked for the displacements and rotation are considered positive.

Example 1: Let us consider an example of a continuous beam shown in Figure 4.10. There are four supports (A, B, C, D). While support D is a fixed support, the rest are on rollers. With reference to Table 4.1, the following observations can be made:

	Degree of Freedom, If Axial Deformation Neglected	
Support	**Restrained**	**Unrestrained**
A	Translation δ_5	Rotation θ_3
B	Translation δ_6	Rotation θ_2
C	Translation δ_7	Rotation θ_1
D	Translation δ_8 Rotation θ_4	Nil

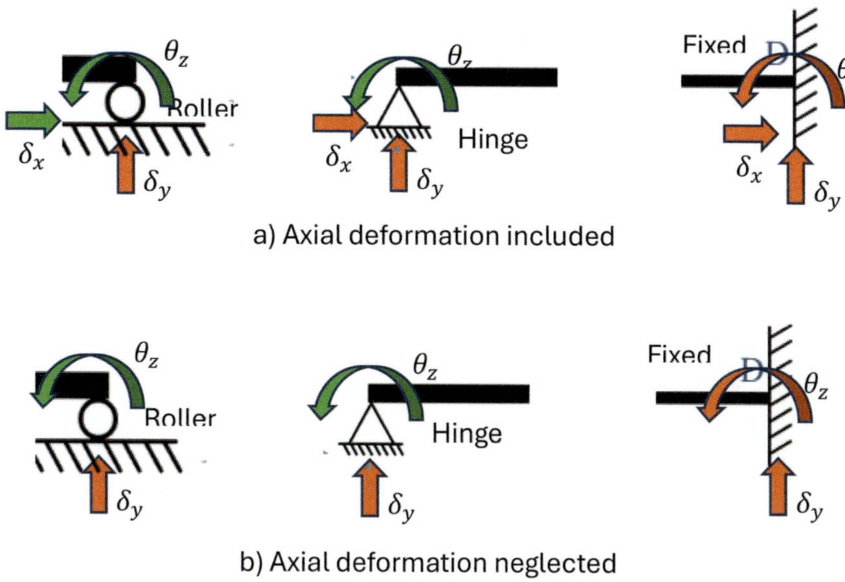

FIGURE 4.9 Types of supports and displacements.

TABLE 4.1
Kinematic Degrees of Freedom at Each Support

	Degree of Freedom, If Axial Deformation Included		Degree of Freedom, If Axial Deformation Neglected	
Support	**Restrained**	**Unrestrained**	**Restrained**	**Unrestrained**
Roller	Translation δ_y	Rotation θ_z Translation δ_x	Translation δ_y	Rotation θ_z
Hinged	Translation δ_x Translation δ_y	Rotation θ_z	Translation δ_y	Rotation θ_z
Fixed	Translation δ_x Translation δ_y Rotation θ_z	Nil	Translation δ_y Rotation θ_z	Nil

The *degree of kinematic indeterminacy* is the total number of unrestrained kinematic degrees of freedom. As seen in the preceding case, the kinematic degree of indeterminacy is three.

Example 2: For the two-span continuous beam shown in what follows, estimate the static and kinematic degree of indeterminacy. Neglect axial deformation in the beam.

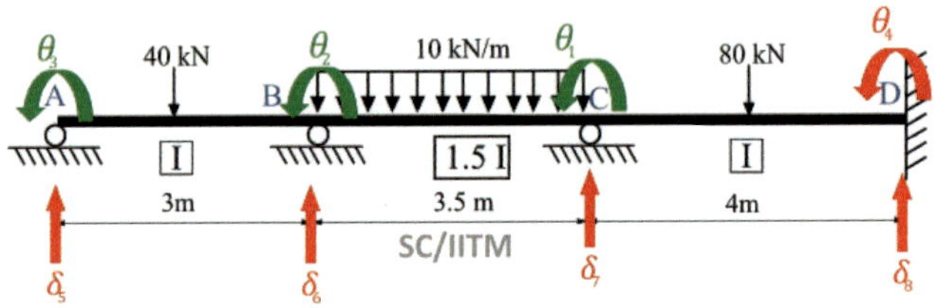

FIGURE 4.10 Continuous beam marked with kinematic degrees of freedom.

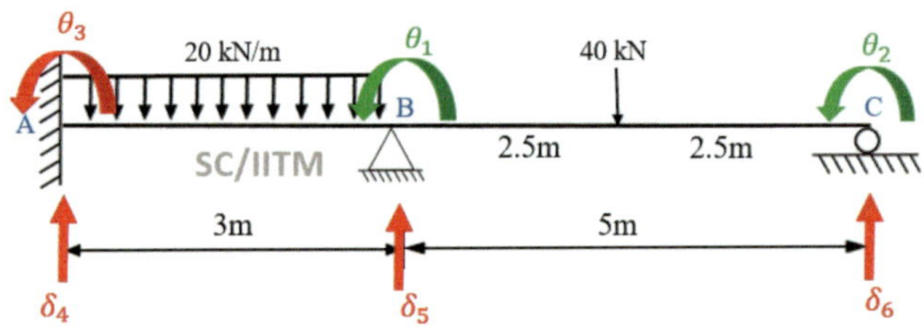

Support	Static Degree of Freedom	Kinematic Degree of Freedom Axial Deformation Neglected: Restrained	Unrestrained
A	2 (V_A, M_A)	Rotation θ_3 Translation δ_4	Nil
B	1 (V_B)	Translation δ_5	Rotation θ_1
C	1 (V_C)	Translation δ_6	Rotation θ_2
Degree of indeterminacy	Static: 4 – 2 = 2	Kinematic: 2 (only the sum of unrestrained degrees of freedom)	

Example 3: For the following frame shown, estimate the kinematic degree of indeterminacy.

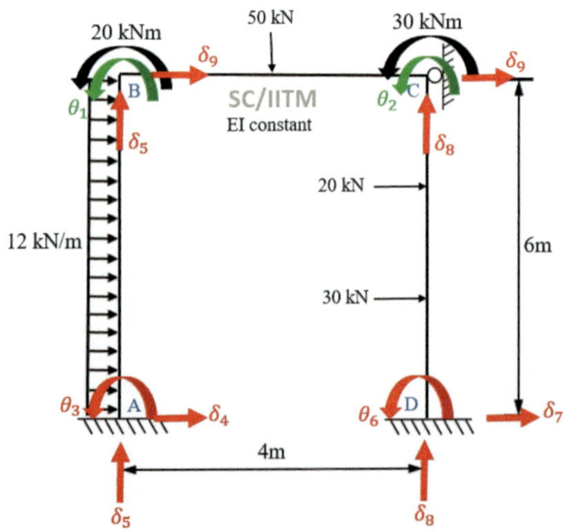

	Kinematic Degree of Freedom Axial Deformation Neglected	
Support	**Restrained**	**Unrestrained**
A	Rotation θ_3 Translation δ_4 Translation δ_5	Nil
B	Translation δ_5 Translation δ_9	Rotation θ_1
C	Translation δ_8 Translation δ_9	Rotation θ_2
D	Rotation θ_6 Translation δ_7 Translation δ_8	Nil
Degree of indeterminacy	Kinematic: 2 (only the sum of unrestrained degrees of freedom)	

Example 4: For the following shown stepped frame, determine the kinematic degree of indeterminacy.

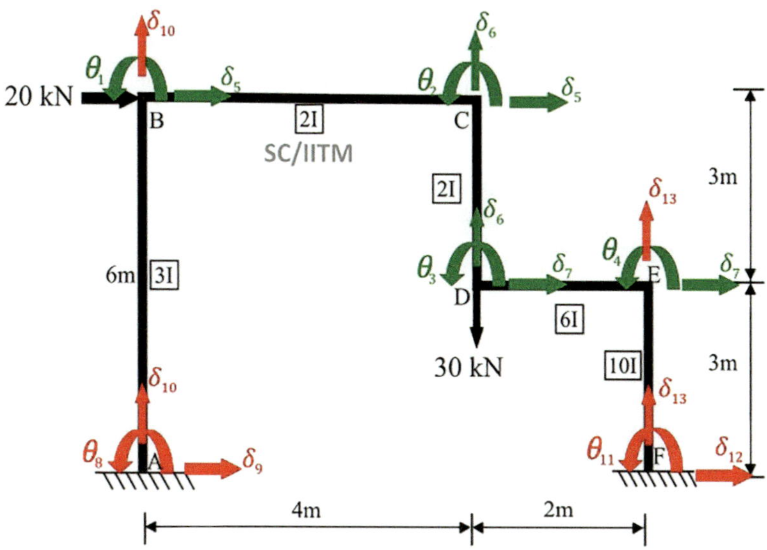

	Kinematic Degree of Freedom Axial Deformation Neglected	
Support	**Restrained**	**Unrestrained**
A	Rotation θ_8 Translation δ_9 Translation δ_{10}	Nil
B	Translation δ_{10}	Rotation θ_1 Translation δ_5
C	Nil	Rotation θ_2 Translation δ_5 Translation δ_6
D	Nil	Rotation θ_3 Translation δ_7 Translation δ_6
E	Translation δ_{13}	Rotation θ_4 Translation δ_7
F	Rotation θ_{11} Translation δ_{12} Translation δ_{13}	Nil
Degree of indeterminacy	Kinematic: 10 (only the sum of unrestrained degrees of freedom)	

In structural analysis, it is essential to use consistent units of measurement. Common units include:

- **Length**: meters (m), millimeters (mm)
- **Force**: newtons (N), kilonewtons (kN)
- **Stress**: pascals (Pa), megapascals (MPa).

4.7 STATIC DETERMINATE STRUCTURES

Static determinate structures are those in which the internal forces and reactions can be determined solely from the equations of static equilibrium. These structures are characterized by having just enough supports and members to maintain stability without any redundancy. The primary advantage of statically determinate structures is the simplicity of analysis, as the forces can be calculated directly using the three fundamental equilibrium equations:

- Sum of vertical forces: ($\sum F_Y = 0$)
- Sum of horizontal forces: ($\sum F_X = 0$)
- Sum of moments: ($\sum M_Z = 0$)

Examples of statically determinate structures include simply supported beams, cantilever beams, and certain types of trusses. These structures are often used in engineering due to their straightforward analysis and predictable behavior under loads.

However, they may be less robust in the event of unexpected loads or support failures, as they lack redundancy.

4.8 STATIC INDETERMINATE STRUCTURES

Static indeterminate structures, also known as *hyperstatic structures*, have more supports or members than are necessary for maintaining equilibrium. As a result, the internal forces and reactions cannot be determined solely from the equations of static equilibrium. Additional compatibility conditions, such as deformation and material properties, must be considered to solve for the unknowns. These structures are often more complex to analyze but offer greater redundancy and resilience. In the event of a support failure or unexpected load, a statically indeterminate structure can redistribute the forces, enhancing its overall stability and safety. Examples include continuous beams, fixed end beams, and certain types of frames and arches.

The analysis of statically indeterminate structures typically involves advanced methods, such as the moment distribution method, slope deflection method, or matrix methods. These techniques account for the additional constraints and compatibility conditions required to solve for the unknown forces and moments.

4.9 ANALYSIS AND DESIGN

While *analysis* deals with determining forces and moments at any section of a member, *design* deals with choosing the appropriate cross section (size, shape, material) to resist the forces. Obviously, design is always carried out to resist the maximum force acting on the structure. Hence, design should employ two checks:

1. Stresses at any section should not exceed permissible limits specified by the design codes.
2. Deflections at any section in the structure should not exceed permissible limits specified in the design codes.

The following table summarizes the difference between analysis and design.

Analysis	Design
Deals with the determination of forces	Deals with checking the stresses at any section by choosing a member shape, size, and material
Carried out using equations (mathematical)	Carried out with the help of design codes given by the Bureau of Indian Standards, New Delhi
Many methods of analysis	Many procedures for the design
Structural analysis broadly divided into: 1. Statically deterministic structure 2. Statically indeterminate structure	No such broad classifications exist; design common to determinate and indeterminate structures
Analysis complicated	Design very simple and easy to carry out

(Continued)

(Continued)

Analysis	Design
Analysis results not a part of the construction drawings	Structural design a part of construction drawings and estimation costing
Software used to do the analysis	Design software available (STAAD Pro, etc.)
To determine reactions at supports; internal forces, such as axial, shear, bending; and torsional moments	To choose appropriate dimensions, shape, size, and material for the structure
To determine the deflection or deformation at any section in a member under the loads	To check that stresses do not exceed permissible values; to check that deflections are within the permissible limits as suggested by the code

4.9.1 Idealization of Structures

To carry out analysis, it is necessary to idealize a structure. During idealization, members are represented by their centroidal axis as lines, while the supports are represented as arrows. Loads acting on the structure are represented as arrows, and continuous arrows for point loads and uniformly distributed loads (UDL). The following figures illustrate the concept of idealization, graphically. Similarly, the connections are also idealized.

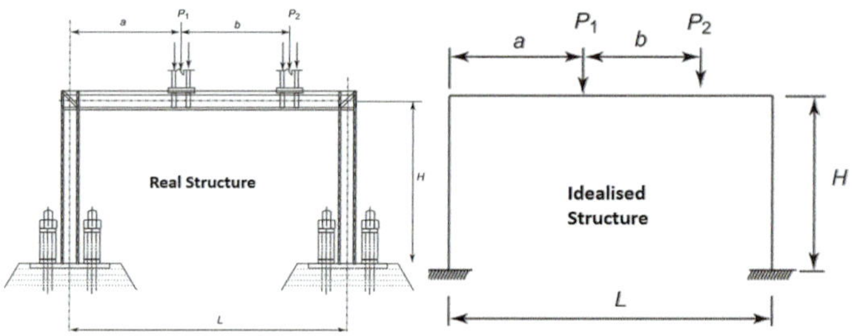

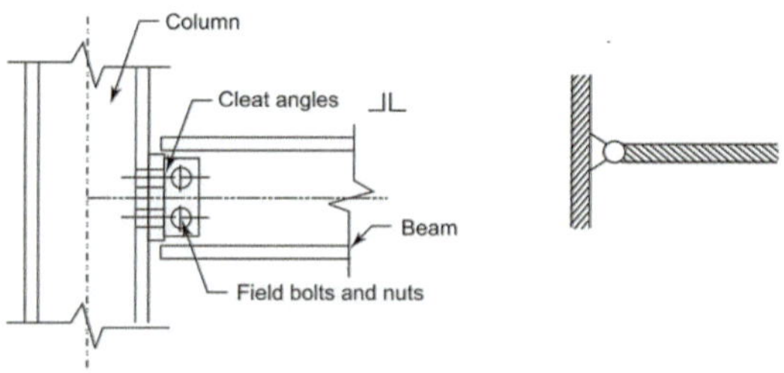

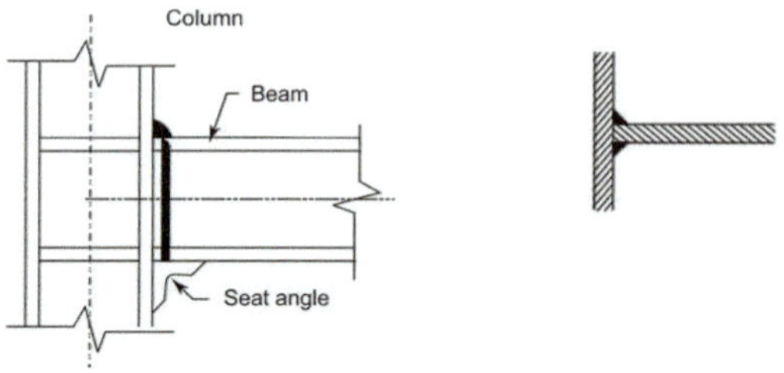

One of the major assumptions made in structural analysis is that the material obeys Hooke's law, which results in linear behavior. This assumption emphasizes that a linear and proportional behavior exists between the load and the deformation caused by the load on the structure. As this assumption makes the analysis simpler, the entire structure is assumed to exhibit a linear behavior. It means that structures will deform/deflect under the action of loads. But upon the removal of the loads, the structure will assume/return to its normal equilibrium position, confirming an elastic linear behavior. Under such a major assumption, the principle of superposition becomes valid for the analysis. The principle of superposition states that *displacements resulting from each force may be added to obtain the total displacements resulting from all the forces acting together.* Conversely, the forces corresponding to each displacement may be added to obtain the total force acting on the structure.

For example, let us consider a cantilever beam loaded as follows.

Let load P_1 cause deflection at points 1, 2, 3, measured as Δ_{11}, Δ_{21}, Δ_{31}. The first subscript refers to the location at which displacements are measured, and the second subscript refers to the location at which the load is applied. Please note that the deflections need not be measured under the load; it can be at different sections.

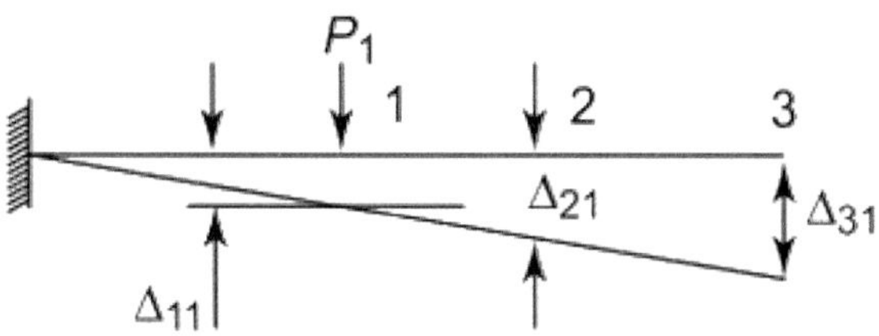

Let P_2 cause deflection at 1, 2, and 3, as shown in what follows:

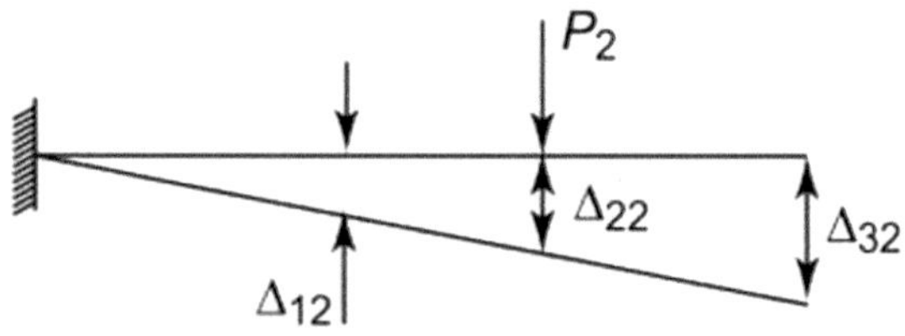

Let P_3 cause deflection at 1, 2, and 3, as shown in what follows:

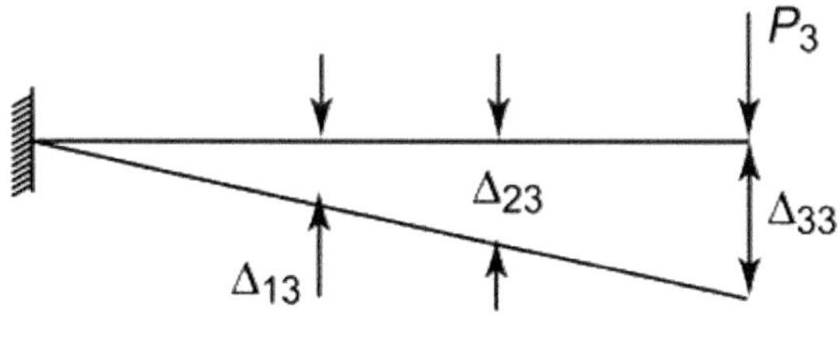

Summation, based on the principle of superposition, is shown in what follows:

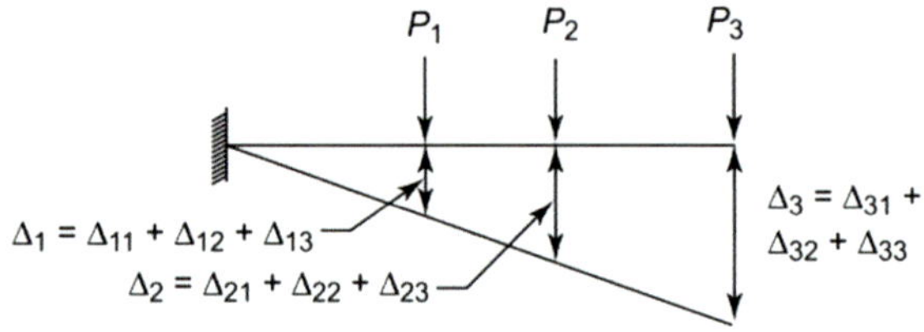

4.10 THEORY OF SIMPLE BENDING

A beam is said to undergo simple bending when it bends under a constant (uniform) bending moment without any shear force. A graphical illustration of simple bending is shown in what follows:

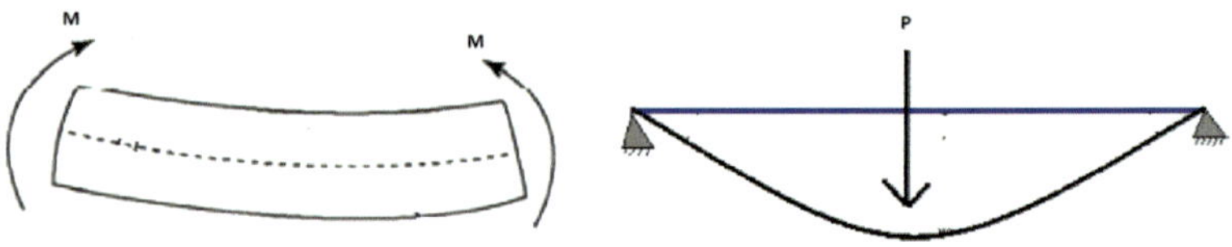

It is important to note the difference between the previously shown figures. In the first case, the beam is undergoing only simple bending, while shear force is absent. But in the second case, where the beam is under the influence of a centrally concentrated load, it also develops shear, which was absent in the first case. The preceding case is not the same as follows. Therefore, the theory of simple bending makes the following assumptions:

- The material of the beam is homogeneous, which means that the material properties are the same throughout its length.
- The material is isotropic, which means that the material possesses the same elastic properties in all directions.
  - Plane sections before bending remain plane after bending, which confirms that the cross-sectional plane does not warp/twist. The cross-sectional plane remains normal to the longitudinal axis of the member before and after bending. Let us now consider a section (1–1), as marked on the following beam. Let the cross section be rectangular (*bxd*) before bending. To measure the characteristics of the cross section, let us consider three fibers (a–b, c–d, e–f), as shown. Even after bending, these fibers are not twisted; they remain perpendicular to the longitudinal axis of the member, as shown.
  - Deflection caused by bending is very small compared to the length of the beam.
  - Shear stress in the plane *xy* (τ_{xy}) is neglected.

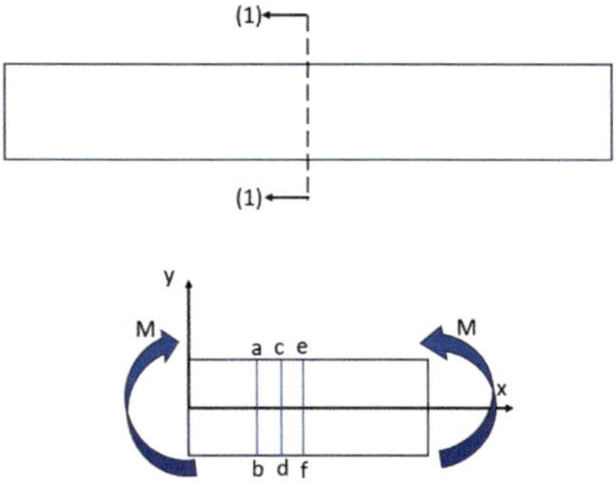

Before bending

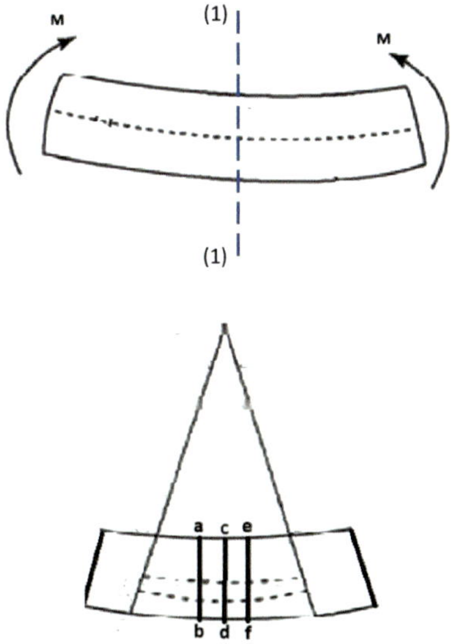

After bending

Consider a beam as shown in the preceding with a rectangular cross section (*bxd*). The beam has two supports (A, B). Let *A* be the jth end, and *B* the kth end of the beam. The X-axis is the longitudinal axis of the member, as indicated. The Y-axis is anticlockwise 90^0 from the positive X-axis. Hence, the bending plane is the X–Y plane, while the bending moment is applied about the Z-axis $\left(M_{zz}\right)$. The moment of inertia about the Z-axis is given as $I_z\left(=\frac{bd^3}{12}\right)$, while about the Y-axis is $I_y\left(=\frac{db^3}{12}\right)$.

The equation of simple bending is given by:

$$\frac{M}{I}=\frac{\sigma}{y}=\frac{E}{R}$$

where *M* is the moment applied, which causes bending on the X–Y plane (in kN-m or N-m); σ is the bending stress (N/mm²); *y* is the distance of any fiber in the cross section, measured from the neutral axis (mm); and *E* is the modulus of elasticity of the material (Young's modulus). For steel, it is 2.1×10^5 N/mm², and for concrete, it is $5000\sqrt{f_{ck}}$, where f_{ck} is the characteristic of compressive strength of concrete mix (in N/mm²). *R* is the radius of curvature of bending beam, measured from the center of curvature to the neutral axis.

Example 1: Compute the bending stress at any typical cross section of a rectangular beam (200 × 400), as shown in the figure. The beam is subjected to a moment of 20 kNm.

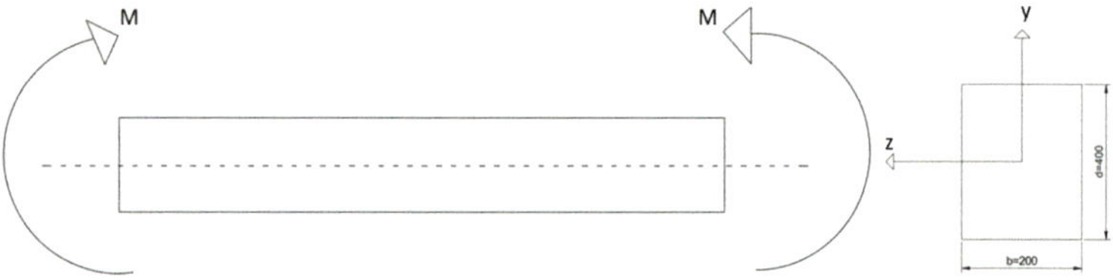

Solution:
M = 20 kNm (given)

$$I_z = \frac{bd^3}{12} = \frac{200x400^3}{12} = 1.067x10^9\ mm^4$$

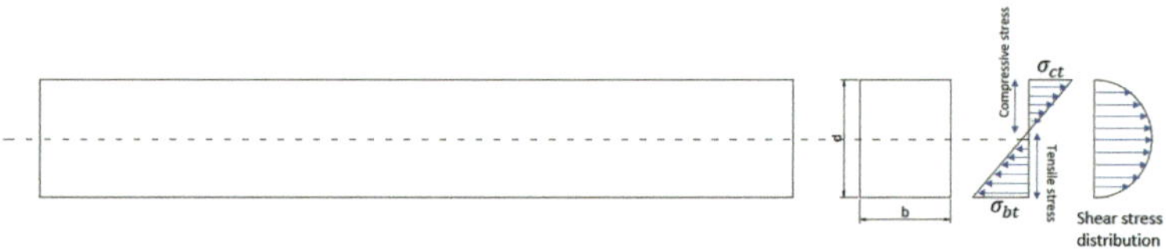

When the beam is under moment *M*, the top fibers shrink and will be in compression, while the bottom fibers will elongate and will be in tension.

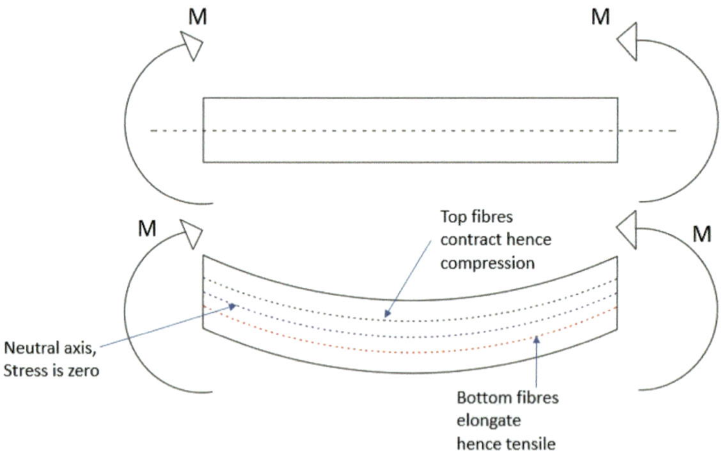

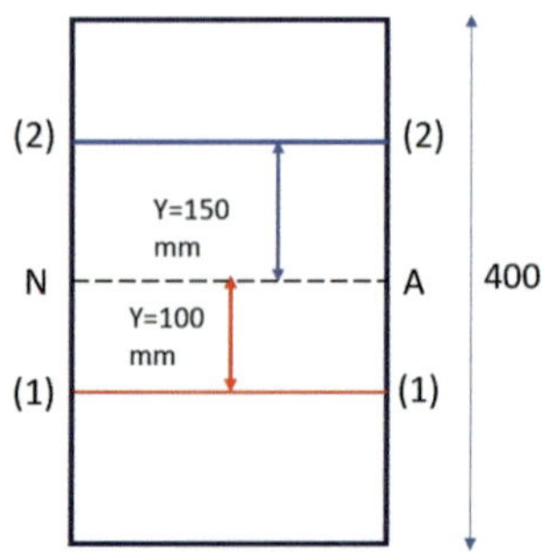

Let us try to compute the bending stress at two sections:
Section (1–1), measured 100 mm below the neutral axis.
Section (2–2), measured 150 mm above the neutral axis.

$$\frac{M}{I} = \frac{\sigma}{y}$$

$$M = \sigma\left(\frac{I}{y}\right)$$

$$M = \sigma(Z)$$

where Z is called section modulus.

$\sigma_{(1)-(1)} = bending\,stess_{(1)-(1)} = \frac{M}{I}y = \frac{2\times10^7}{1.067\times10^9}\times100 = 1.874$ N/mm^2 (tensile), as the section (1–1) is below the neutral axis and the beam is subjected to sagging *BM*.

$\sigma_{(2)-(2)} = bending\,stess_{(2)-(2)} = \frac{M}{I}y = \frac{2\times10^7}{1.067\times10^9}\times150 = -2.812$ N/mm^2 (compressive), as the section is located above the neutral axis and the beam is subjected to sagging *BM*.

If one is interested in finding the maximum bending stress, please choose y as the maximum.

For example, maximum compression bending stress will be:

$$\sigma_{bc} = \sigma_{(Compression)_{Max}} = \frac{M}{I}200 = \frac{2\times10^7}{1.067\times10^9}\times200 = 3.749 \text{ N/mm}^2 \text{ (compressive)}$$

Maximum tensile bending stress is the same as that of compressive, as the NA coincides with the centroidal axis.

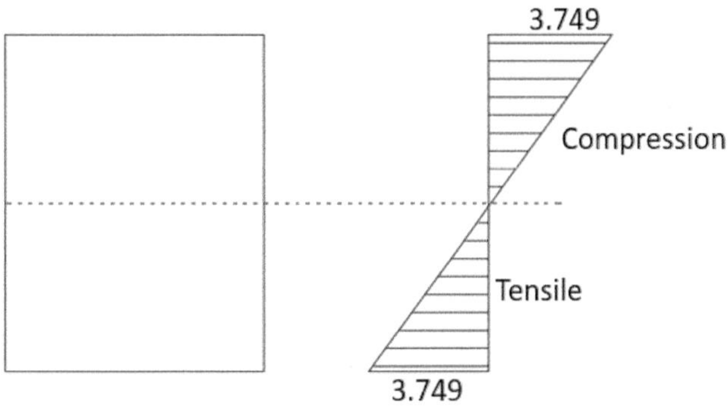

Example 2: Compute the bending stress at any typical cross section of a circular bar (300 dia), as shown in the figure. The bar is subjected to a moment of 100 kNm.

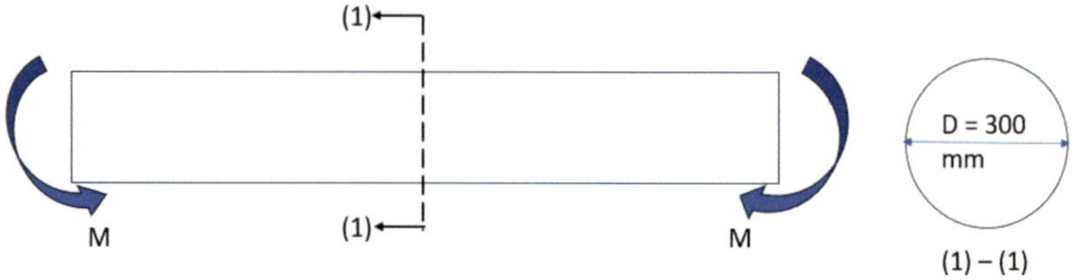

$$M = 100 \text{ kNm}$$

The material is steel, $E_{steel} = 2.1 \times 10^5 \dfrac{N}{mm^2}$.

Compute the maximum bending stress in the cross section under uniform bending.

$$\frac{M}{I} = \frac{\sigma}{y} = \frac{E}{R}$$

$$M = 100 \text{ KNM} = 100 \times 10^6 = 1 \times 10^8 N - mm$$

$$I_z = \frac{\pi d^4}{64} = \frac{\pi \times 300^4}{64} = 3.97 \times 10^8 mm^4$$

$$\sigma_{bt} = \frac{M}{I} y = \frac{1 \times 10^8}{3.97 \times 10^8} \times 150 = 37.78 \frac{N}{mm^2}$$

4.11 ANALYSIS OF BEAMS

Beams are fundamental structural elements that primarily resist loads applied laterally to their axis. They are designed to withstand bending moments, shear forces, and sometimes, axial forces. Beams can be made from various materials, including steel, concrete, timber, and composites. They are used in a wide range of structures, from simple residential buildings to complex bridges and skyscrapers.

Static and kinematic indeterminacies of any structural system are indicators of the amount or extent of computational efforts required to analyze the structural system either using the flexibility approach or the stiffness approach. The unknowns in each method of analysis or approach are those released to convert the structural system into statically or kinematic determinate system so that standard equations can be applied to evaluate the unknowns.

Differences between the flexibility and the stiffness methods are summarized as follows:

Flexibility Approach	Stiffness Approach
This deals with static indeterminacy.	This deals with kinematic indeterminacy.
Unknowns are actions such as shear force, axial force, bending moments, etc.	Unknowns are joint displacements, such as rotational and translational displacements.
Problem formulation converts the structural system into statically determinate structure.	Problem formulation converts the structural system into kinematic determinate structure.

If the problem formulation reduces the number of unknowns, then it is the best formulation attempted by a mathematician or an engineer. It is important to note that the number of unknowns in a system of equations depends on the choice of the method demanded. For the computer method of structural analysis, one should keep in mind that the method recommended should be more or less generic and not problem-specific. The degrees of static and kinematic indeterminacies of a few example structures are arrived at with explanations.

4.11.1 Continuous Beam

Consider a three-span continuous beam as shown in Figure 4.11. The system has one hinged support and three roller supports, whose reactions are marked as R1, R2, R3, and R4 in Figure 4.11. Displacement unknowns neglecting the axial deformations are designated as θ1, θ2, θ3, and θ4. The degree of static indeterminacy is the number of unknown reactions: system of standard equations = (5 − 3) = 2. The degree of kinematic indeterminacy is the number of unrestrained displacements, which is 4.

4.11.2 Fixed Beam

Let us now consider a fixed beam, as shown in Figure 4.12. Support reactions are R1, R2, and R3 at fixed end A, and R4, R5, and R6 at fixed end B, respectively. The fixed beam at both the supports does not permit any unrestrained displacements. Thus, the degree of static indeterminacy is 3 = 6 − 3, and the degree of kinematic indeterminacy is NIL.

4.11.3 Simply Supported Beam

Let us consider a simply supported beam as shown in Figure 4.13. Reaction components R1, R2, and R3 at the supports are shown in the figure, while the unrestrained

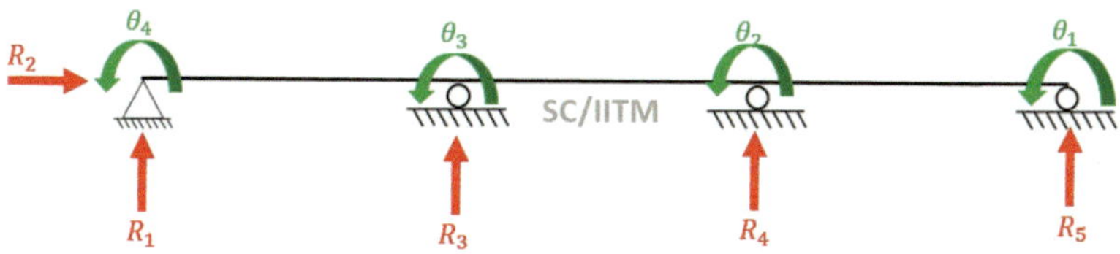

FIGURE 4.11 Continuous beam.

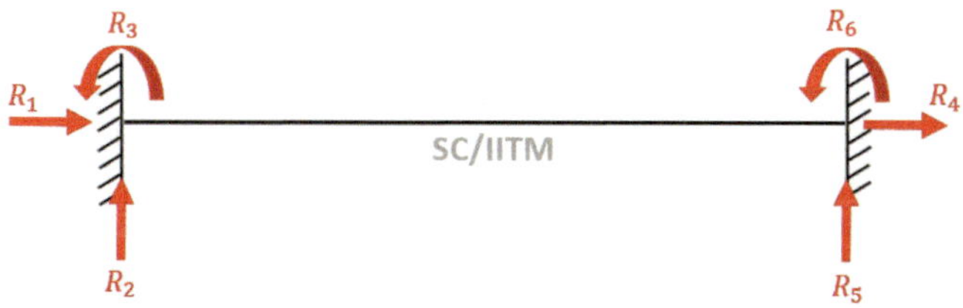

FIGURE 4.12 Fixed beam.

FIGURE 4.13 Simply supported beam.

displacements are θ1 and θ2. Thus, the degree of static indeterminacy is NIL = 3 – 3, and the degree of kinematic indeterminacy is 2, as there are two unrestrained displacements, as shown in the figure.

4.11.4 Frame

Let us consider a single-story frame with a single bay as shown in Figure 4.14. The frame has one end fixed and another end on roller support. Unknown reactions are R1, R2, R3, R4, and R5, as marked in the figure. Independent unrestrained displacements are θ1, θ2, δ3, and θ4 by neglecting axial deformation. It is important to note that even though axial deformation is neglected in computing the degrees of indeterminacies, overall sway displacement is an indication of rigid-body motion, which will be present. Hence, the degree of static indeterminacy is 2 = 5 – 3, and the degree of kinematic indeterminacy is 4.

4.11.4.1 Choice of the Method

From the preceding examples, it is seen that the number of unknowns are dependent upon the choice of the method. The following observations are noteworthy:

- There are essentially two methods to solve statically indeterminate structures, namely, the flexibility method and the stiffness method. The choice of the method depends on computational convenience.
- For the flexibility method, there are several alternatives to choose the redundant or the unknowns. Thus, the choice of redundant has significant effect on the computational effort in the flexibility method.
- On the other hand, there is no choice of unknown quantities in the stiffness method because there is only one possible restrained structure. Therefore, this leads to a set of standard procedure.
- One of the main objectives of the computer methods of structural analysis is that the choice of the method should not be geometry-specific but be more generic and repetitive in nature.

Thus, the stiffness method is a better choice, fulfilling all the aforementioned requirements. In this book, the direct stiffness method is used for elaborating the application procedures. During problem formulation using the stiffness method, variables in a

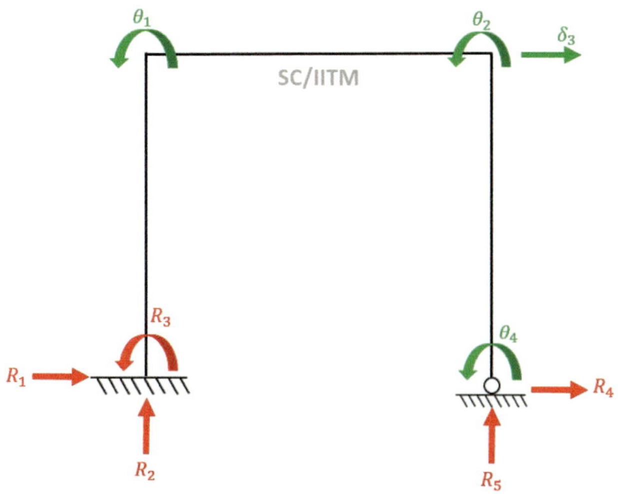

FIGURE 4.14 Single-story frame with single bay.

given structural system, such as rotational or translational displacements for every joint, will result in the formation of a set of linear equations. The following section throws some light on the discussions of linear equations.

4.12 STIFFNESS MATRIX OF A STANDARD BEAM ELEMENT

A *beam element* is one of the basic elements to be used in the structural analysis of planar orthogonal frames. There are some sign conventions that need to be followed before deriving the stiffness matrix of the beam element.

- Anticlockwise end moment, joint rotation, and joint moments are taken as positive.
- Upward force or displacement of the joint is positive.
- Force toward the right or axial displacement toward right of the joint is positive.
- Upward end shear at ends of the beam is positive.
- Right direction force at the ends of the beam is positive.

Let us limit the discussion to planar orthogonal structures in the following section. Planar orthogonal frames have members aligned either horizontal or vertical, and the included angle between them is 90°.

A *fixed beam* is considered as a basic module for the analysis. Consider a fixed beam undergoing deformation due to bending, neglecting the axial deformation. The standard fixed beam is shown in Figure 4.15. The two joints of the beam element are j and k and the length of the member is L_i. In terms of mathematical conditions, the considered beam element is fixed at nodes j and k. It has constant EI over its entire length. The left end of the beam is designated as the jth node, and the right end of

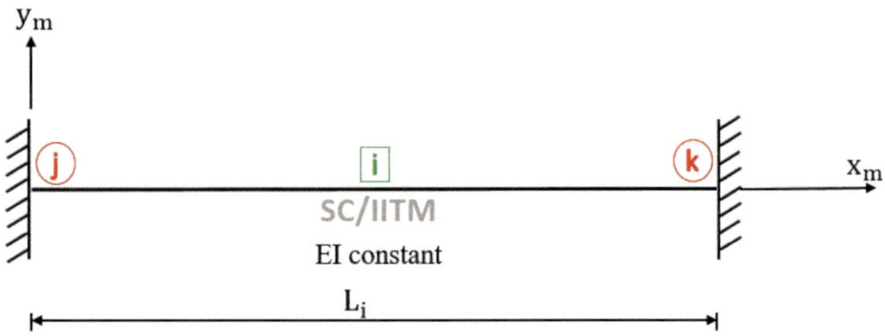

FIGURE 4.15 Reference beam element.

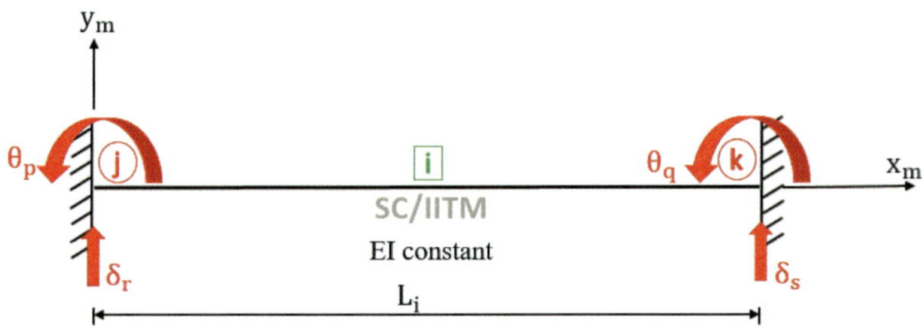

FIGURE 4.16 Rotational and translational displacements of a beam element.

the beam is designated as the kth node. The member is designated as the ith member. (x_m, y_m) are local axes of the member. It is very important to note the axis system. The axis system is such that it has its origin at the jth end; x_m is directed toward the kth end. y_m is counter-clockwise 90° to the x_m axis. Therefore, the (x_m, y_m) plane defines the plane of bending of the beam element.

Let us neglect the axial deformation. For the stiffness method, one should identify possible displacements, both translational and rotational, at each end of the beam. Possible rotational and translational displacements are shown in Figure 4.16. Suitable subscripts are used for denoting the rotational and translational moments. Note the order by which the moments are marked, which can make the readers understand the computer programs easily. Displacements at the jth end and the kth end are $\left(\theta_p, \delta_r\right)$ and $\left(\theta_q, \delta_s\right)$, respectively. All these displacements occur and are mapped to the (x_m, y_m) plane; out-of-plane bending is not considered.

Now, let us derive the stiffness coefficient, k_{ij}. By definition, k_{ij} is the force in the ith degree of freedom by imposing unit displacement jth degree of freedom, which can be either in translational or rotational, by restraining all other degrees of freedom. Let us give unit rotation at the jth end, keeping all other degrees of freedom restrained, as shown in Figure 4.17. This will invoke members with end forces $k^i_{pp}, k^i_{qp}, k^i_{rp}, k^i_{sp}$. For example, k^i_{pp} is the force in the pth degree of freedom by giving

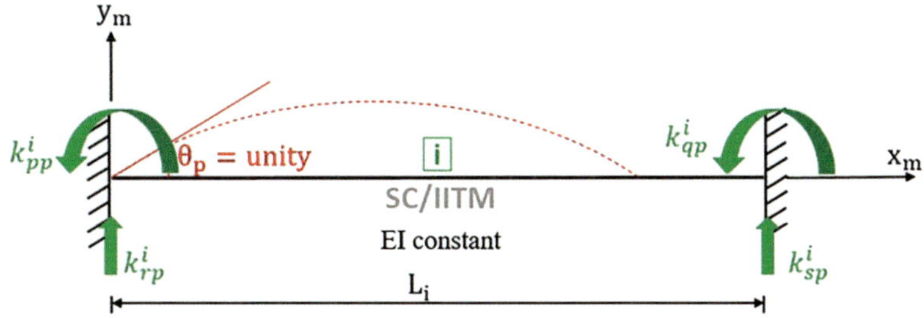

FIGURE 4.17 Fixed beam element with unit rotation at the *j*th end.

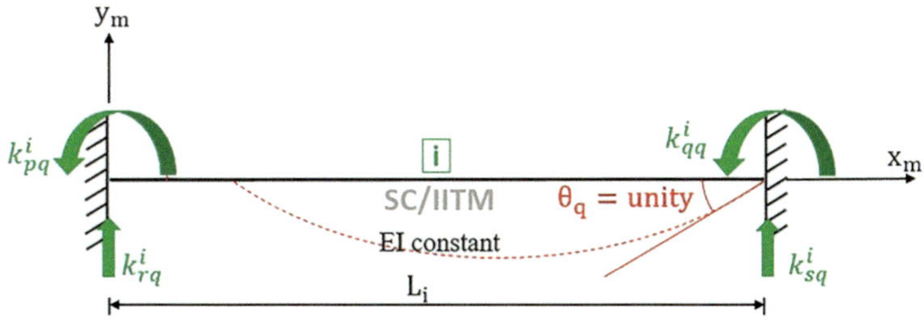

FIGURE 4.18 Fixed beam element with unit rotation at the *k*th end.

unit displacement in the *p*th degree of freedom of the *i*th member. Similarly, k^i_{qp} is the force in the *q*th degree of freedom by giving unit displacement in the *p*th degree of freedom of the *i*th member.

The second subscript in all the notations is common, which is *p*. It indicates that the unit displacement is given in the *p*th degree. By this manner, it is seen that stiffness coefficients are generated column-wise, as the second subscript in all the force components is the same, indicating the same column. Thus, the obtained stiffness coefficients correspond to the first column of the stiffness matrix. Similarly, by applying unit rotation at the *k*th end of the member, stiffness coefficients are derived, as shown in Figure 4.18. The stiffness coefficients in this case will be $k^i_{pq}, k^i_{qq}, k^i_{rq}, k^i_{sq}$. Figures 4.19 and 4.20 show the stiffness coefficients by imposing unit displacements at the *j*th and the *k*th ends, respectively. The stiffness coefficients, by applying unit displacement at the *j*th end, are $k^i_{pr}, k^i_{qr}, k^i_{rr}, k^i_{sr}$. By applying unit displacement at the *k*th end, the following stiffness coefficients are obtained: $k^i_{ps}, k^i_{qs}, k^i_{rs}, k^i_{ss}$. A tangent can be drawn by connecting the deflected position of the beam at which the unit rotation is applied and the initial position of the beam at the other end. It can be seen that the beam has undergone a rotation of 1/Li, where *Li* is the length of the member. Hence, rotation at the ends of the beam is equal to 1/Li.

For the *i*th beam element experiencing arbitrary end displacements, namely, (θ_p, δ_r) and (θ_q, δ_s), the corresponding end reactions (moment, shear) are required to

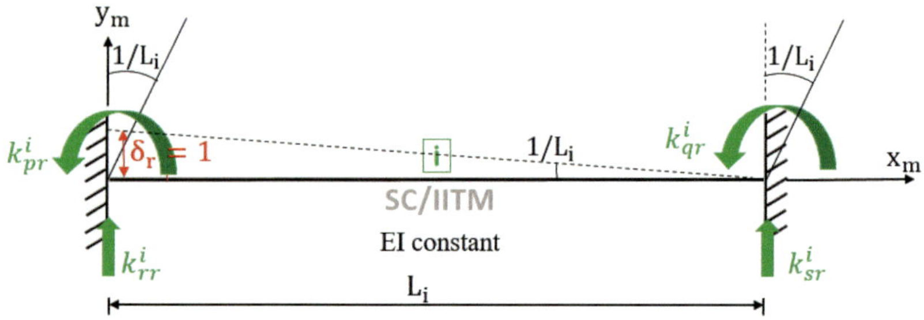

FIGURE 4.19 Fixed beam element with unit displacement at the *j*th end.

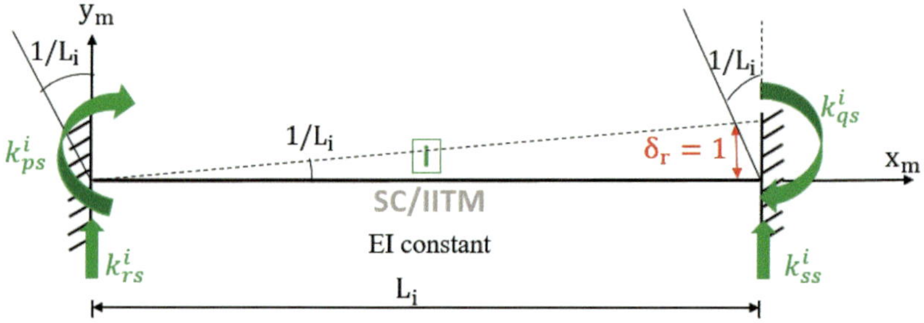

FIGURE 4.20 Fixed beam element with unit displacement at the *k*th end.

be estimated. They need to be estimated by maintaining equilibrium of the restrained member. The governing equations are as follows:

$$m_p^i = k_{pp}^i \theta_p + k_{pq}^i \theta_q + k_{pr}^i \delta_r + k_{ps}^i \delta_s$$

It can be seen form the preceding equation that the first subscript corresponds to the end at which the unit rotation or displacement is applied.

$$m_q^i = k_{qp}^i \theta_p + k_{qq}^i \theta_q + k_{qr}^i \delta_r + k_{qs}^i \delta_s$$

Similarly, force equations are written as follows:

$$p_r^i = k_{rp}^i \theta_p + k_{rq}^i \theta_q + k_{rr}^i \delta_r + k_{rs}^i \delta_s$$
$$p_s^i = k_{sp}^i \theta_p + k_{sq}^i \theta_q + k_{sr}^i \delta_r + k_{ss}^i \delta_s$$

The preceding equations give the end moments and end shear forces for arbitrary displacements $\theta_p, \theta_q, \delta_r$, and δ_s, which are unity at the respective degrees of freedom. These equations can be generalized as follows:

$$\{m_i\} = [k]_i \{\delta_i\}$$

Where:

$$\{m_i\} = \begin{Bmatrix} m_p \\ m_q \\ p_r \\ p_s \end{Bmatrix}, \{\delta_i\} = \begin{Bmatrix} \theta_p \\ \theta_q \\ \delta_r \\ \delta_s \end{Bmatrix} \text{ and}$$

$$[k]_i = \begin{bmatrix} k_{pp} & k_{pq} & k_{pr} & k_{ps} \\ k_{qp} & k_{qq} & k_{qr} & k_{qs} \\ k_{rp} & k_{rq} & k_{rr} & k_{rs} \\ k_{sp} & k_{sq} & k_{sr} & k_{ss} \end{bmatrix}$$

We need to evaluate only a set of rotational coefficients in the stiffness matrix, and the other coefficients can be easily written in terms of the same rotational coefficients. These rotational coefficients are $k^i_{pp}, k^i_{pq}, k^i_{qp}, k^i_{qq}$.

For example, in order to evaluate the end shear, the following relationship holds good:

$$k^i_{rp} = \frac{k^i_{pp} + k^i_{qp}}{L_i}$$

$$k^i_{sp} = -\frac{k^i_{pp} + k^i_{qp}}{L_i}$$

The negative sign in the preceding equation is due to the fact that the direction of k^i_{sp} is opposite to that of the end shear developed by the restraining moments, as shown in Figure 4.21. The upward shear is positive, and the downward shear is negative, from which the coefficients k_{rp} and k_{sp} can be obtained. It can be seen that the second subscript in the stiffness coefficients indicates the end at which the unit rotation is applied, and the first subscript indicates the respective forces in the degrees

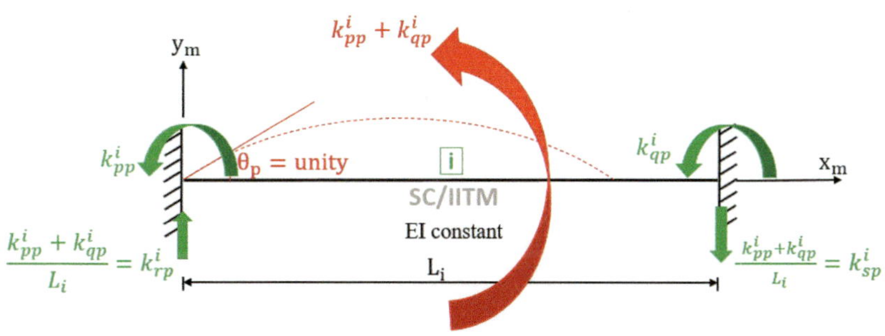

FIGURE 4.21 Fixed beam element showing rotation coefficients.

of freedom. In the same manner, the remaining stiffness coefficients can also be expressed in terms of the rotation coefficients $k^i_{pp}, k^i_{pq}, k^i_{qp}, k^i_{qq}$. Thus, out of 16 coefficients in the stiffness matrix, we need to evaluate only 4 rotational coefficients.

Referring to Figure 4.18, where unit rotation is applied at the kth end, the following relationship holds good:

$$k^i_{rq} = \frac{k^i_{pq} + k^i_{qq}}{L_i}$$

$$k^i_{sq} = -\frac{k^i_{pq} + k^i_{qq}}{L_i}$$

By referring to Figure 4.19, where unit displacement is applied at the jth end, the following relationship holds good:

$$k^i_{pr} = \frac{k^i_{pp} + k^i_{pq}}{L_i}$$

$$k^i_{qr} = -\frac{k^i_{qp} + k^i_{qq}}{L_i}$$

$$k^i_{rr} = \frac{k^i_{pr} + k^i_{qr}}{L_i} = \left[\frac{k^i_{pp} + k^i_{pq}}{(L_i)^2}\right] + \left[\frac{k^i_{qp} + k^i_{qq}}{(L_i)^2}\right]$$

Thus:

$$k^i_{rr} = \frac{k^i_{pp} + k^i_{pq} + k^i_{qp} + k^i_{qq}}{(L_i)^2}$$

Similarly:

$$k^i_{sr} = -\frac{k^i_{pp} + k^i_{pq} + k^i_{qp} + k^i_{qq}}{(L_i)^2}$$

By referring to Figure 4.20, where unit displacement is applied at the kth end, the following relationship holds good:

$$k^i_{ps} = -\frac{k^i_{pp} + k^i_{pq}}{L_i}$$

$$k^i_{qs} = -\frac{k^i_{qp} + k^i_{qq}}{L_i}$$

$$k_{rs}^{i} = -\frac{k_{ps}^{i} + k_{qs}^{i}}{L_i} = -\frac{k_{pp}^{i} + k_{pq}^{i} + k_{qp}^{i} + k_{qq}^{i}}{(L_i)^2}$$

$$k_{ss}^{i} = -\frac{k_{pp}^{i} + k_{pq}^{i} + k_{qp}^{i} + k_{qq}^{i}}{(L_i)^2}$$

Form the preceding equations, it can be seen that the end shear coefficients are expressed in terms of rotational coefficients. Thus, by evaluating the four rotational coefficients, a complete stiffness matrix with 16 stiffness coefficients can be framed, as given in the following:

$$[k] = \begin{bmatrix} k_{pp} & k_{pq} & \frac{k_{pp} + k_{pq}}{L} & -\left(\frac{k_{pp} + k_{pq}}{L}\right) \\ k_{qp} & k_{qq} & \frac{k_{qp} + k_{qq}}{L} & -\left(\frac{k_{qp} + k_{qq}}{L}\right) \\ \frac{k_{pp} + k_{pq}}{L} & \frac{k_{pq} + k_{qq}}{L} & \frac{k_{pp} + k_{pq} + k_{qp} + k_{qq}}{L^2} & \frac{k_{pp} + k_{pq} + k_{qp} + k_{qq}}{L^2} \\ -\left(\frac{k_{pp} + k_{pq}}{L}\right) & -\left(\frac{k_{pq} + k_{qq}}{L}\right) & -\left(\frac{k_{pp} + k_{pq} + k_{qp} + k_{qq}}{L^2}\right) & -\left(\frac{k_{pp} + k_{pq} + k_{qp} + k_{qq}}{L^2}\right) \end{bmatrix}$$

Flexibility and stiffness matrices are related as follows:

$$[\mathrm{D}][\mathrm{K}] = [\mathrm{I}]$$

Let $\begin{bmatrix} \delta_{jj} & \delta_{jk} \\ \delta_{kj} & \delta_{kk} \end{bmatrix}$ be the flexibility coefficients of a member with nodes j and k

And $\begin{bmatrix} k_{pp} & k_{pq} \\ k_{qp} & k_{qq} \end{bmatrix}$ be the stiffness coefficients.

The following relationship holds good:

$$\begin{bmatrix} S_{jj} & S_{jk} \\ S_{kj} & S_{kk} \end{bmatrix} \begin{bmatrix} k_{pp} & k_{pq} \\ k_{qp} & k_{qq} \end{bmatrix} = \begin{bmatrix} 1 & 0 \\ 0 & 1 \end{bmatrix}$$

Hence, the stiffness matrix k can be evaluated using the following relationship:

$$[\mathrm{K}][\mathrm{D}]^{-1} = [\mathrm{I}]$$

4.12.1 Stiffness Coefficients from Flexibility Coefficients

Consider a simply supported beam as shown in Figure 4.22. Unit rotation is applied at the jth end, and the flexibility coefficients are determined. Similarly, flexibility

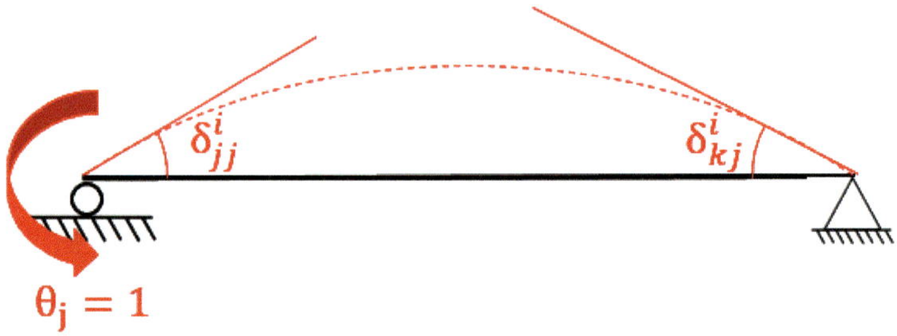

FIGURE 4.22 Simply supported beam—unit rotation at *j*.

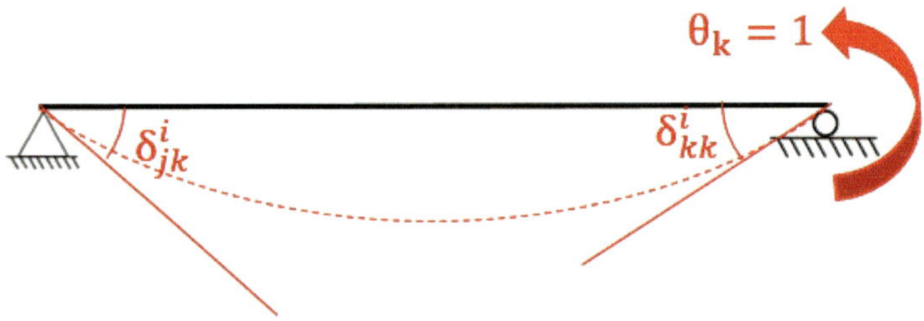

FIGURE 4.23 Simply supported beam—unit rotation at *k*.

coefficients are determined by applying unit displacement at the *k*th end, as shown in Figure 4.23. The flexibility coefficients $\left(\delta^i_{jj}, \delta^i_{kj}\right)$ define rotations at ends *j* and *k*, respectively, of the *i*th member, caused due to unit moment applied at the *j*th end. Similarly, flexibility coefficients $\left(\delta^i_{jk}, \delta^i_{kk}\right)$ define rotations at the *j*th and the *k*th ends of the *i*th member due to the unit moment applied at the *k*th end.

Let us consider a beam fixed at the end *q* and imposed by unit rotation at the end *p*, as shown in Figure 4.24. Similarly, assume the beam fixed at end *p* and imposed by unit rotation at end *q*, as shown in Figure 4.25. The stiffness coefficients $\left(k^i_{pp}, k^i_{qp}\right)$ define the end moments required at the *j*th and the *k*th ends to maintain equilibrium when the *j*th end is subjected to unit rotation, while the *k*th end is restrained. Similarly, the stiffness coefficients $\left(k^i_{pq}, k^i_{qq}\right)$ define the end moments required at the *j*th and the *k*th ends to maintain equilibrium, when the *k*th end is subjected to unit rotation and the *j*th end is restrained.

Thus:

$$\begin{bmatrix} S_{jj} & S_{jk} \\ S_{kj} & S_{kk} \end{bmatrix} \begin{bmatrix} k_{pp} & k_{pq} \\ k_{qp} & k_{qq} \end{bmatrix} = \begin{bmatrix} 1 & 0 \\ 0 & 1 \end{bmatrix}$$

Expanding the preceding equation:

$$k^i_{pp}\delta^i_{jj} + k^i_{qp}\delta^i_{jk} = 1$$

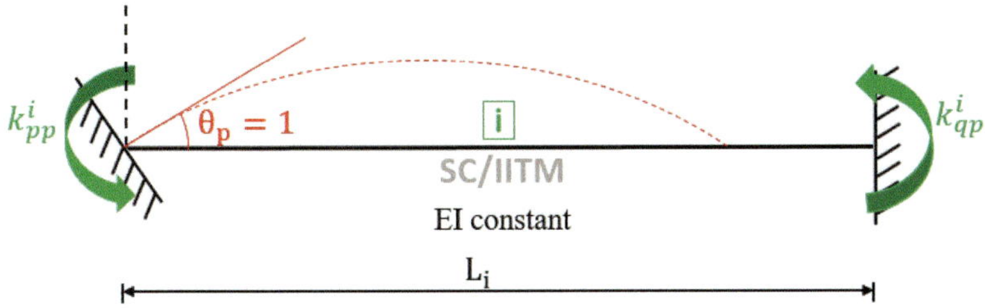

FIGURE 4.24 Fixed beam—unit rotation at *p*.

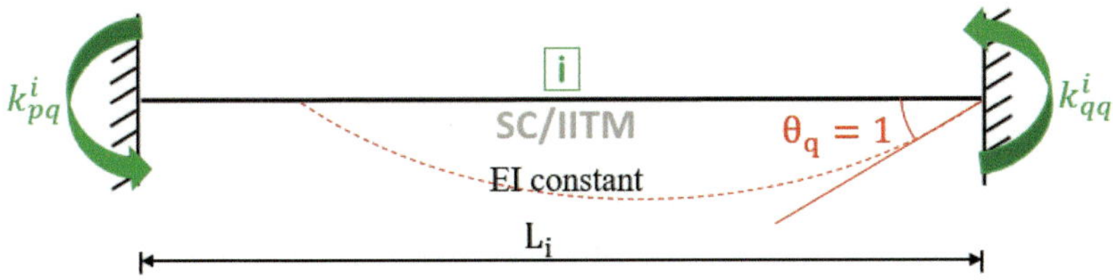

FIGURE 4.25 Fixed beam—unit rotation at *q*.

The preceding equation is required at the *j*th and the *k*th ends to maintain equilibrium, when the *k*th end is subjected to unit rotation and *j*th end $k^i_{pp}\delta^i_{kj} + k^i_{qp}\delta^i_{kk} = 1$.

$$k^i_{pq}\delta^i_{jj} + k^i_{qq}\delta^i_{jk} = 1$$

$$k^i_{pq}\delta^i_{kj} + k^i_{qq}\delta^i_{kk} = 1$$

Let us denote the flexibility matrix as D_r, and stiffness matrix as k_r. The subscript r stands for rotational degrees of freedom. In order to estimate the flexibility matrix for the beam element, assume the simply supported beam with unit moment at the *j*th end, as shown in Figure 4.26. The anticlockwise moment is balanced by the clockwise couple created by the forces. The bending moment diagram is also shown with tension at the top and compression at the bottom.

Let us replace the loading diagram with a conjugate beam, as shown in Figure 4.27. Taking moment about A:

$$V_B = \left[\left\{\frac{1}{2}L_i\left(\frac{1}{EI}\right)\right\}\frac{1}{3}L_i\right]\frac{1}{L_i} = \frac{L_i}{6EI} \text{ (downward)}$$

$$V_A = \left\{\frac{1}{2}L_i\left(\frac{1}{EI}\right)\right\} - \frac{L_i}{6EI} = \frac{L_i}{3EI} \text{ (upward)}$$

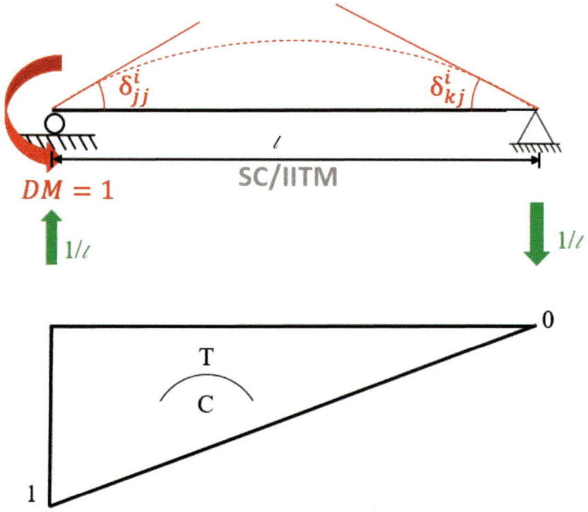

FIGURE 4.26 Simply supported beam.

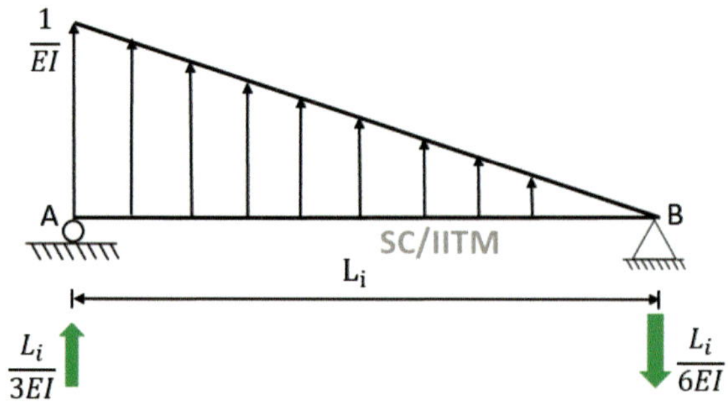

FIGURE 4.27 Conjugate beam.

The same procedure is followed for the other case to derive the following flexibility matrix:

$$D_r = \begin{bmatrix} \frac{L}{3EI} & -\frac{L}{6EI} \\ -\frac{L}{6EI} & \frac{L}{3EI} \end{bmatrix}$$

Therefore:

$$k_r = [D_r]^{-1} = \frac{12(EI)^2}{L^2}\begin{bmatrix} \frac{L}{3EI} & \frac{L}{6EI} \\ \frac{L}{6EI} & \frac{L}{3EI} \end{bmatrix} = \begin{bmatrix} \frac{4EI}{L} & \frac{2EI}{L} \\ \frac{2EI}{L} & \frac{4EI}{L} \end{bmatrix}$$

Thus, from the preceding four rotational coefficients, the whole stiffness matrix can be derived.

$$K_i = \begin{bmatrix} \frac{4EI}{l} & \frac{2EI}{l} & \frac{6EI}{l^2} & -\frac{6EI}{l^2} \\ \frac{2EI}{l} & \frac{4EI}{l} & \frac{6EI}{l^2} & -\frac{6EI}{l^2} \\ \frac{6EI}{l^2} & \frac{6EI}{l^2} & \frac{12EI}{l^3} & -\frac{12EI}{l^3} \\ -\frac{6EI}{l^2} & -\frac{6EI}{l^2} & -\frac{12EI}{l^3} & \frac{12EI}{l^3} \end{bmatrix}$$

4.13 STIFFNESS METHOD OF ANALYSIS FOR BEAMS AND PLANAR ORTHOGONAL STRUCTURES

The stiffness method is more generic and easily programmable compared to that of the flexibility method. The stiffness method is also not problem-specific. The stiffness matrix of a fixed beam element can be easily developed only with the rotational coefficients. Let us apply this method to a planar orthogonal structure. Consider a single-bay, single-story frame, as shown in Figure 4.28. Both the ends of the frame are fixed, and the frame is subjected to some arbitrary loading. The height of the

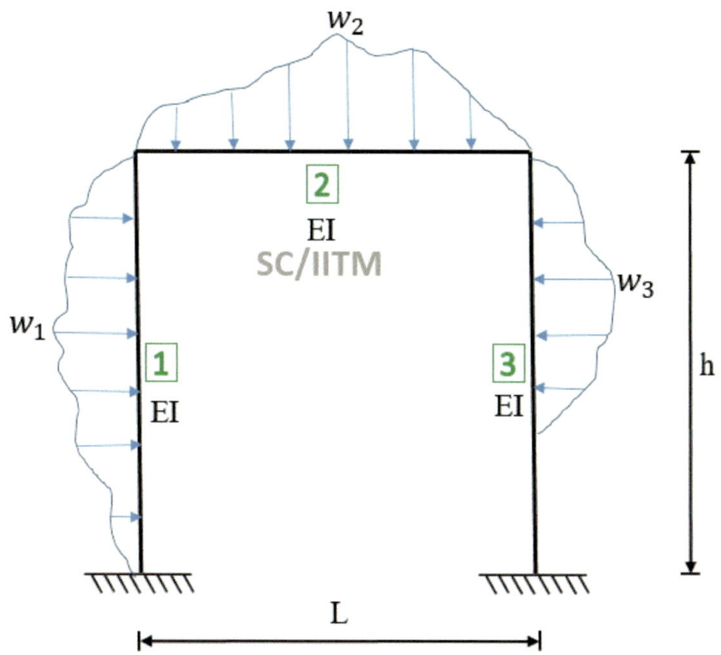

FIGURE 4.28 Planar orthogonal structure.

frame is taken as h, and flexural rigidity is EI. There are some basic steps to formulate a stiffness method of analysis for solving this problem. The unknowns in the stiffness method are displacements, which can be translational as well as rotational. Thus, the stiffness method is a generic method which can be applied to any frame under arbitrary loading, because the unknowns in the analysis are not actions but displacements. It is more or less a well-defined procedure. The members are numbered-indexed using square brackets, as shown in the figure. There are three members in the frame considered for the analysis. Let the joints be numbered in a sequence using circles. There are four joints in the frame.

For every member, it is now important to identify the joint numbering in accordance to the standard fixed beam. Certain literature use the transformation matrix for solving the problem, where the orientation of the member with respect to the origin axis becomes important. But we will handle this following a different method without any transformation matrix. So the elemental member will be a fixed beam, irrespective of the boundary conditions of the original beam considered for the analysis.

STEP 1: MARK THE DEGREES OF FREEDOM

The elemental member should have both the ends fixed with four degrees of freedom in translation and rotation by neglecting axial deformation, as shown in Figure 4.29. We will follow the same notations and order for all the problems.

STEP 2: IDENTIFY UNRESTRAINED AND RESTRAINED DISPLACEMENTS AT EACH JOINT

While numbering the displacements, first label the unrestrained displacements of all the joints. Then, label restrained displacements of all the joints. This is called grouping, which helps to do cross-partitioning of the matrix. This will make the analysis more simple, closed-form, and very easy.

For example, consider a frame with both ends fixed, as shown in Figure 4.30. The frame has three members and four joints. Joints 2 and 3 are free to rotate and displace, which means the frame can sway. The unrestrained displacements are marked first: θ_1, θ_2 and δ_3, neglecting the axial deformation. Thus, the stiffness matrix in unrestrained degrees of freedom is a 3×3 matrix. Joints 1 and 4 are fixed, not allowing any rotation and displacement. The restrained displacements are δ_4, δ_5, θ_6, θ_7, δ_8, and δ_9. Thus, the total number of degrees of freedom of the frame is 9. It shows that the size of the stiffness matrix will be 9×9. But the stiffness matrix of the whole frame can be partitioned.

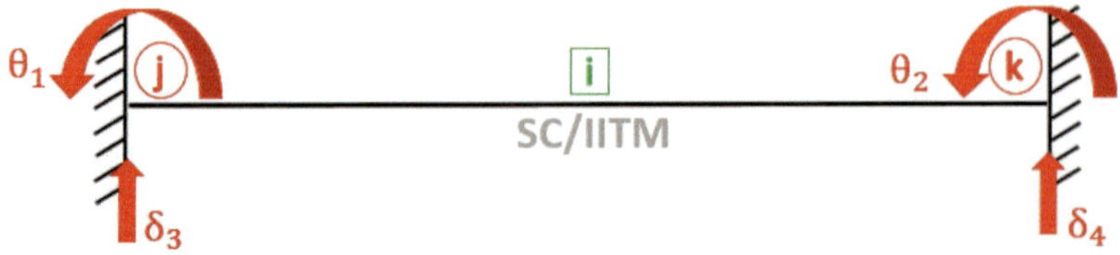

FIGURE 4.29 Elemental fixed beam with degrees of freedom.

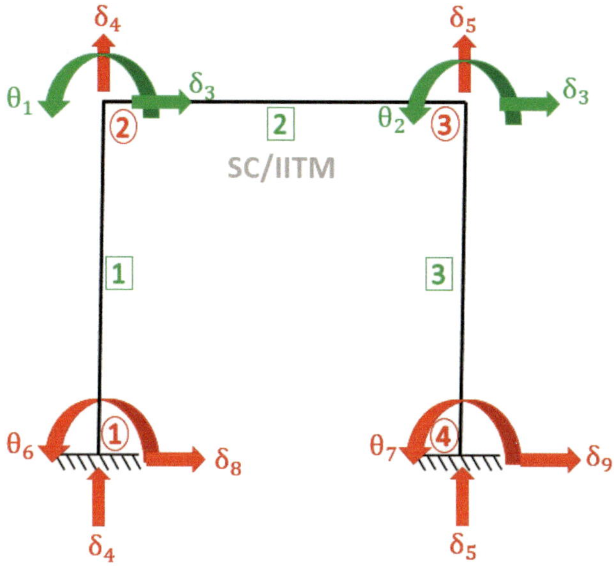

FIGURE 4.30 Single-bay, single-story frame.

The stiffness matrix in unrestrained degrees of freedom is denoted by k_{uu}, and the stiffness matrix in restrained degrees of freedom is denoted by k_{rr}, as shown in the following.

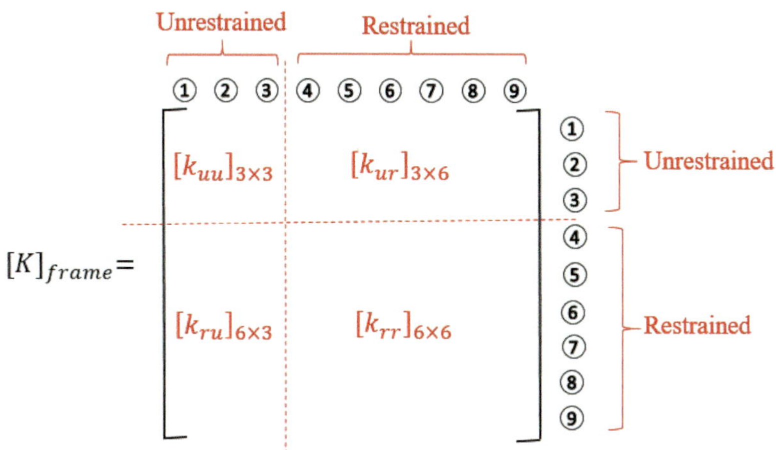

STEP 3: DETERMINE THE UNRESTRAINED DISPLACEMENTS

Let us consider the stiffness matrix in terms of unrestrained and restrained stiffness matrices. Let the unrestrained and restrained displacements be denoted as Δ_u and Δ_r, respectively. The product of the stiffness matrix and the displacement will give the sum of the joint load vector and partial reaction vector, which can again be partitioned into unrestrained and restrained load vectors and reaction vectors, respectively. The partitioned stiffness matrix, partitioned displacement vector, partitioned joint load vector, and partitioned reaction vector are given in the following equation:

$$\begin{bmatrix} [K_{uu}] & [K_{ur}] \\ [K_{ru}] & [K_{rr}] \end{bmatrix} \begin{Bmatrix} \Delta_u \\ \Delta_r \end{Bmatrix} = \begin{Bmatrix} J_{Lu} \\ J_{Lr} \end{Bmatrix} + \begin{Bmatrix} [0] \\ R_r \end{Bmatrix}$$

Let us expand the preceding equation:

$$[k_{uu}]\{\Delta_u\} = \{J_L\}_u$$
$$\{\Delta_u\} = [k_{uu}]^{-1}\{J_L\}_u$$

Thus, Δ_u can be computed from the previous equation. To compute Δ_u, we need to know the joint load vector also.

STEP 4: ESTIMATE THE JOINT LOAD VECTOR (J_L)

The joint load vector is derived based on the fixed end moments generated from the applied loads on each member. The fixed end moments for different loading conditions are given in Table 4.3.

Joint load will be simply the reversal of the fixed end moments. End moments due to displacements can also be found out as given in Table 4.4.

TABLE 4.3
Fixed End Moments

Sl. No.	Beam	M_F^{AB}	M_F^{BA}
1	Fixed beam with centrally concentrated load 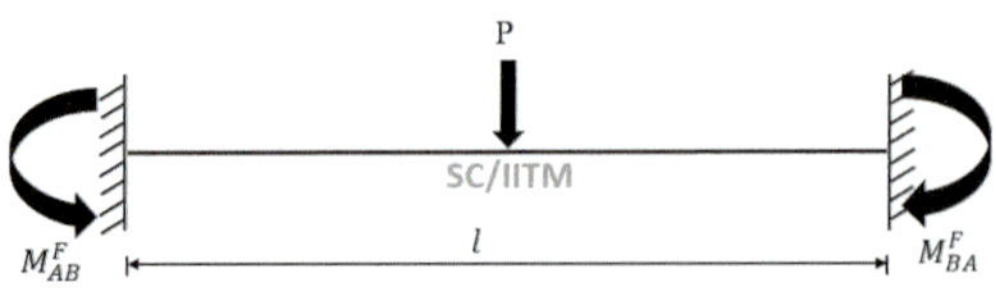	$+\frac{Pl}{8}$	$-\frac{Pl}{8}$
2	Fixed beam with uniformly distributed load 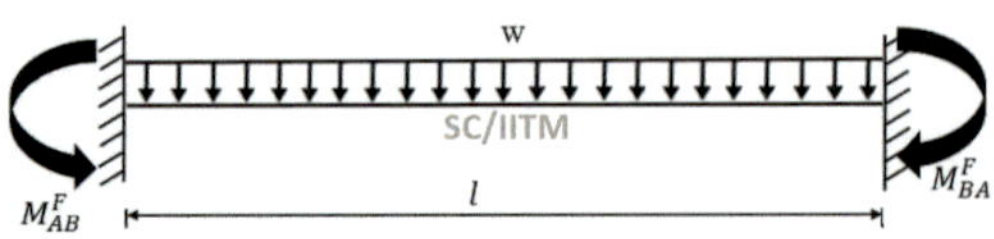	$+\frac{wl^2}{12}$	$-\frac{wl^2}{12}$
3	Fixed beam with eccentric concentrated load 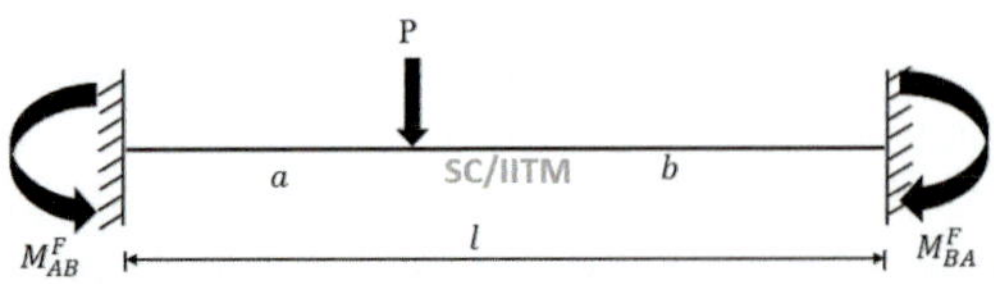	$+\frac{pab^2}{l^2}$	$-\frac{pba^2}{l^2}$

TABLE 4.3 (*Continued*)

Fixed End Moments

Sl. No.	Beam	M_F^{AB}	M_F^{BA}
4	Fixed bean with uniformly varying load 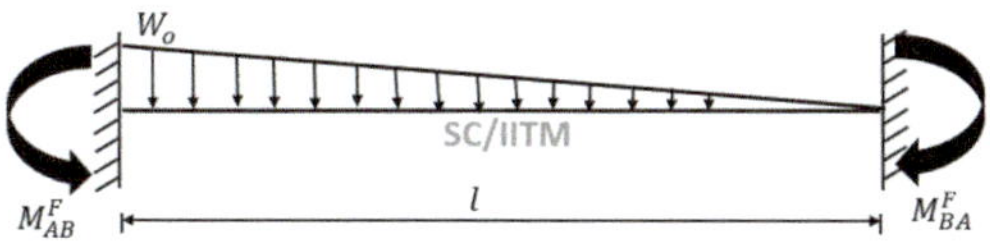	$+\frac{W_o l^2}{20}$	$-\frac{W_o l^2}{20}$
5	Fixed beam with triangular loading 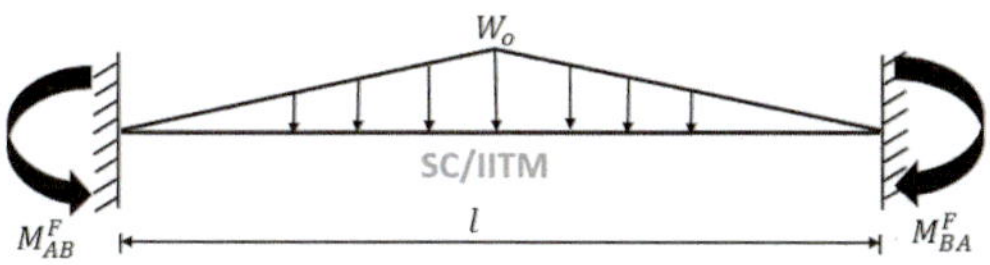	$+\frac{5}{96}W_o l^2$	$-\frac{5}{96}W_o l^2$
6	Fixed beam with eccentric anticlockwise moment 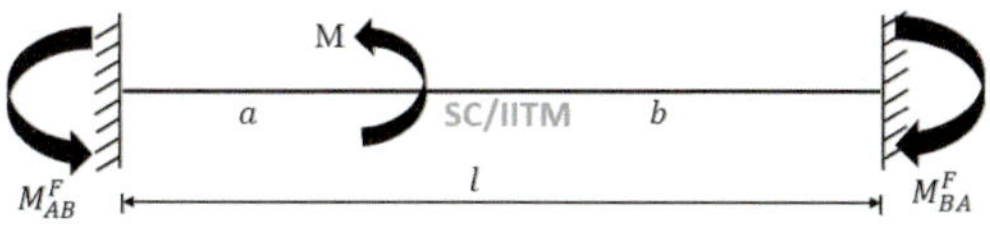	$+\frac{M}{l^2}b(b-2a)$	$+\frac{M}{l^2}a(2b-a)$
7	Fixed beam with central clockwise moment 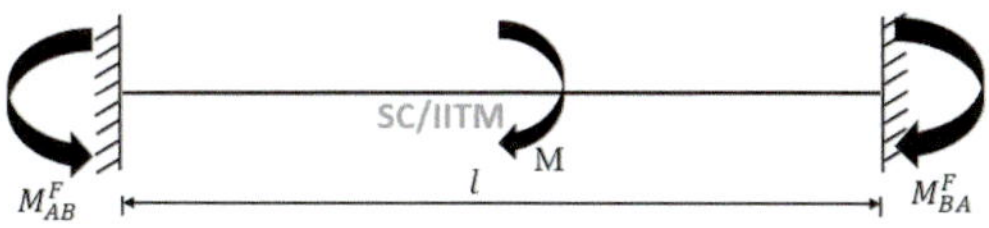	$-\frac{M}{4}$	$+\frac{M}{4}$

Note: Sign convention: Anticlockwise moments are positive.

TABLE 4.4

End Moments Duet to Displacements

Sl. No.	Beam	M_F^{AB}	M_F^{BA}
1	Beam with rotation at the *j*th end 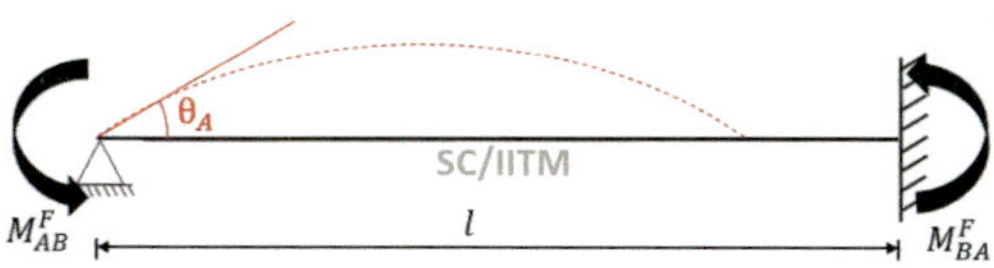	$+\frac{4EI\theta_A}{l}$	$+\frac{2EI\theta_A}{l}$
2	Beam with rotation at the *k*th end 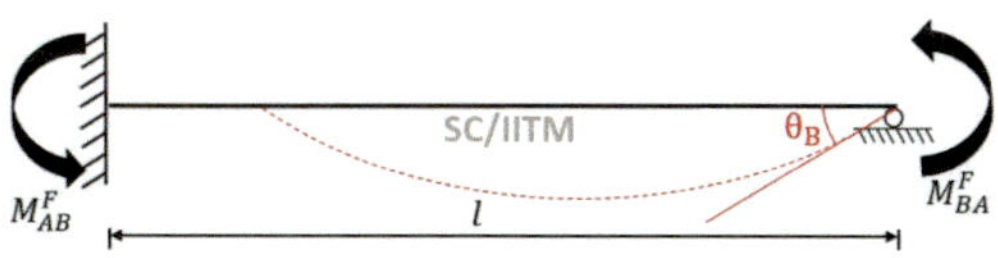	$+\frac{2EI\theta_B}{l}$	$+\frac{4EI\theta_B}{l}$

(*Continued*)

TABLE 4.4 (*Continued*)

End Moments Duet to Displacements

Sl. No.	Beam	M_F^{AB}	M_F^{BA}
3	Simply supported beam with rotation 	$+\frac{3EI\theta_A}{l}$	NIL
4	Beam with settlement of support 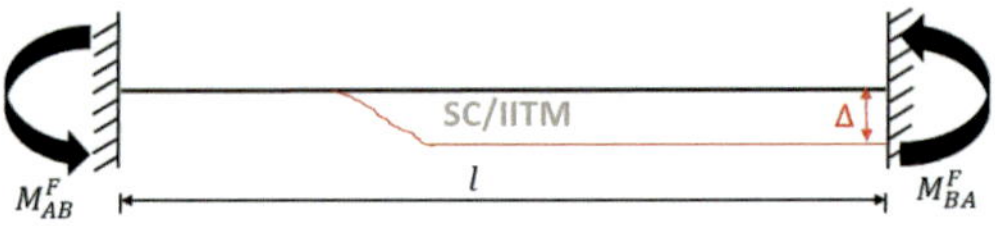	$+\frac{6EI\Delta}{l^2}$	$+\frac{6EI\Delta}{l^2}$
5	Beam with settlement of support 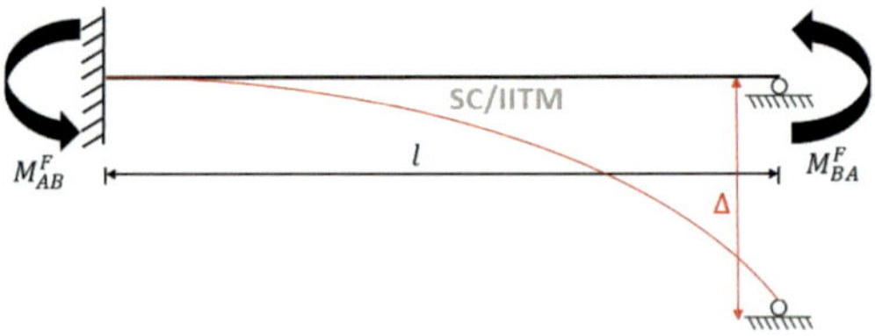	$+\frac{3EI\Delta}{l^2}$	NIL

Thus, these are the standard formats by which one can find the end moments either caused by the load on the beam or by the displacement on the joint or rotation on the joint. By reversing this, joint loads can be obtained, from which the unrestrained displacements can be computed.

4.13.1 Example Problem 1

Analyze the continuous beam and find the reactions and end moments of the beam shown in the following:

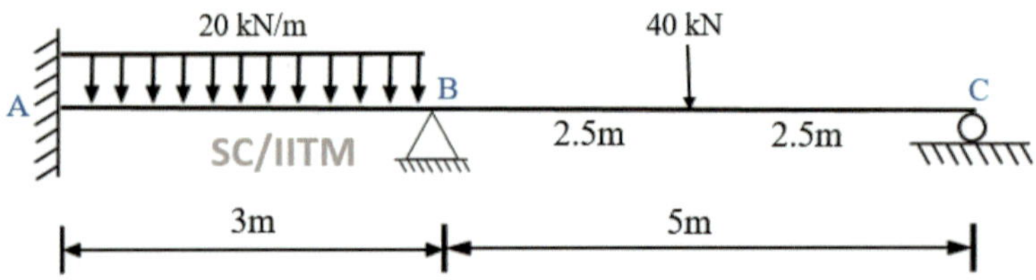

4.13.2 Solution

4.13.2.1 Marking Unrestrained and Restrained Degrees of Freedom

The unrestrained and restrained degrees of freedom are marked on the continuous beam, as shown in what follows:

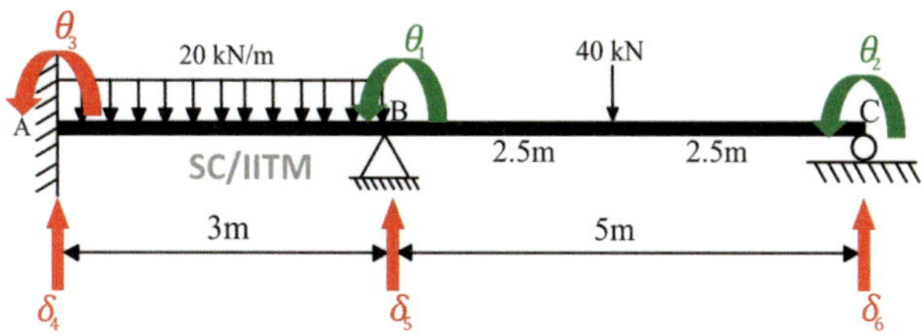

Thus, there are, in total, six degrees of- freedom, with two unrestrained and four restrained degrees of freedom.

$$\text{Unrestrained degrees of freedom} = 2\left[\theta_1, \theta_2\right]$$

$$\text{Restrained degrees of freedom} = 4\left[\theta_3, \delta_4, \delta_5, \delta_6\right]$$

Thus, the unrestrained sub-matrix will be of size 2 × 2. The size of the total stiffness matrix will be 6 × 6. The element stiffness matrix will be 4 × 4. Let us compare the given problem with a standard fixed beam to obtain the labels. Even though support B is simply supported, the basic element is a fixed beam.

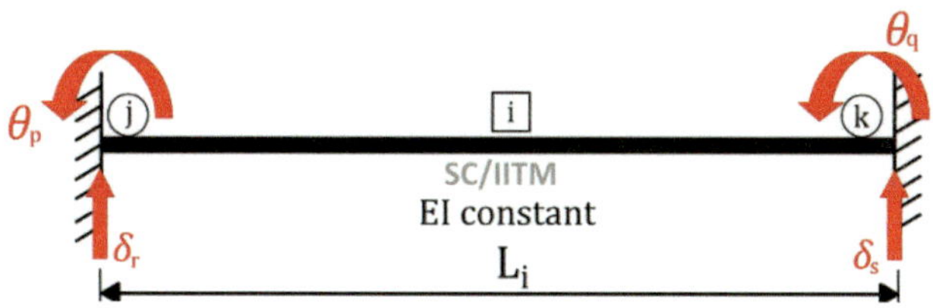

Thus, the labels are:

$$\begin{aligned} AB &= [3, 1, 4, 5] \\ BC &= [1, 2, 5, 6] \end{aligned}$$

The labels are ordered in such a manner that the rotations are followed by translations.

4.13.2.2 Formulation of the Stiffness Matrix

We already know the formulation of the stiffness matrix for a standard fixed beam element. The stiffness matrix is given by:

$$K_i = \begin{bmatrix} \frac{4EI}{l} & \frac{2EI}{l} & \frac{6EI}{l^2} & -\frac{6EI}{l^2} \\ \frac{2EI}{l} & \frac{4EI}{l} & \frac{6EI}{l^2} & -\frac{6EI}{l^2} \\ \frac{6EI}{l^2} & \frac{6EI}{l^2} & \frac{12EI}{l^3} & -\frac{12EI}{l^3} \\ -\frac{6EI}{l^2} & -\frac{6EI}{l^2} & -\frac{12EI}{l^3} & \frac{12EI}{l^3} \end{bmatrix}$$

Thus, the size of the stiffness matrix is 4 × 4, neglecting axial deformation. For this problem, the element stiffness matrices are given by:

$$K_{AB} = EI\begin{bmatrix} 1.333 & 0.667 & 0.667 & -0.667 \\ 0.667 & 1.333 & 0.667 & -0.667 \\ 0.667 & 0.667 & 0.445 & -0.445 \\ -0.667 & -0.667 & -0.445 & 0.445 \end{bmatrix} \begin{matrix} ③ \\ ① \\ ④ \\ ⑤ \end{matrix}$$

(columns: ③ ① ④ ⑤)

$$K_{BC} = EI\begin{bmatrix} 0.80 & 0.40 & 0.24 & -0.24 \\ 0.40 & 0.80 & 0.24 & -0.24 \\ 0.24 & 0.24 & 0.096 & -0.096 \\ -0.24 & -0.24 & -0.096 & 0.096 \end{bmatrix} \begin{matrix} ① \\ ② \\ ⑤ \\ ⑥ \end{matrix}$$

(columns: ① ② ⑤ ⑥)

Thus, the total stiffness matrix will be written as follows:

$$[K] = \begin{bmatrix} [k_{uu}]_{2\times2} & [k_{ur}]_{2\times4} \\ [k_{ru}]_{4\times2} & [k_{rr}]_{4\times4} \end{bmatrix}$$

We need only the stiffness matrix of unrestrained degrees of freedom.

$$K_{UU} = EI\begin{bmatrix} 2.133 & 0.4 \\ 0.4 & 0.8 \end{bmatrix} \begin{matrix} ① \\ ② \end{matrix}$$

(columns: ① ②)

Then:

$$K_{UU}^{-1} = \frac{1}{1.546EI}\begin{bmatrix} 2.133 & 0.4 \\ 0.4 & 0.8 \end{bmatrix}$$

4.13.2.3 Calculation of Fixed End Moments

Now, the fixed end moments are calculated to find the unrestrained displacements. The fixed end moments and reactions of the members are shown in the following:

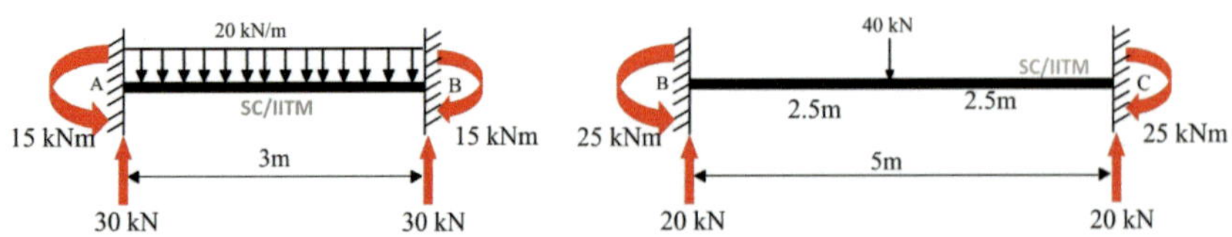

$$M_{AB}^{F} = \frac{wl^2}{12} = \frac{20 \times 3^2}{12} = +15\ kNm$$

$$M_{BA}^{F} = -\frac{wl^2}{12} = -\frac{20 \times 3^2}{12} = -15\ kNm$$

$$V_A = +30\ \text{kN}$$

$$V_B = +30\ \text{kN} \quad M_{BC}^{F} = \frac{pl}{8} = \frac{40 \times 5}{8} = +25\ kNm$$

$$M_{CB}^{F} = -\frac{pl}{8} = -\frac{40 \times 5}{8} = -25\ kNm$$

$$V_B = +20\ \text{kN}$$

$$V_C = +20\ \text{kN}$$

Thus, for the whole beam, the fixed end moments are shown as thus:

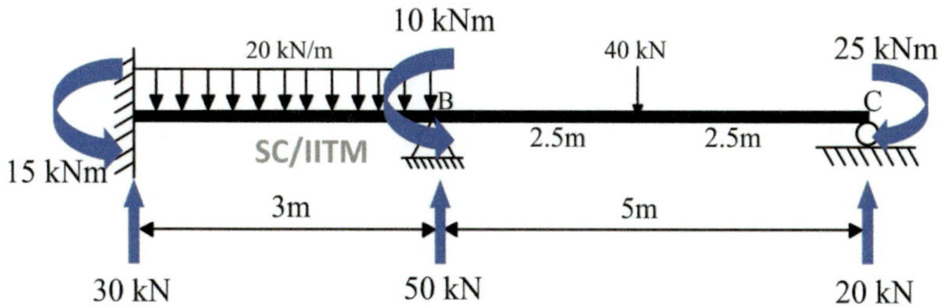

4.13.2.4 Calculation of Joint Load Vectors

The joint loads will be the reversal of the fixed end moments and the reactions. The joint load for the continuous beam is shown in what follows:

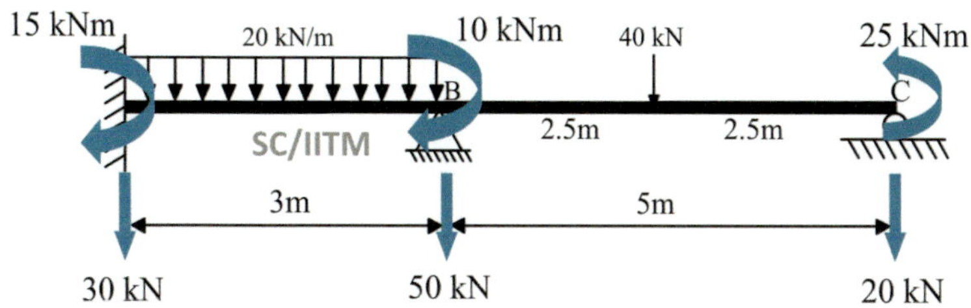

Now, the joint load vector is given by:

$$J_L = \begin{Bmatrix} J_{LU} \\ --- \\ J_{LR} \end{Bmatrix}$$

$$J_L = \begin{Bmatrix} -10 \\ +25 \\ -15 \\ -30 \\ -50 \\ -20 \end{Bmatrix} \begin{matrix} ① \\ ② \\ ③ \\ ④ \\ ⑤ \\ ⑥ \end{matrix}$$

The unrestrained joint load vector is given by:

$$J_{LU} = \begin{Bmatrix} -10 \\ +25 \end{Bmatrix} \begin{matrix} ① \\ ② \end{matrix}$$

4.13.2.5 Calculation of Displacements

Now:

$$\{\Delta_u\} = [k_{uu}]^{-1}\{J_L\}_u$$

$$\begin{Bmatrix} \theta_1 \\ \theta_2 \end{Bmatrix} = \frac{1}{1.546\,EI}\begin{bmatrix} 0.8 & -0.4 \\ -0.4 & 2.133 \end{bmatrix}\begin{Bmatrix} -10 \\ +25 \end{Bmatrix} = \begin{Bmatrix} -18/1.546\,EI \\ 57.325/1.546\,EI \end{Bmatrix}$$

The vector of restrained displacements will be zero.

$$\begin{Bmatrix} \theta_3 \\ \delta_4 \\ \delta_5 \\ \delta_6 \end{Bmatrix} = 0$$

4.13.2.6 Calculation of End Moment and Shear

The general equation to find the end moment of the ith beam is:

$$[M]_i = k_i\delta_i + (FEM)_i$$

The equation can be rewritten element-wise as follows:

$$M_{AB} = [k]_{AB}\{\delta\} + \{FEM\}$$

$$\begin{Bmatrix} M_3 \\ M_1 \\ V_4 \\ V_5 \end{Bmatrix} = EI\begin{bmatrix} 1.333 & 0.667 & 0.667 & -0.667 \\ 0.667 & 1.333 & 0.667 & -0.667 \\ 0.667 & 0.667 & 0.445 & -0.445 \\ -0.667 & -0.667 & -0.445 & 0.445 \end{bmatrix}\begin{Bmatrix} \theta_3 \\ \theta_1 \\ \delta_4 \\ \delta_5 \end{Bmatrix} + \begin{Bmatrix} -15 \\ -15 \\ +30 \\ +30 \end{Bmatrix} = \begin{Bmatrix} 7.234 \\ -30.52 \\ 22.234 \\ 37.766 \end{Bmatrix}$$

$$M_{BC} = [k]_{BC}\{\delta\} + \{FEM\}$$

$$\begin{Bmatrix} M_1 \\ M_2 \\ V_5 \\ V_6 \end{Bmatrix} = EI \begin{bmatrix} 0.8 & 0.4 & 0.24 & -0.24 \\ 0.4 & 0.8 & 0.24 & -0.24 \\ 0.24 & 0.24 & 0.096 & -0.096 \\ -0.24 & -0.24 & -0.096 & 0.096 \end{bmatrix} \begin{Bmatrix} \theta_1 \\ \theta_2 \\ \delta_5 \\ \delta_6 \end{Bmatrix} + \begin{Bmatrix} 25 \\ -25 \\ 20 \\ 20 \end{Bmatrix} = \begin{Bmatrix} 30.517 \\ 0 \\ 26.105 \\ 13.895 \end{Bmatrix}$$

The member end moments and shear are shown in what follows:

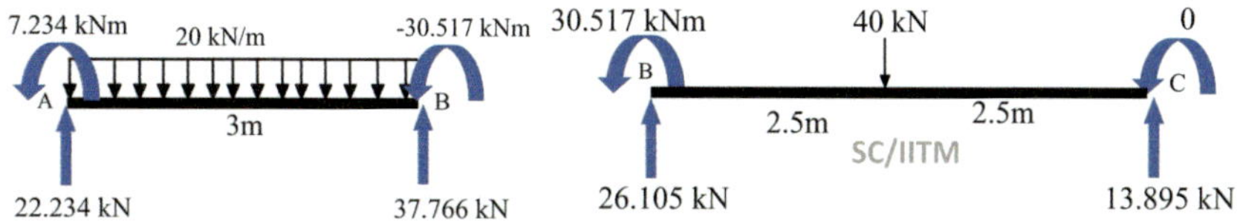

Check for span AB:

$$\sum V = 0,$$

$$22.234 + 37.766 = 20 \times 3 = 60 \text{ kN}$$

$$MA = (20 \times 3 \times 1.5) + 30.52 - 37.766(3) = 7.234 \text{ kNm}$$

Hence, okay.

Check for span BC:

$$\sum V = 0,$$

$$(26.105 + 13.895) = 40 \text{ kN}$$

$$M_B = (40 \times 2.5) - (13.895 \times 5) = 30.52 \text{ kNm}$$

Hence, okay.

By superimposing the end moments of the members into the continuous beam, the final end moments and shear are obtained as shown in the following:

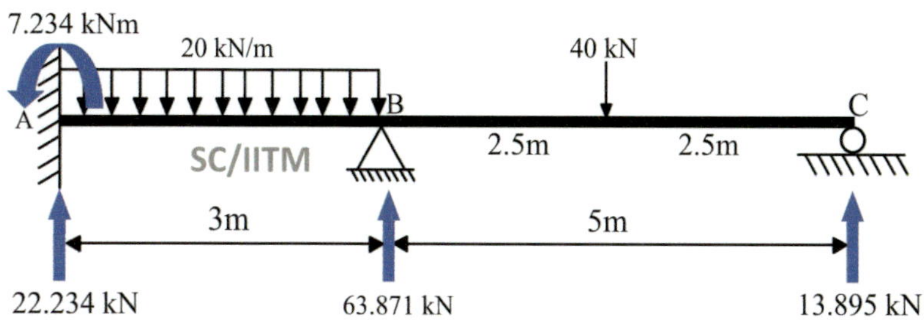

CHECK

We can say that:

$$[k_{ru}]\{\Delta_u\} - \{J_l\}_r = \{R_r\}$$

$$[k_{ru}] = EI\begin{bmatrix} 0.667 & 0 \\ 0.667 & 0 \\ -0.427 & 0.24 \\ -0.24 & -0.24 \end{bmatrix}$$

$$\{\Delta_u\}\frac{1}{1.546EI}\begin{Bmatrix} -18 \\ 57.325 \end{Bmatrix}$$

$$\{U_l\}_r = \begin{Bmatrix} -15 \\ -30 \\ -50 \\ -20 \end{Bmatrix} \begin{matrix} ③ \\ ④ \\ ⑤ \\ ⑥ \end{matrix}$$

Thus:

$$\{R_r\} = \begin{Bmatrix} M_3 \\ V_4 \\ V_5 \\ V_6 \end{Bmatrix} = \begin{Bmatrix} 7.234 \\ 22.234 \\ 63.81 \\ 13.895 \end{Bmatrix}. \text{ Hence, okay.}$$

MATLAB PROGRAM

```
%% stiffness matrix method
% Input
clc;
clear;
n = 2; % number of members
I = [1 1]; %Moment of inertias in m4
L = [3 5]; % length in m
uu = 2; % Number of unrestrained degrees of freedom
ur = 4; % Number of restrained degrees of freedom
uul = [1 2]; % global labels of unrestrained dof
url = [3 4 5 6]; % global labels of restrained dof
l1 = [3 1 4 5]; % Global labels for member 1
l2 = [1 2 5 6]; % Global labels for member 2
l = [l1; l2];
dof = uu+ur;
Ktotal = zeros (dof);
fem1 = [15 -15 30 30]; % Local Fixed end moments of
member 1
fem2 = [25 -25 20 20]; % Local Fixed end moments of
member 2

%% Creation of joint load vector
```

```
jl = [-10; 25; -15; -30; -50; -20]; % values given in kN
or kNm
jlu = [-10; 25]; % load vector in unrestrained dof

%% rotation coefficients for each member
rc1 = 4.*I./L;
rc2 = 2.*I./L;
%% stiffness matrix 4 by 4 (axial deformation neglected)
for i = 1:n
Knew = zeros (dof);
k1 = [rc1(i); rc2(i); (rc1(i)+rc2(i))/L(i); (-(rc1(i)+
rc2(i))/L(i))];
k2 = [rc2(i); rc1(i); (rc1(i)+rc2(i))/L(i); (-(rc1(i)+
rc2(i))/L(i))];
k3 = [(rc1(i)+rc2(i))/L(i); (rc1(i)+rc2(i))/L(i);
(2*(rc1(i)+rc2(i))/(L(i)^2)); (-2*(rc1(i)+rc2(i))/
(L(i)^2))];
k4 = -k3;
K = [k1 k2 k3 k4];
fprintf ('Member Number =');
disp (i);
fprintf ('Local Stiffness matrix of member, [K] = \n');
disp (K);
  for p = 1:4
    for q = 1:4
      Knew((l(i,p)),(l(i,q))) = K(p,q);
    end
  end
  Ktotal = Ktotal + Knew;
  if i == 1
      Kg1 = K;
  else
      Kg2 = K;
end
end
fprintf ('Stiffness Matrix of complete structure,
[Ktotal] = \n');
disp (Ktotal);
Kunr = zeros(uu);
forx = 1:uu
  fory = 1:uu
    Kunr(x,y) = Ktotal(x,y);
  end
end
fprintf ('Unrestrained Stiffness sub-matrix, [Kuu] = \n');
disp (Kunr);
KuuInv = inv(Kunr);
```

```
fprintf ('Inverse of Unrestrained Stiffness sub-matrix,
[KuuInverse] = \n');
disp (KuuInv);
%% Calculation of displacements
delu = KuuInv*jlu;
fprintf ('Joint Load vector, [Jl] = \n');
disp (jl');
fprintf ('Unrestrained displacements, [DelU] = \n');
disp (delu');
delr = zeros (ur,1);
del = [delu; delr];
deli= zeros (4,1);
for i = 1:n
  for p = 1:4
    deli(p,1) = del((l(i,p)),1);
  end
  if i == 1
    delbar1 = deli;
    mbar1 = (Kg1 * delbar1)+fem1';
    fprintf ('Member Number =');
    disp (i);
    fprintf ('Global displacement matrix [DeltaBar] = \n');
    disp (delbar1');
    fprintf ('Global End moment matrix [MBar] = \n');
    disp (mbar1');
  else
    delbar2 = deli;
    mbar2 = (Kg2 * delbar2)+fem2';
    fprintf ('Member Number =');
    disp (i);
    fprintf ('Global displacement matrix [DeltaBar] = \n');
    disp (delbar2');
    fprintf ('Global End moment matrix [MBar] = \n');
    disp (mbar2');
end
end
%% check
mbar = [mbar1'; mbar2'];
jf = zeros(dof,1);
for a = 1:n
for b = 1:4 % size of k matrix
d = l(a,b);
jfnew = zeros(dof,1);
jfnew(d,1)= mbar(a,b);
jf = jf+jfnew;
end
end
```

```
fprintf ('Joint forces = \n');
disp (jf');
```

MATLAB OUTPUT

```
Member Number = 1
Local Stiffness matrix of member, [K] =
   1.3333    0.6667    0.6667   -0.6667
   0.6667    1.3333    0.6667   -0.6667
   0.6667    0.6667    0.4444   -0.4444
  -0.6667   -0.6667   -0.4444    0.4444
Member Number = 2
Local Stiffness matrix of member, [K] =
   0.8000    0.4000    0.2400   -0.2400
   0.4000    0.8000    0.2400   -0.2400
   0.2400    0.2400    0.0960   -0.0960
  -0.2400   -0.2400   -0.0960    0.0960
Stiffness Matrix of complete structure, [Ktotal] =
   2.1333    0.4000    0.6667    0.6667   -0.4267   -0.2400
   0.4000    0.8000         0         0    0.2400   -0.2400
   0.6667         0    1.3333    0.6667   -0.6667         0
   0.6667         0    0.6667    0.4444   -0.4444         0
  -0.4267    0.2400   -0.6667   -0.4444    0.5404   -0.0960
  -0.2400   -0.2400         0         0   -0.0960    0.0960
Unrestrained Stiffness sub-matix, [Kuu] =
  2.1333  0.4000
  0.4000  0.8000
Inverse of Unrestrained Stiffness sub-matix,
[KuuInverse] =
   0.5172  -0.2586
  -0.2586   1.3793
Joint Load vector, [Jl] =
  -10  25  -15  -30  -50  -20
Unrestrained displacements, [DelU] =
  -11.6379   37.0690
Member Number = 1
Global displacement matrix [DeltaBar] =
         0  -11.6379       0  0
Global End moment matrix [MBar] =
    7.2414  -30.5172  22.2414  37.7586
Member Number = 2
Global displacement matrix [DeltaBar] =
  -11.6379   37.0690        0        0
Global End moment matrix [MBar] =
   30.5172   -0.0000  26.1034  13.8966
Joint forces =
         0   -0.0000   7.2414  22.2414  63.8621  13.8966
```

4.13.3 Example Problem 2: Orthogonal Frame

Analyze the single-bay, single-story frame shown in what follows:

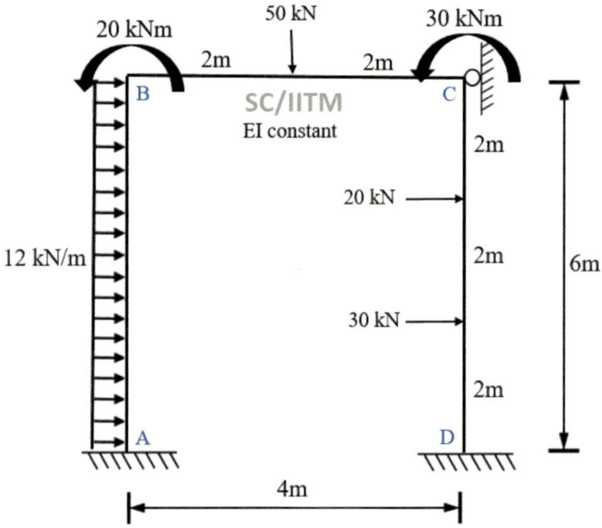

4.13.4 Solution

4.13.4.1 Marking Unrestrained and Restrained Degrees of Freedom

The unrestrained and restrained degrees of freedom are marked on the continuous beam as shown.

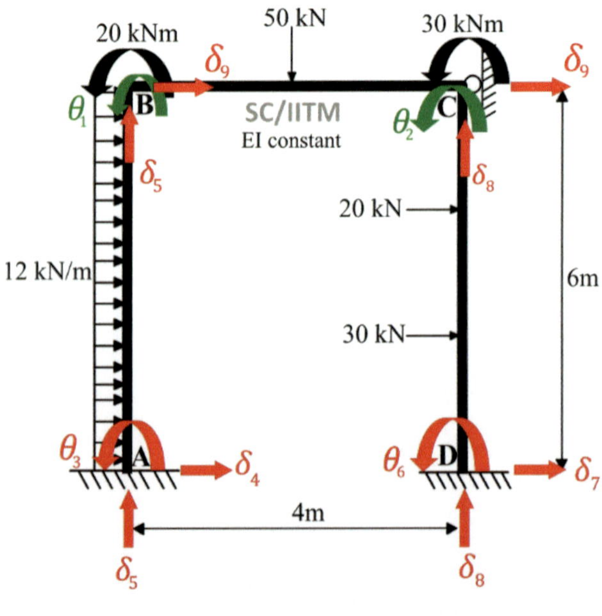

Thus, the size of the stiffness matrix will be of size 9 × 9.

Unrestrained degrees of freedom $= 2[\theta_1, \theta_2]$.

Restrained degrees of freedom $= 7[\theta_3, \theta_4, \delta_4, \delta_5, \delta_6, \delta_8, \delta_9]$.

Thus, the unrestrained sub-matrix will be of size 2 × 2. The size of total stiffness matrix will be 9 × 9. The element stiffness matrix will be 4 × 4, neglecting axial deformation.

$$[K] = \begin{bmatrix} [k_{uu}]_{2\times2} & [k_{ur}]_{2\times7} \\ [k_{ru}]_{7\times2} & [k_{rr}]_{7\times7} \end{bmatrix}$$

The orientation and the labels of the members are given in the following table.

Member	*j*th End	*k*th End	Degrees of Freedom Labels
BA	B	A	1, 3, 9, 4
BC	B	C	1, 2, 5, 8
CD	C	D	2, 6, 9, 7

Thus:

$$\{\Delta\}_{9\times1} = \begin{Bmatrix} \theta_1 \\ \theta_2 \\ \theta_3 \\ \delta_4 \\ \delta_5 \\ \delta_6 \\ \delta_7 \\ \delta_8 \\ \delta_9 \end{Bmatrix} = \begin{Bmatrix} [\Delta_u]_{2\times1} \\ [\Delta_r]_{7\times1} \end{Bmatrix}$$

4.13.4.2 Formulation of Stiffness Matrix

The stiffness matrix is given by:

$$K_i = \begin{bmatrix} \frac{4EI}{l} & \frac{2EI}{l} & \frac{6EI}{l^2} & -\frac{6EI}{l^2} \\ \frac{2EI}{l} & \frac{4EI}{l} & \frac{6EI}{l^2} & -\frac{6EI}{l^2} \\ \frac{6EI}{l^2} & \frac{6EI}{l^2} & \frac{12EI}{l^3} & -\frac{12EI}{l^3} \\ -\frac{6EI}{l^2} & -\frac{6EI}{l^2} & -\frac{12EI}{l^3} & \frac{12EI}{l^3} \end{bmatrix}$$

Thus, the size of the stiffness matrix is 4 × 4, neglecting axial deformation. For this problem, the element stiffness matrices are given by:

$$\begin{array}{cc} & \begin{array}{cccc} \textcircled{1} & \textcircled{3} & \textcircled{9} & \textcircled{4} \end{array} \\ K_{AB} = EI & \begin{bmatrix} 1.333 & 0.667 & 0.333 & -0.333 \\ 0.667 & 1.333 & 0.333 & -0.333 \\ 0.333 & 0.333 & 0.111 & -0.111 \\ -0.333 & -0.333 & -0.111 & 0.111 \end{bmatrix} \begin{array}{c} \textcircled{1} \\ \textcircled{3} \\ \textcircled{9} \\ \textcircled{4} \end{array} \end{array}$$

$$\begin{array}{cc} & \begin{array}{cccc} \textcircled{1} & \textcircled{2} & \textcircled{5} & \textcircled{8} \end{array} \\ K_{BC} = EI & \begin{bmatrix} 1 & 0.5 & 0.375 & -0.375 \\ 0.5 & 1 & 0.375 & -0.375 \\ 0.375 & 0.375 & 0.188 & -0.188 \\ -0.375 & -0.375 & -0.188 & 0.188 \end{bmatrix} \begin{array}{c} \textcircled{1} \\ \textcircled{2} \\ \textcircled{5} \\ \textcircled{8} \end{array} \end{array}$$

$$\begin{array}{cc} & \begin{array}{cccc} \textcircled{2} & \textcircled{6} & \textcircled{9} & \textcircled{7} \end{array} \\ K_{CD} = EI & \begin{bmatrix} 1.33 & 0.667 & 0.333 & -0.333 \\ 0.667 & 1.333 & 0.333 & -0.333 \\ 0.333 & 0.333 & 0.111 & -0.111 \\ -0.333 & -0.333 & -0.111 & 0.111 \end{bmatrix} \begin{array}{c} \textcircled{2} \\ \textcircled{6} \\ \textcircled{9} \\ \textcircled{7} \end{array} \end{array}$$

We need only the stiffness matrix of unrestrained degrees of freedom.

$$\begin{array}{cc} & \begin{array}{cc} \textcircled{1} & \textcircled{2} \end{array} \\ K_{UU} = EI & \begin{bmatrix} 2.333 & 0.5 \\ 0.5 & 2.333 \end{bmatrix} \begin{array}{c} \textcircled{1} \\ \textcircled{2} \end{array} \end{array}$$

Then:

$$K_{UU}^{-1} = \frac{1}{5.193EI} \begin{bmatrix} 2.333 & -0.5 \\ -0.5 & 2.333 \end{bmatrix}$$

4.13.4.3 Calculation of Fixed End Moments

The fixed end moments and reactions of the members are given in the following.

For member AB (Figure 4.36):

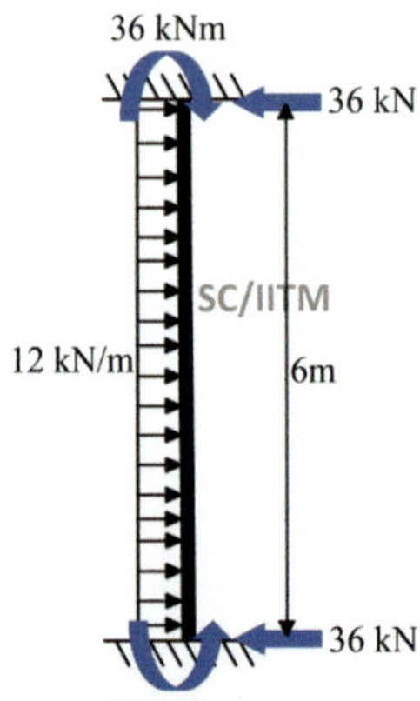

$$M_{AB}^{F} = -\frac{wl^2}{12} = -\frac{12\times 6^2}{12} = -36 \text{ KNm}$$
$$M_{BA}^{F} = +36 \text{ KNm}$$
$$V_A = -36 \text{ kN}$$
$$V_B = -36 \text{ kN}$$

For member *BC* (Figure 4.37):

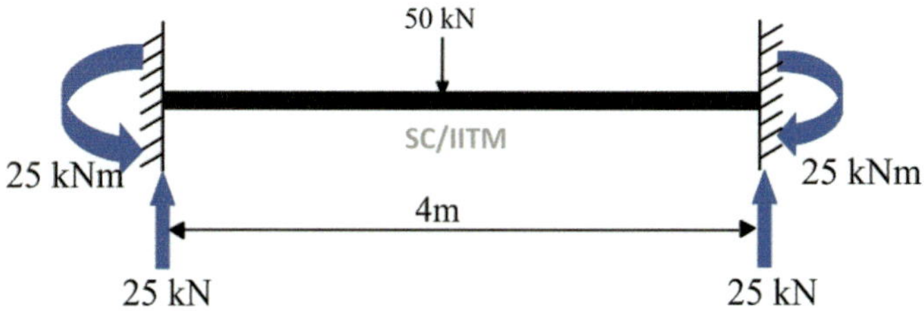

$$M_{BC}^{F} = \frac{pl}{8} = \frac{50\times 4}{8} = +25 \text{ KNm}$$
$$M_{CB}^{F} = -25 \text{ KNm}$$
$$V_B = +25 \text{ kN}$$
$$V_C = +25 \text{ kN}$$

For member *CD* (Figure 4.38):

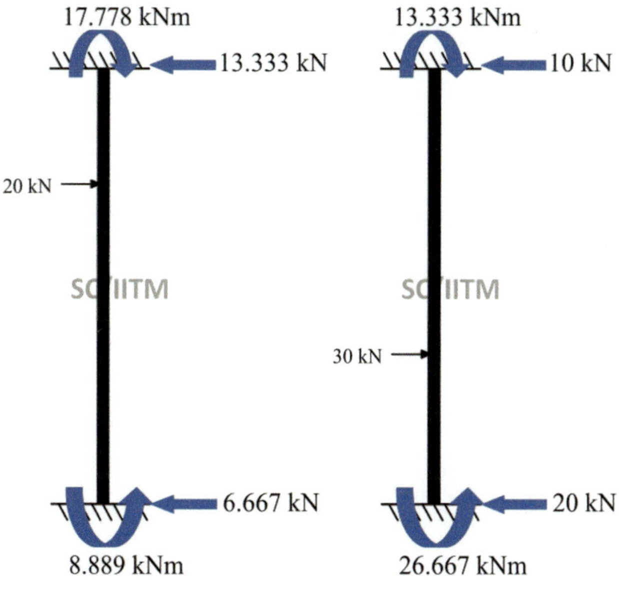

For a 20 kN load:

$$M_{CD}^{F} = \frac{pab^2}{l^2} = \frac{20\times 2\times 4^2}{6^2} = +17.778 \text{ kNm}$$

$$M_{DC}^{F} = \frac{pa^2b}{l^2} = \frac{20\times 2^2\times 4}{6^2} = -8.889 \text{ kNm}$$

$$V_C = +13.333 \text{ kN}$$

$$V_D = +6.667 \text{ kN}$$

For a 30 kN load:

$$M_{CD}^{F} = \frac{pab^2}{l^2} = \frac{30\times 4\times 2^2}{6^2} = +13.333 \text{ kNm}$$

$$M_{DC}^{F} = \frac{pa^2b}{l^2} = \frac{30\times 2\ \ \times 4^2}{6^2} = -26.667 \text{ kNm}$$

$$V_C = +10 \text{ kN}$$

$$V_D = +20 \text{ kN}$$

Thus, for the member *CD*, the fixed end moments and reactions are given by:

$$M_{CD}^{F} = -17.778 - 13.333 = -31.111 \text{ kNm}$$

$$M_{DC}^{F} = +8.889 + 26.667 = +35.556 \text{ kNm}$$

$$V_C = -13.333 - 10 = -23.333 \text{ kN}$$

$$V_D = -6.667 - 20 = -26.667 \text{ kN}$$

Thus, for the whole beam, the fixed end moments are written as follows:

$$AB = \begin{Bmatrix} -36 \\ +36 \\ -36 \\ -36 \end{Bmatrix},\ BC = \begin{Bmatrix} +25 \\ -25 \\ +25 \\ +25 \end{Bmatrix},\ CD = \begin{Bmatrix} -31.111 \\ +35.556 \\ -23.333 \\ -26.667 \end{Bmatrix}$$

4.13.4.4 Calculation of Joint Load Vectors

The size of the joint load vector is 9 × 1.

$$\{J_L\}_{9\times 1} = \begin{Bmatrix} +31 \\ 86.111 \\ \hdashline -36 \\ +36 \\ -25 \\ -35.556 \\ +26.667 \\ -25.000 \\ +59.333 \end{Bmatrix} \begin{matrix} (1) \\ (2) \\ (3) \\ (4) \\ (5) \\ (6) \\ (7) \\ (8) \\ (9) \end{matrix}$$

The joint load vector can be split into unrestrained and restrained degrees of freedom.

$$\{J_L\} = \begin{Bmatrix} (J_L)_{\substack{u \\ 2\times1}} \\ --- \\ (J_L)_{\substack{r \\ 7\times1}} \end{Bmatrix}_{9\times1}$$

$$\text{Thus, } J_{Lu} = \begin{Bmatrix} +31 \\ 86.111 \end{Bmatrix} \begin{matrix} ① \\ ② \end{matrix}$$

4.13.4.5 Calculation of Displacements

Now:

$$\{\Delta_u\} = [k_{uu}]^{-1}\{J_L\}_u$$

$$\begin{Bmatrix} \theta_1 \\ \theta_2 \end{Bmatrix} = \frac{1}{5.193EI}\begin{bmatrix} 2.333 & -0.5 \\ -0.5 & 2.333 \end{bmatrix}\begin{Bmatrix} 31 \\ 86.111 \end{Bmatrix} = \frac{1}{5.193EI}\begin{Bmatrix} 29.268 \\ 185.397 \end{Bmatrix} radians$$

4.13.4.6 Calculation of End Moment and Shear

The general equation to find the end moment of the ith beam is:

$$[M]_i = k_i\delta_i + (FEM)_i$$

The equation can be rewritten element-wise as follows:

$$M_{AB} = [k]_{AB}\{\delta\} + \{FEM\}$$

$$\begin{Bmatrix} M_1 \\ M_3 \\ V_9 \\ V_4 \end{Bmatrix} = EI\begin{bmatrix} 1.333 & 0.667 & 0.333 & -0.333 \\ 0.667 & 1.333 & 0.333 & -0.333 \\ 0.333 & 0.333 & 0.111 & -0.111 \\ -0.333 & -0.333 & -0.111 & 0.111 \end{bmatrix}\begin{Bmatrix} \theta_1 \\ \theta_3 \\ \delta_9 \\ \delta_4 \end{Bmatrix} + \begin{Bmatrix} -36 \\ +36 \\ -36 \\ +36 \end{Bmatrix}$$

We know that $\theta_1 = \dfrac{29.268}{5.193EI}$, $\theta_3 = \delta_9 = \delta_4 = 0$.

$$\text{Thus, } \begin{Bmatrix} M_1 \\ M_3 \\ V_9 \\ V_4 \end{Bmatrix} = \begin{Bmatrix} -28.487 \\ 39.759 \\ 34.123 \\ -34.123 \end{Bmatrix}$$

$$M_{BC} = [k]_{BC}\{\delta\} + \{FEM\}$$

$$\begin{Bmatrix} M_1 \\ M_2 \\ V_5 \\ V_8 \end{Bmatrix} = EI \begin{bmatrix} 1 & 0.5 & 0.375 & -0.375 \\ 0.5 & 1 & 0.375 & -0.375 \\ 0.375 & 0.375 & 0.188 & -0.188 \\ -0.375 & -0.375 & -0.188 & 0.188 \end{bmatrix} \begin{Bmatrix} \theta_1 \\ \theta_2 \\ \delta_5 \\ \delta_8 \end{Bmatrix} + \begin{Bmatrix} 25 \\ -25 \\ 25 \\ 25 \end{Bmatrix} = \begin{Bmatrix} 48.487 \\ 13.519 \\ 40.502 \\ 9.498 \end{Bmatrix}$$

Similarly, for the member *CD*:

$$M_{CD} = [k]_{CD}\{\delta\} + \{FEM\}$$

$$\begin{Bmatrix} M_2 \\ M_6 \\ V_9 \\ V_7 \end{Bmatrix} = EI \begin{bmatrix} 1.333 & 0.667 & 0.333 & -0.333 \\ 0.667 & 1.333 & 0.333 & -0.333 \\ 0.333 & 0.333 & 0.111 & -0.111 \\ -0.333 & -0.333 & -0.111 & 0.111 \end{bmatrix} \begin{Bmatrix} \theta_2 \\ \theta_6 \\ \delta_9 \\ \delta_7 \end{Bmatrix} + \begin{Bmatrix} -31.111 \\ 35.556 \\ -23.333 \\ -26.667 \end{Bmatrix} = \begin{Bmatrix} 16.479 \\ 59.369 \\ -11.444 \\ -38.556 \end{Bmatrix}$$

The member end moments and shear are shown in Figure 4.39, and the final moments in Figure 4.40.

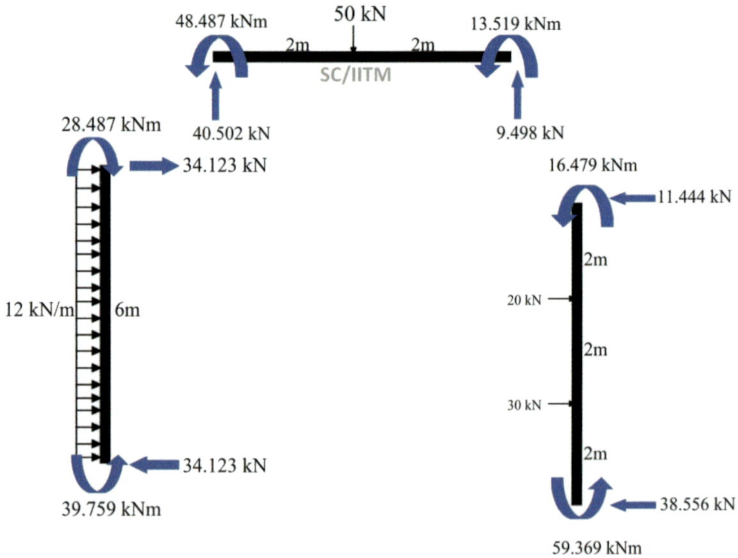

CHECK

In member *AB*:

$$\sum F_H = 0$$

$$(12 \times 6) - 34.123 - 37.877 = 0$$

$$\sum M_A = 0$$

$$+28.487 + (12 \times 6 \times 3) - (34.123 \times 6) + 39.75 = 0$$

In member *BC*:

$$\sum F_V = 0$$
$$40.502 + 9.798 - 50 = 0$$
$$\sum M_B = 0$$
$$-13.159 - (9.498 \times 4) + (50 \times 2) + 98.489 = 0$$

In member *CD*:

$$\sum F_H = 0$$
$$20 + 30 - 11.444 - 38.556 = 0$$
$$\sum M_D = 0$$
$$+16.479 + (11.444 \times 6) - (20 \times 4) - (30 \times 2) + 59.369 = 0$$

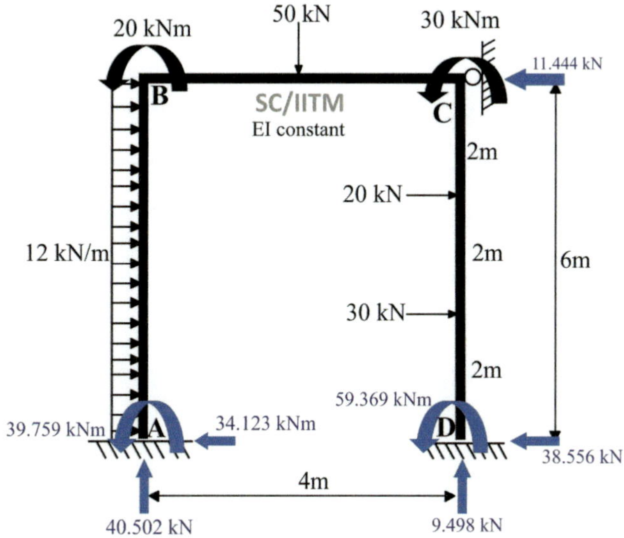

MATLAB PROGRAM

```
%% stiffness matrix method
% Input
clc;
clear;
n = 3; % number of members
I = [2 1 2]; %Moment of inertis in m4
L = [6 4 6]; % length in m
uu = 2; % Number of unrestrained degrees of freedom
ur = 7; % Number of restrained degrees of freedom
uul = [1 2]; % global labels of unrestained dof
```

```
url = [3 4 5 6 7 8 9]; % global labels of restained dof
l1 = [1 3 9 4]; % Global labels for member 1
l2 = [1 2 5 8]; % Global labels for member 2
l3 = [2 6 9 7]; % Global labels for member 3
l = [l1; l2; l3];
dof = uu+ur;
Ktotal = zeros (dof);
fem1 = [-36 36 -36 -36]; % Local Fixed end moments of
member 1
fem2 = [25 -25 25 25]; % Local Fixed end moments of
member 2
fem3 = [-31.111 35.556 -23.333 -26.667]; % Local Fixed
end moments of member 3
%% rotation coefficients for each member
rc1 = 4.*I./L;
rc2 = 2.*I./L;
%% stiffness matrix 4 by 4 (axial deformation neglected)
for i = 1:n
  Knew = zeros (dof);
  k1 = [rc1(i); rc2(i); (rc1(i)+rc2(i))/L(i);
  (-(rc1(i)+rc2(i))/L(i))];
  k2 = [rc2(i); rc1(i); (rc1(i)+rc2(i))/L(i);
  (-(rc1(i)+rc2(i))/L(i))];
  k3 = [(rc1(i)+rc2(i))/L(i); (rc1(i)+rc2(i))/L(i);
  (2*(rc1(i)+rc2(i))/(L(i)^2)); (-2*(rc1(i)+rc2(i))/
  (L(i)^2))];
  k4 = -k3;
  K = [k1 k2 k3 k4];
  fprintf ('Member Number =');
  disp (i);
  fprintf ('Local Stiffness matrix of member, [K] = \n');
  disp (K);
  for p = 1:4
    for q = 1:4
      Knew((l(i,p)),(l(i,q))) = K(p,q);
    end
  end
  Ktotal = Ktotal + Knew;
  if i == 1
    Kg1 = K;
  elseif i == 2
    Kg2 = K;
  else
    Kg3 = K;
  end
end
```

```
fprintf ('Stiffness Matrix of complete structure,
[Ktotal] = \n');
disp (Ktotal);
Kunr = zeros(uu);
for x = 1:uu
  for y = 1:uu
    Kunr(x,y) = Ktotal(x,y);
  end
end
fprintf ('Unrestrained Stiffness sub-matix, [Kuu] = \n');
disp (Kunr);
KuuInv = inv(Kunr);
fprintf ('Inverse of Unrestrained Stiffness sub-matix,
[KuuInverse] = \n');
disp (KuuInv);
%% Creation of joint load vector
jl = [31; 86.111; -36; 36; -25; -35.556; 26.665; -25;
59.333]; % values given in kN or kNm
jlu = [31; 86.111]; % load vector in unrestrained dof
delu = KuuInv*jlu;
fprintf ('Joint Load vector, [Jl] = \n');
disp (jl');
fprintf ('Unrestrained displacements, [DelU] = \n');
disp (delu');
delr = zeros (ur,1);
del = [delu; delr];
deli = zeros (4,1);
for i = 1:n
  for p = 1:4
    deli(p,1) = del((l(i,p)),1) ;
  end
  if i == 1
    delbar1 = deli;
    mbar1 = (Kg1 * delbar1)+fem1';
    fprintf ('Member Number =');
    disp (i);
    fprintf ('Global displacement matrix [DeltaBar] = \n');
    disp (delbar1');
    fprintf ('Global End moment matrix [MBar] = \n');
    disp (mbar1');
  elseif i == 2
    delbar2 = deli;
    mbar2 = (Kg2 * delbar2)+fem2';
    fprintf ('Member Number =');
    disp (i);
    fprintf ('Global displacement matrix [DeltaBar] = \n');
```

```
    disp (delbar2');
    fprintf ('Global End moment matrix [MBar] = \n');
    disp (mbar2');
  else
    delbar3 = deli;
    mbar3 = (Kg3 * delbar3)+fem3';
    fprintf ('Member Number =');
    disp (i);
    fprintf ('Global displacement matrix [DeltaBar] =
    \n');
    disp (delbar3');
    fprintf ('Global End moment matrix [MBar] = \n');
    disp (mbar3');
  end
end
%% check
mbar = [mbar1'; mbar2'; mbar3'];
jf = zeros(dof,1);
for a = 1:n
  for b = 1:4 % size of k matrix
    d = l(a,b);
    jfnew = zeros(dof,1);
    jfnew(d,1) = mbar(a,b);
    jf = jf+jfnew;
  end
end
fprintf ('Joint forces = \n');
disp (jf');
```

MATLAB OUTPUT

```
Member Number = 1
Local Stiffness matrix of member, [K] =
   1.3333   0.6667   0.3333  -0.3333
   0.6667   1.3333   0.3333  -0.3333
   0.3333   0.3333   0.1111  -0.1111
  -0.3333  -0.3333  -0.1111   0.1111
Member Number = 2
Local Stiffness matrix of member, [K] =
   1.0000   0.5000   0.3750  -0.3750
   0.5000   1.0000   0.3750  -0.3750
   0.3750   0.3750   0.1875  -0.1875
  -0.3750  -0.3750  -0.1875   0.1875
Member Number = 3
Local Stiffness matrix of member, [K] =
  1.3333   0.6667   0.3333  -0.3333
  0.6667   1.3333   0.3333  -0.3333
```

```
  0.3333   0.3333   0.1111  -0.1111
 -0.3333  -0.3333  -0.1111   0.1111
Stiffness Matrix of complete structure, [Ktotal] =
   2.3333   0.5000   0.6667  -0.3333   0.3750        0
        0  -0.3750   0.3333
   0.5000   2.3333        0        0   0.3750   0.6667
  -0.3333  -0.3750   0.3333
   0.6667        0   1.3333  -0.3333        0        0
        0        0   0.3333
  -0.3333        0  -0.3333   0.1111        0        0
        0        0  -0.1111
   0.3750   0.3750        0        0   0.1875        0
        0  -0.1875        0
        0   0.6667        0        0        0   1.3333
  -0.3333        0   0.3333
        0  -0.3333        0        0        0  -0.3333
   0.1111        0  -0.1111
  -0.3750  -0.3750        0        0  -0.1875        0
        0   0.1875        0
   0.3333   0.3333   0.3333  -0.1111        0   0.3333
  -0.1111        0   0.2222
Unrestrained Stiffness sub-matix, [Kuu] =
   2.3333   0.5000
   0.5000   2.3333
Inverse of Unrestrained Stiffness sub-matix, [KuuInverse] =
   0.4492  -0.0963
  -0.0963   0.4492
Joint Load vector, [Jl] =
   31.0000   86.1110  -36.0000   36.0000  -25.0000
  -35.5560   26.6650  -25.0000   59.3330
Unrestrained displacements, [DelU] =
    5.6364   35.6969
Member Number = 1
Global displacement matrix [DeltaBar] =
    5.6364         0         0         0
Global End moment matrix [MBar] =
  -28.4848   39.7576  -34.1212  -37.8788
Member Number = 2
Global displacement matrix [DeltaBar] =
    5.6364   35.6969         0         0
Global End moment matrix [MBar] =
   48.4848   13.5151   40.5000    9.5000
Member Number = 3
Global displacement matrix [DeltaBar] =
   35.6969         0         0         0
Global End moment matrix [MBar] =
   16.4849   59.3539  -11.4340  -38.5660
```

```
Joint forces =
   20.0000    30.0000   39.7576  -37.8788   40.5000
   59.3539  -38.5660    9.5000  -45.5552
```

4.13.5 Solution

4.13.5.1 Marking Unrestrained and Restrained Degrees of Freedom

The unrestrained and restrained degrees of freedom are marked on the continuous beam as shown in the following:

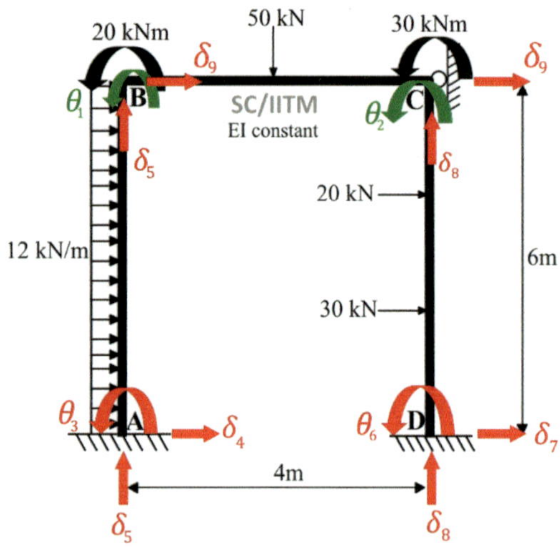

Thus, the size of the stiffness matrix will be 9 × 9.
Unrestrained degrees of freedom = 2 [θ_1, θ_2].

Restrained degrees of freedom = 7 [θ_3, θ_4, δ_4, δ_5, δ_6, δ_8, δ_9].

Thus, the unrestrained sub-matrix will be of size 2 × 2. The size of the total stiffness matrix will be 9 × 9. The element stiffness matrix will be 4 × 4, neglecting axial deformation.

$$[K] = \begin{bmatrix} [k_{uu}]_{2\times2} & [k_{ur}]_{2\times7} \\ [k_{ru}]_{7\times2} & [k_{rr}]_{7\times7} \end{bmatrix}$$

The orientation and the labels of the members are given in the following table.

Member	*j*th End	*k*th End	Degrees of Freedom Labels
BA	B	A	1, 3, 9, 4
BC	B	C	1, 2, 5, 8
CD	C	D	2, 6, 9, 7

Thus:

$$\{\Delta\}_{9\times1}=\begin{Bmatrix}\theta_1\\\theta_2\\\theta_3\\\delta_4\\\delta_5\\\delta_6\\\delta_7\\\delta_8\\\delta_9\end{Bmatrix}=\begin{Bmatrix}[\Delta_u]_{2\times1}\\[\Delta_r]_{7\times1}\end{Bmatrix}$$

4.13.5.2 Formulation of the Stiffness Matrix

The stiffness matrix is given by:

$$K_i=\begin{bmatrix}\frac{4EI}{l} & \frac{2EI}{l} & \frac{6EI}{l^2} & -\frac{6EI}{l^2}\\ \frac{2EI}{l} & \frac{4EI}{l} & \frac{6EI}{l^2} & -\frac{6EI}{l^2}\\ \frac{6EI}{l^2} & \frac{6EI}{l^2} & \frac{12EI}{l^3} & -\frac{12EI}{l^3}\\ -\frac{6EI}{l^2} & -\frac{6EI}{l^2} & -\frac{12EI}{l^3} & \frac{12EI}{l^3}\end{bmatrix}$$

Thus, the size of the stiffness matrix is 4 × 4, neglecting axial deformation. For this problem, the element stiffness matrices are given by:

$$\begin{array}{cc} & \begin{array}{cccc}\textcircled{1} & \textcircled{3} & \textcircled{9} & \textcircled{4}\end{array}\\ K_{AB}=EI & \begin{bmatrix}1.333 & 0.667 & 0.333 & -0.333\\ 0.667 & 1.333 & 0.333 & -0.333\\ 0.333 & 0.333 & 0.111 & -0.111\\ -0.333 & -0.333 & -0.111 & 0.111\end{bmatrix}\begin{array}{c}\textcircled{1}\\\textcircled{3}\\\textcircled{9}\\\textcircled{4}\end{array}\end{array}$$

$$\begin{array}{cc} & \begin{array}{cccc}\textcircled{1} & \textcircled{2} & \textcircled{5} & \textcircled{8}\end{array}\\ K_{BC}=EI & \begin{bmatrix}1 & 0.5 & 0.375 & -0.375\\ 0.5 & 1 & 0.375 & -0.375\\ 0.375 & 0.375 & 0.188 & -0.188\\ -0.375 & -0.375 & -0.188 & 0.188\end{bmatrix}\begin{array}{c}\textcircled{1}\\\textcircled{2}\\\textcircled{5}\\\textcircled{8}\end{array}\end{array}$$

$$\begin{array}{cc} & \begin{array}{cccc}\textcircled{2} & \textcircled{6} & \textcircled{9} & \textcircled{7}\end{array}\\ K_{CD}=EI & \begin{bmatrix}1.33 & 0.667 & 0.333 & -0.333\\ 0.667 & 1.333 & 0.333 & -0.333\\ 0.333 & 0.333 & 0.111 & -0.111\\ -0.333 & -0.333 & -0.111 & 0.111\end{bmatrix}\begin{array}{c}\textcircled{2}\\\textcircled{6}\\\textcircled{9}\\\textcircled{7}\end{array}\end{array}$$

We need only the stiffness matrix of unrestrained degrees of freedom.

$$K_{UU} = EI \begin{bmatrix} 2.333 & 0.5 \\ 0.5 & 2.333 \end{bmatrix} \begin{matrix} (1) \\ (2) \end{matrix}$$

with columns ① ② and rows ① ②.

Then:

$$K_{UU}^{-1} = \frac{1}{5.193EI} \begin{bmatrix} 2.333 & -0.5 \\ -0.5 & 2.333 \end{bmatrix}$$

4.13.5.3 Calculation of Fixed End Moments

The fixed end moments and reactions of the members are given in the following:

For member AB:

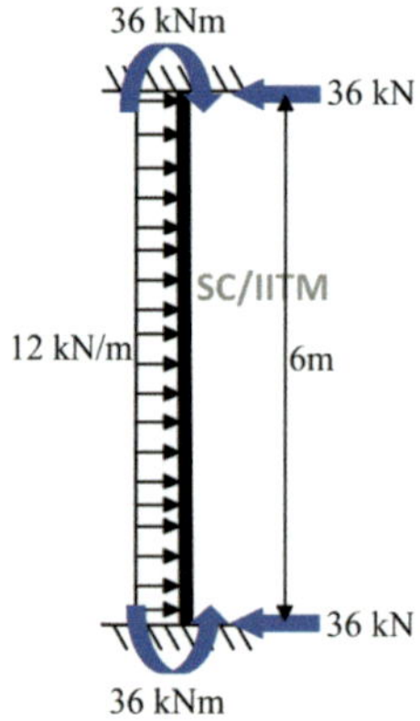

$$M_{AB}^{F} = -\frac{wl^2}{12} = -\frac{12 \times 6^2}{12} = -36 \text{ KNM}$$

$$M_{BA}^{F} = +-36 \text{ KNM}$$

$$V_A = -36 \text{ kN}$$

$$V_B = -36 \text{ kN}$$

For member BC:

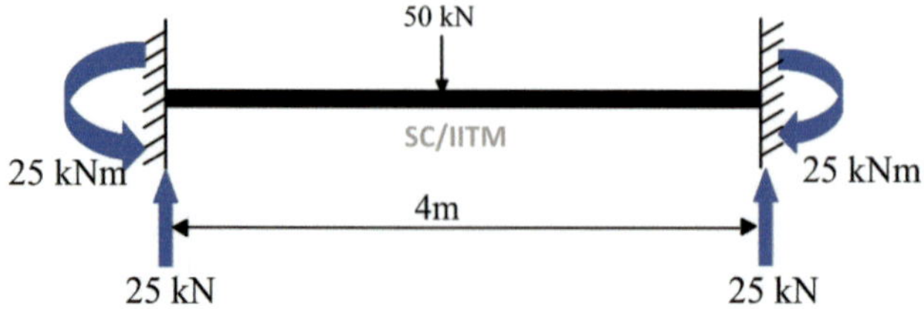

$$M_{BC}^{F} = \frac{pl}{8} = \frac{50 \times 4}{8} = +25 \text{ KM}$$

$$M_{CB}^{F} = -25 \text{ KNM}$$

$$V_B = +25 \text{ kN}$$

$$V_C = +25 \text{ kN}$$

For member *CD*:

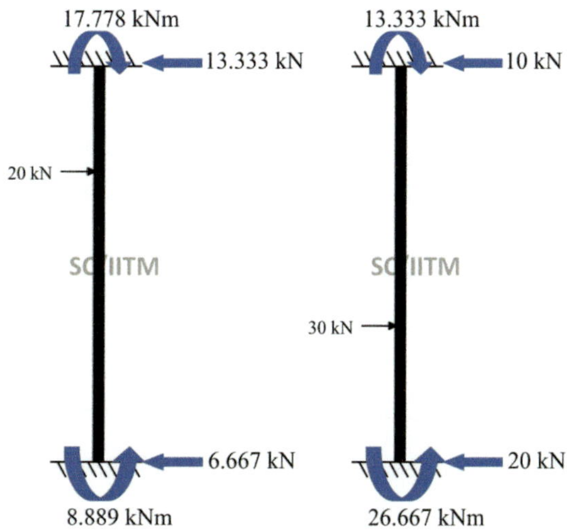

For a 20 kN load:

$$M_{CD}^{F} = \frac{pab^2}{l^2} = \frac{20 \times 2 \times 4^2}{6^2} = +17.778 \text{ KNM}$$

$$M_{DC}^{F} = \frac{pa^2 b}{l^2} = \frac{20 \times 2^2 \times 4}{6^2} = -8.889 \text{ KNM}$$

$$V_C = +13.333 \text{ kN}$$

$$V_D = +6.667 \text{ kN}$$

For a 30 kN load:

$$M_{CD}^{F} = \frac{pab^2}{l^2} = \frac{30 \times 4 \times 2^2}{6^2} = +13.333 \text{ KNM}$$

$$M_{DC}^{F} = \frac{pa^2 b}{l^2} = \frac{30 \times 2 \ \times 4^2}{6^2} = VC = +10 \text{ kN}$$

$$V_C = +10 \text{ kN}$$

$$V_D = +20 \text{ kN}$$

Thus, for the member *CD*, the fixed end moments and reactions are given by:

$$M_{CD}^{F} = -17.778 - 13.333 = -31.111 \text{ KNM}$$
$$M_{DC}^{F} = +8.889 + 26.667 = +35.556 \text{ KNM}$$
$$V_{C} = -13.333 - 10 = -23.333 \text{ kN}$$
$$V_{D} = -6.667 - 20 = -26.667 \text{ kN}$$

Thus, for the whole beam, the fixed end moments are written as follows:

$$AB = \begin{Bmatrix} -36 \\ +36 \\ -36 \\ -36 \end{Bmatrix}, \; BC = \begin{Bmatrix} +25 \\ -25 \\ +25 \\ +25 \end{Bmatrix}, \; CD = \begin{Bmatrix} -31.111 \\ +35.556 \\ -23.333 \\ -26.667 \end{Bmatrix}$$

4.13.5.4 Calculation of Joint Load Vectors

The size of the joint load vector is 9 × 1.

$$\{J_L\}_{9\times1} = \begin{Bmatrix} +31 \\ 86.111 \\ \hline -36 \\ +36 \\ -25 \\ -35.556 \\ +26.667 \\ -25.000 \\ +59.333 \end{Bmatrix} \begin{matrix} ① \\ ② \\ ③ \\ ④ \\ ⑤ \\ ⑥ \\ ⑦ \\ ⑧ \\ ⑨ \end{matrix}$$

The joint load vector can be split into unrestrained and restrained degrees of freedom.

$$\{J_L\} = \begin{Bmatrix} (J_L)_{u\,2\times1} \\ --- \\ (J_L)_{r\,7\times1} \end{Bmatrix}_{9\times1}$$

$$\text{Thus, } J_{Lu} = \begin{Bmatrix} +31 \\ 86.111 \end{Bmatrix} \begin{matrix} ① \\ ② \end{matrix}$$

4.13.5.5 Calculation of Displacements

Now:

$$\{\Delta_u\} = [k_{uu}]^{-1} \{J_L\}_u$$

$$\begin{Bmatrix} \theta_1 \\ \theta_2 \end{Bmatrix} = \frac{1}{5.193EI} \begin{bmatrix} 2.333 & -0.5 \\ -0.5 & 2.333 \end{bmatrix} \begin{Bmatrix} 31 \\ 86.111 \end{Bmatrix} = \frac{1}{5.193EI} \begin{Bmatrix} 29.268 \\ 185.397 \end{Bmatrix} radians$$

4.13.5.6 Calculation of End Moment and Shear

The general equation to find the end moment of the *i*th beam is:

$$[M]_i = k_i \delta_i + (FEM)_i$$

The equation can be rewritten element-wise as follows:

$$M_{AB} = [k]_{AB} \{\delta\} + \{FEM\}$$

$$\begin{Bmatrix} M_1 \\ M_3 \\ V_9 \\ V_4 \end{Bmatrix} = EI \begin{bmatrix} 1.333 & 0.667 & 0.333 & -0.333 \\ 0.667 & 1.333 & 0.333 & -0.333 \\ 0.333 & 0.333 & 0.111 & -0.111 \\ -0.333 & -0.333 & -0.111 & 0.111 \end{bmatrix} \begin{Bmatrix} \theta_1 \\ \theta_3 \\ \delta_9 \\ \delta_4 \end{Bmatrix} + \begin{Bmatrix} -36 \\ +36 \\ -36 \\ +36 \end{Bmatrix}$$

We know that $\theta_1 = \dfrac{29.268}{5.193EI}$, $\theta_3 = \delta_9 = \delta_4 = 0$.

$$\text{Thus,} \begin{Bmatrix} M_1 \\ M_3 \\ V_9 \\ V_4 \end{Bmatrix} = \begin{Bmatrix} -28.487 \\ 39.759 \\ 34.123 \\ -34.123 \end{Bmatrix}$$

$$M_{BC} = [k]_{BC} \{\delta\} + \{FEM\}$$

$$\begin{Bmatrix} M_1 \\ M_2 \\ V_5 \\ V_8 \end{Bmatrix} = EI \begin{bmatrix} 1 & 0.5 & 0.375 & -0.375 \\ 0.5 & 1 & 0.375 & -0.375 \\ 0.375 & 0.375 & 0.188 & -0.188 \\ -0.375 & -0.375 & -0.188 & 0.188 \end{bmatrix} \begin{Bmatrix} \theta_1 \\ \theta_2 \\ \delta_5 \\ \delta_8 \end{Bmatrix} + \begin{Bmatrix} 25 \\ -25 \\ 25 \\ 25 \end{Bmatrix} = \begin{Bmatrix} 48.487 \\ 13.519 \\ 40.502 \\ 9.498 \end{Bmatrix}$$

Similarly, for the member CD:

$$M_{CD} = [k]_{CD}\{\delta\} + \{FEM\}$$

$$\begin{Bmatrix} M_2 \\ M_6 \\ V_9 \\ V_7 \end{Bmatrix} = EI \begin{bmatrix} 1.333 & 0.667 & 0.333 & -0.333 \\ 0.667 & 1.333 & 0.333 & -0.333 \\ 0.333 & 0.333 & 0.111 & -0.111 \\ -0.333 & -0.333 & -0.111 & 0.111 \end{bmatrix} \begin{Bmatrix} \theta_2 \\ \theta_6 \\ \delta_9 \\ \delta_7 \end{Bmatrix} + \begin{Bmatrix} -31.111 \\ 35.556 \\ -23.333 \\ -26.667 \end{Bmatrix} = \begin{Bmatrix} 16.479 \\ 59.369 \\ -11.444 \\ -38.556 \end{Bmatrix}$$

The member end moments, shear, and final moments are shown in what follows:

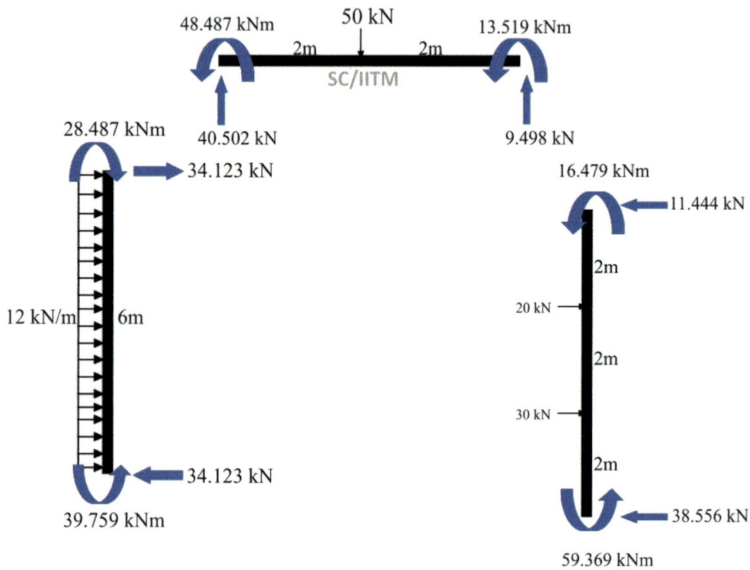

CHECK

In member AB:

$$\sum F_H = 0$$
$$(12 \times 6) - 34.123 - 37.877 = 0$$
$$\sum M_A = 0$$
$$+28.487 + (12 \times 6 \times 3) - (34.123 \times 6) + 39.75 = 0$$

In member BC:

$$\sum F_V = 0$$
$$40.502 + 9.798 - 50 = 0$$
$$\sum M_B = 0$$
$$-13.159 - (9.498 \times 4) + (50 \times 2) + 98.489 = 0$$

In member *CD*:

$$\sum F_H = 0$$

$$20 + 30 - 11.444 - 38.556 = 0$$

$$\sum M_D = 0$$

$$+16.479 + (11.444 \times 6) - (20 \times 4) - (30 \times 2) + 59.369 = 0$$

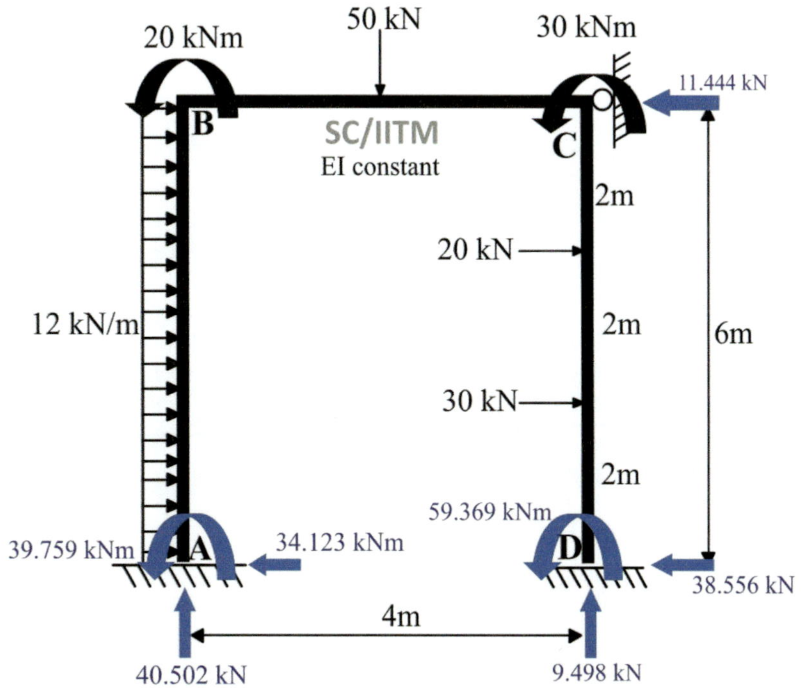

MATLAB PROGRAM

```
%% stiffness matrix method
% Input
clc;
clear;
n = 3; % number of members
I = [2 1 2]; %Moment of inertis in m4
L = [6 4 6]; % length in m
uu = 2; % Number of unrestrained degrees of freedom
ur = 7; % Number of restrained degrees of freedom
uul = [1 2]; % global labels of unrestained dof
url = [3 4 5 6 7 8 9]; % global labels of restained dof
l1 = [1 3 9 4]; % Global labels for member 1
l2 = [1 2 5 8]; % Global labels for member 2
l3 = [2 6 9 7]; % Global labels for member 3
```

```
l = [l1; l2; l3];
dof = uu+ur;
Ktotal = zeros (dof);
fem1 = [-36 36 -36 -36]; % Local Fixed end moments of
member 1
fem2 = [25 -25 25 25]; % Local Fixed end moments of
member 2
fem3 = [-31.111 35.556 -23.333 -26.667]; % Local Fixed
end moments of member 3
%% rotation coefficients for each member
rc1 = 4.*I./L;
rc2 = 2.*I./L;
%% stiffness matrix 4 by 4 (axial deformation neglected)
for i = 1:n
  Knew = zeros (dof);
  k1 = [rc1(i); rc2(i); (rc1(i)+rc2(i))/L(i);
  (-(rc1(i)+rc2(i))/L(i))];
  k2 = [rc2(i); rc1(i); (rc1(i)+rc2(i))/L(i);
  (-(rc1(i)+rc2(i))/L(i))];
  k3 = [(rc1(i)+rc2(i))/L(i); (rc1(i)+rc2(i))/L(i);
  (2*(rc1(i)+rc2(i))/(L(i)^2)); (-2*(rc1(i)+rc2(i))/
  (L(i)^2))];
  k4 = -k3;
  K = [k1 k2 k3 k4];
  fprintf ('Member Number =');
  disp (i);
  fprintf ('Local Stiffness matrix of member, [K] =
  \n');
  disp (K);
  for p = 1:4
    for q = 1:4
      Knew((l(i,p)),(l(i,q))) = K(p,q);
    end
  end
  Ktotal = Ktotal + Knew;
  if i == 1
    Kg1 = K;
  elseif i == 2
    Kg2 = K;
  else
    Kg3 = K;
  end
end
fprintf ('Stiffness Matrix of complete structure,
[Ktotal] = \n');
disp (Ktotal);
```

```
Kunr = zeros(uu);
for x = 1:uu
  for y = 1:uu
    Kunr(x,y) = Ktotal(x,y);
  end
end
fprintf ('Unrestrained Stiffness sub-matix, [Kuu] =
\n');
disp (Kunr);
KuuInv = inv(Kunr);
fprintf ('Inverse of Unrestrained Stiffness sub-matix,
[KuuInverse] = \n');
disp (KuuInv);
%% Creation of joint load vector
jl = [31; 86.111; -36; 36; -25; -35.556; 26.665; -25;
59.333]; % values given in kN or kNm
jlu = [31; 86.111]; % load vector in unrestrained dof
delu = KuuInv*jlu;
fprintf ('Joint Load vector, [Jl] = \n');
disp (jl');
fprintf ('Unrestrained displacements, [DelU] = \n');
disp (delu');
delr = zeros (ur,1);
del = [delu; delr];
deli = zeros (4,1);
for i = 1:n
  for p = 1:4
    deli(p,1) = del((l(i,p)),1) ;
  end
  if i == 1
    delbar1 = deli;
    mbar1 = (Kg1 * delbar1)+fem1';
    fprintf ('Member Number =');
    disp (i);
    fprintf ('Global displacement matrix [DeltaBar] =
    \n');
    disp (delbar1');
    fprintf ('Global End moment matrix [MBar] = \n');
    disp (mbar1');
    delbar2 = deli;
    mbar2 = (Kg2 * delbar2)+fem2';
    fprintf ('Member Number =');
    disp (i);
    fprintf ('Global displacement matrix [DeltaBar] = \n');
    disp (delbar2');
    fprintf ('Global End moment matrix [MBar] = \n');
```

```
    disp (mbar2');
  else
    delbar3 = deli;
    mbar3 = (Kg3 * delbar3)+fem3';
    fprintf ('Member Number =');
    disp (i);
    fprintf ('Global displacement matrix [DeltaBar] = \n');
    disp (delbar3');
    fprintf ('Global End moment matrix [MBar] = \n');
    disp (mbar3');
  end
end
%% check
mbar = [mbar1'; mbar2'; mbar3'];
jf = zeros(dof,1);
for a = 1:n
  for b = 1:4 % size of k matrix
    d = l(a,b);
    jfnew = zeros(dof,1);
    jfnew(d,1) = mbar(a,b);
    jf = jf+jfnew;
  end
end
fprintf ('Joint forces = \n');
disp (jf');
```

MATLAB OUTPUT

```
Member Number = 1
Local Stiffness matrix of member, [K] =
   1.3333   0.6667   0.3333  -0.3333
   0.6667   1.3333   0.3333  -0.3333
   0.3333   0.3333   0.1111  -0.1111
  -0.3333  -0.3333  -0.1111   0.1111
Member Number = 2
Local Stiffness matrix of member, [K] =
   1.0000   0.5000   0.3750  -0.3750
   0.5000   1.0000   0.3750  -0.3750
   0.3750   0.3750   0.1875  -0.1875
  -0.3750  -0.3750  -0.1875   0.1875
Member Number = 3
Local Stiffness matrix of member, [K] =
   1.3333   0.6667   0.3333  -0.3333
   0.6667   1.3333   0.3333  -0.3333
   0.3333   0.3333   0.1111  -0.1111
```

```
  -0.3333  -0.3333  -0.1111   0.1111
Stiffness Matrix of complete structure, [Ktotal] =
   2.3333   0.5000   0.6667  -0.3333   0.3750        0
        0  -0.3750   0.3333
   0.5000   2.3333        0        0   0.3750   0.6667
  -0.3333  -0.3750   0.3333
   0.6667        0   1.3333  -0.3333        0        0
        0        0   0.3333
  -0.3333        0  -0.3333   0.1111        0        0
        0        0  -0.1111
   0.3750   0.3750        0        0   0.1875        0
        0  -0.1875        0
        0   0.6667        0        0        0   1.3333
  -0.3333        0   0.3333
        0  -0.3333        0        0        0  -0.3333
   0.1111        0  -0.1111
  -0.3750  -0.3750        0        0  -0.1875        0
        0   0.1875        0
   0.3333   0.3333   0.3333  -0.1111        0   0.3333
  -0.1111        0   0.2222
Unrestrained Stiffness sub-matix, [Kuu] =
   2.3333   0.5000
   0.5000   2.3333
Inverse of Unrestrained Stiffness sub-matix,
[KuuInverse] =
   0.4492  -0.0963
  -0.0963   0.4492
Joint Load vector, [Jl] =
   31.0000   86.1110  -36.0000   36.0000  -25.0000
  -35.5560   26.6650  -25.0000   59.3330
Unrestrained displacements, [DelU] =
    5.6364   35.6969
Member Number = 1
Global displacement matrix [DeltaBar] =
    5.6364         0         0         0
Global End moment matrix [MBar] =
  -28.4848  -39.7576  -34.1212  -37.8788
Member Number = 2
Global displacement matrix [DeltaBar] =
    5.6364   35.6969         0         0
Global End moment matrix [MBar] =
   48.4848   13.5151   40.5000    9.5000
Member Number = 3
Global displacement matrix [DeltaBar] =
   35.6969         0         0         0
```

```
Global End moment matrix [MBar] =
   16.4849   59.3539  -11.4340  -38.5660
Joint forces =
   20.0000   30.0000   39.7576  -37.8788   40.5000
   59.3539  -38.5660    9.5000  -45.5552
```

4.14 PLANAR TRUSS SYSTEM

A *planar truss* is a two-dimensional framework composed of straight members that are joined together at nodes (joints) to form a rigid structure. These trusses are typically used in structures like bridges, roofs, and towers to efficiently transfer loads. They work by ensuring that all members experience either tension or compression, minimizing the risk of bending, which allows the structure to be lightweight and material-efficient.

Figure 4.31 shows the different types of steel roof truss, highlighting the geometric design with respect to the span of the truss.

Figure 4.32 shows details of a north light roofing. Following are a few important classifications of steel trusses:

Pratt Truss: It is characterized by diagonals slanting toward the center, making it effective for handling tension in the diagonals and compression in the verticals.

Howe Truss: It is the opposite of the Pratt, with diagonals that slope away from the center, allowing compression in the diagonals and tension in the verticals.

Warren Truss: It features a series of equilateral triangles, distributing both tension and compression evenly among members.

King Post and Queen Post Trusses: These are simpler forms, often used in short-span structures, like small bridges or residential roofs. The king post truss has a central vertical member, while the queen post truss has two vertical members.

It is therefore obvious from the construction drawings that, in a truss, joints are considered to be pinned; no moment is transferred through them. As a result, they can only resist axial forces and experience axial deformations. Consequently, each node or joint in the truss can have only two independent components of translation relative to the reference axes. Similarly, in the local axis system, each joint is limited to two translation movements. Therefore, no rotations occur at the ends. In architecture, planar trusses are particularly important because they enable large spans and open spaces without needing many interior supports.

The primary action in planar trusses involves the efficient distribution of loads through tension and compression modes only. When a load is applied, each member of the truss either stretches under tension or compression. This action allows the truss to handle significant external loads while minimizing material usage. The force distribution depends on the type of load applied (dead loads, like the weight

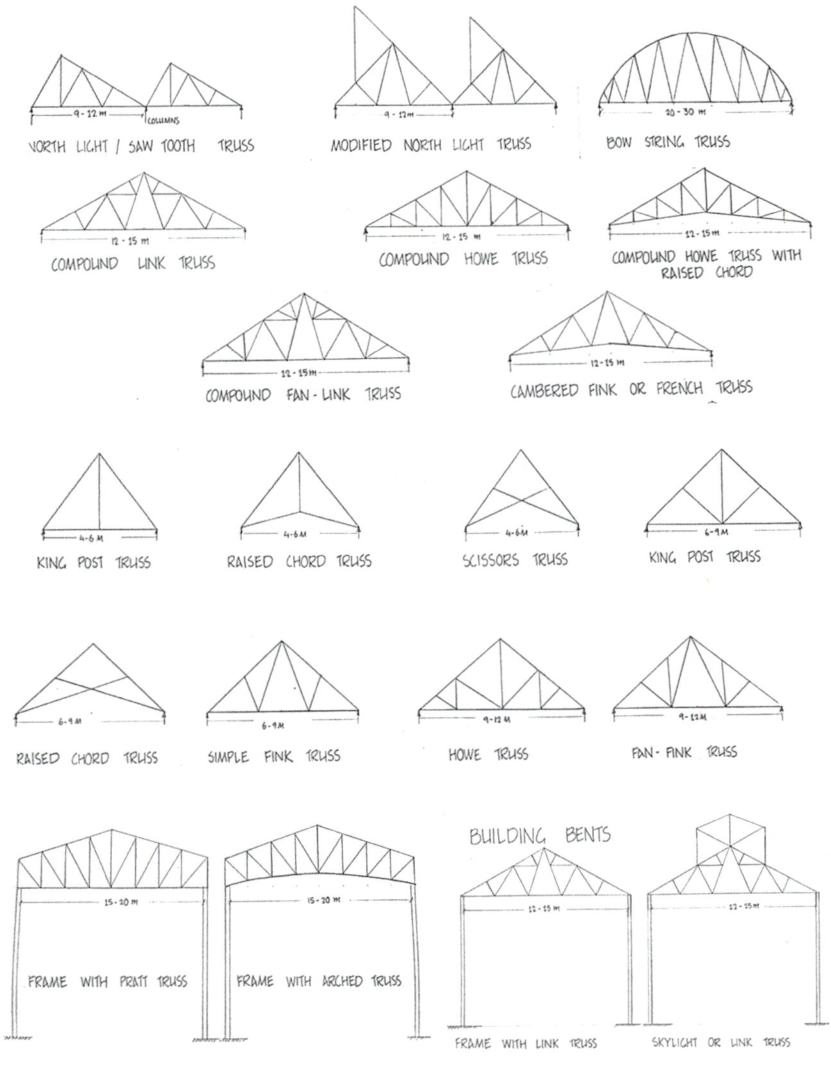

FIGURE 4.31 Types of steel trusses and span limitations.

of the roof, or live loads, like wind or snow), and trusses are designed to ensure that members can handle the specific forces. Figure 4.33 illustrates a basic truss and its force distribution.

Planar trusses are vital in architectural design due to their ability to cover large spans with minimal materials, reducing both weight and cost. They provide flexibility in design, allowing architects to create open, airy spaces without obstructive columns or supports. In large structures, such as auditoriums, sports stadiums, or

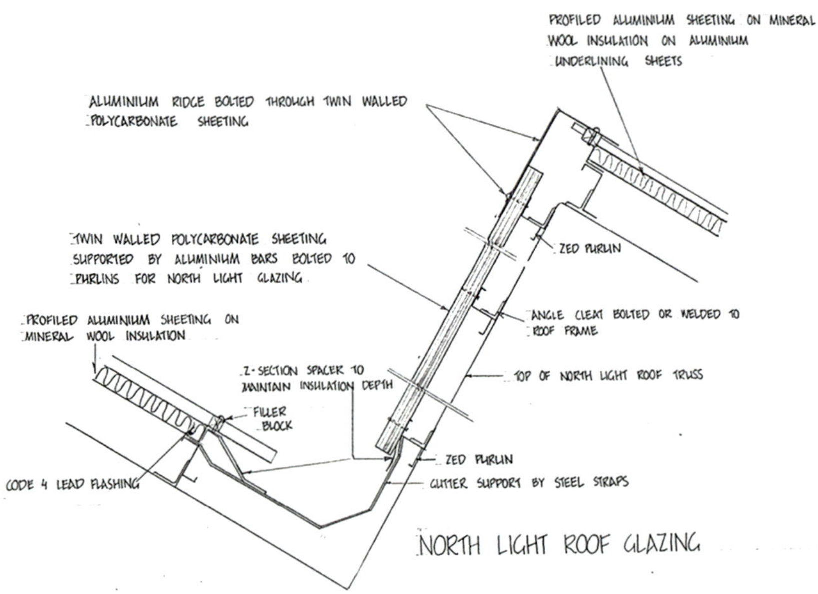

FIGURE 4.32 North light roofing details.

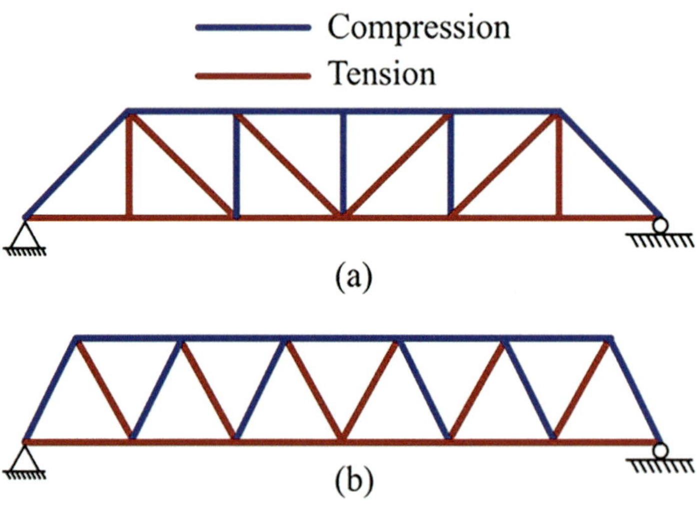

FIGURE 4.33 Axial loads on members: (a) Pratt truss; (b) Warren truss.

exhibition halls, planar trusses enable the construction of roofs that span significant distances while maintaining structural integrity. Moreover, trusses enhance aesthetic appeal, as their geometric designs can be exposed to add architectural beauty, blending function and form. Their lightweight nature also makes them ideal for prefabrication and quick installation on-site, improving construction efficiency. Planar trusses play a fundamental role in architecture by allowing the creation of efficient, material-saving structures that cover large areas without interior support. They are a perfect example of how structural design and aesthetic appeal can work together to achieve both functional and beautiful architectural spaces.

4.15 ANALYSIS OF TRUSS MEMBERS

Trusses are structural frameworks composed of interconnected members, typically arranged in triangular units. The primary load actions on truss members include thus:

- **Tension:** This occurs when a member is subjected to pulling forces, causing it to elongate. In a truss, members that are primarily in tension are typically the diagonal members and the bottom chords.
- **Compression:** This action occurs when a member is subjected to pushing forces, leading to shortening. Members that usually experience compression include the top chords and some diagonal members in specific configurations.
- **Shear:** Although trusses primarily resist axial loads (tension and compression), shear forces can also act at the joints where members connect, particularly under lateral loads, or when the truss is not symmetrically loaded.

Understanding these load actions is essential for analyzing truss structures effectively.

The analysis of a truss involves determining the internal forces in each member due to external loads and support reactions. The two primary methods for analyzing trusses are:

1. Method of joints
2. Stiffness matrix method

4.15.1 Method of Joints

The method of joints focuses on each joint in the truss, applying the principles of static equilibrium:

1. *Calculate support reactions.* Begin by determining the reactions at the supports using static equilibrium equations ($\sum F_x = 0; \sum F_y = 0$).
2. *Isolate joints.* Examine each joint where members connect. Ideally, start with a joint that has only two unknown member forces, as this simplifies calculations.
3. *Draw free-body diagrams.* For each joint, create a free-body diagram showing all forces acting on it. Assume all member forces are either tensile or compressive.
4. *Apply equilibrium equations.* Use the equations of equilibrium to solve for unknown forces at the joint:

   $\sum F_x = 0; \sum F_y = 0$

5. *Repeat for adjacent joints.* Move to adjacent joints with known forces, and repeat the process until all member forces are determined.
6. *Identify zero members.* Some members may have no force acting on them (zero members), often found in configurations where two non-collinear members meet at a joint without external loading.

Analyzing trusses through these methods allows engineers to ensure structural integrity under various loading conditions. Understanding how different load actions affect truss members is crucial for effective design and safety assessments in engineering applications.

4.16 STIFFNESS METHOD FOR PLANAR TRUSS

Figure 4.34 explains the transformation between the local and the global axes. Consider a truss member inclined to an arbitrary angle of θ. This angle is measured with respect to the global axes (X–Y) in an anticlockwise direction, which is considered as positive in the analysis. There will be independent translations happening along the X and the Y directions (Chandrasekaran, 2018).

Refer to the following:

$$C_x = \cos\theta,\ C_y = \sin\theta.$$

In order to convert the local responses with respect to the reference axes responses, the transformation matrix can be written as follows:

$$\begin{Bmatrix} p_r \\ p_s \\ p_t \\ p_h \end{Bmatrix} = \begin{bmatrix} C_x & 0 & -C_y & 0 \\ 0 & C_x & 0 & -C_y \\ C_y & 0 & C_x & 0 \\ 0 & C_y & 0 & C_x \end{bmatrix} \begin{Bmatrix} \overline{p}_r \\ \overline{p}_s \\ \overline{p}_t \\ \overline{p}_h \end{Bmatrix}$$

$$\{p\}_i = [T_T]_i \{\overline{p}\}_i$$

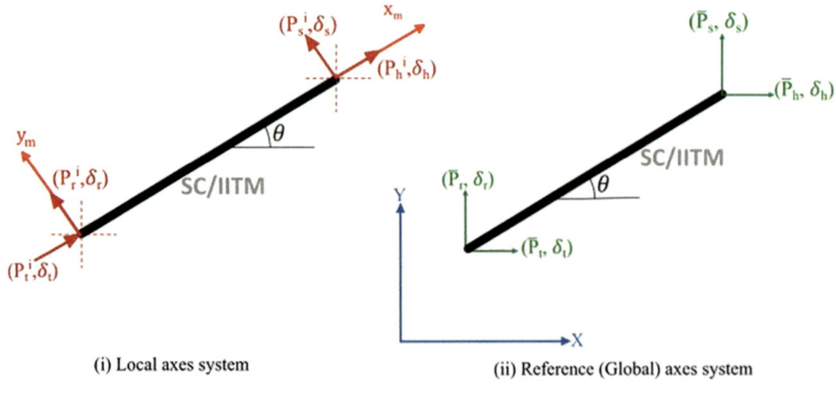

FIGURE. 4.34 Transformation between local and global axes.

Hence:

$$\{\delta_T\}_i = [T_T]_i \{\bar{\delta}_T\}_i$$
$$\{\bar{\delta}_T\}_i = [T_T]_i^T \{\delta_T\}_i$$

Where:

$$\{\delta_T\}_i = \begin{Bmatrix} \delta_r \\ \delta_s \\ \delta_t \\ \delta_h \end{Bmatrix}, \{\bar{\delta}_T\}_i = \begin{Bmatrix} \bar{\delta}_r \\ \bar{\delta}_s \\ \bar{\delta}_t \\ \bar{\delta}_h \end{Bmatrix}$$

4.17 STIFFNESS MATRIX

Now, the standard stiffness matrix of the truss element without end rotations is simply given as follows:

$$[K_T]_i = \begin{bmatrix} 0 & 0 & 0 & 0 \\ 0 & 0 & 0 & 0 \\ 0 & 0 & \frac{AE}{l} & -\frac{AE}{l} \\ 0 & 0 & -\frac{AE}{l} & \frac{AE}{l} \end{bmatrix}$$

Now, the global stiffness matrix of the truss member can be obtained using the following equation:

$$[\bar{K}_T]_i = [T_T]_i^T [K_T]_i [T_T]_i$$

The responses of the truss member in the reference axes system is given by:

$$[\bar{P}_T]_i = [\bar{K}_T]_i \{\bar{\delta}_T\}_i + \{\bar{F}_p\}_i$$

Where:

$$[\bar{P}_T]_i = \begin{Bmatrix} \bar{p}_r \\ \bar{p}_s \\ \bar{p}_s \\ \bar{p}_h \end{Bmatrix}, \{\bar{F}_p\}_i = \begin{Bmatrix} \bar{F}_{ps} \\ \bar{F}_{pr} \\ \bar{F}_{pt} \\ \bar{F}_{ph} \end{Bmatrix}_i$$

Hence, the planar truss problem is much simpler than the orthogonal frame analysis

4.17.1 Example Problem 1: Planar Truss

Analyze the planar truss system shown in Figure using the stiffness method. *E* is constant for all the members:

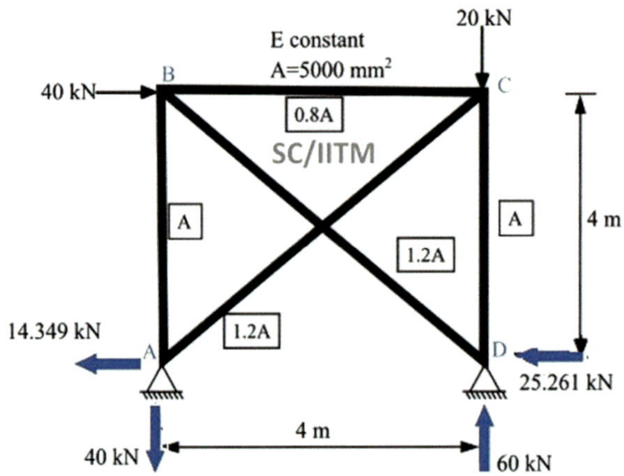

4.17.2 Solution

4.17.2.1 Marking Restrained and Unrestrained Degrees of Freedom

$$\text{Unrestrained degrees of freedom} = 4\left(\delta_1, \delta_2, \delta_3, \delta_4\right) \text{(marked in green color)}$$
$$\text{Restrained degrees of freedom} = 4\left(\delta_5, \delta_6, \delta_7, \delta_8\right) \text{(marked in red color)}$$

The global axes system and the local axes for every member should be marked to find θ.

The unrestrained and restrained degrees of freedom are marked. The local and global axes system are also shown in what follows:

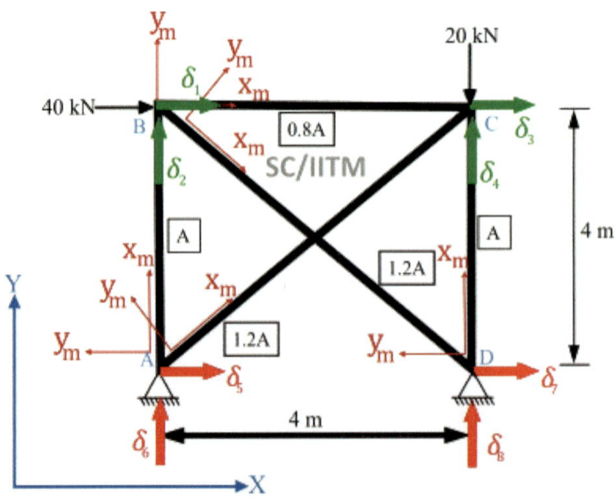

	Ends						
Member	**j**	**k**	**Length (m)**	**θ (Degrees)**	**C_x**	**C_y**	**Global Labels**
AB	A	B	4	90	0	1	(6, 2, 5, 1)
BC	B	C	4	0	1	0	(2, 4, 1, 3)
CD	D	C	4	90	0	1	(8, 4, 7, 3)
BD	B	D	5.646	−45	0.707	−0.707	(2, 8, 1, 7)
AC	A	C	5.646	45	0.707	−0.707	(6, 4, 5, 3)

4.17.2.2 Calculation of Local Stiffness Matrix

The local stiffness matrix for the truss element is given by:

$$[K_T]_i = \begin{bmatrix} 0 & 0 & 0 & 0 \\ 0 & 0 & 0 & 0 \\ 0 & 0 & \frac{AE}{l} & -\frac{AE}{l} \\ 0 & 0 & -\frac{AE}{l} & \frac{AE}{l} \end{bmatrix}$$

$$K_{AB} = E \times 10^{-3} \begin{array}{c} \begin{array}{cccc} ⑥ & ② & ⑤ & ① \end{array} \\ \left[\begin{array}{cccc} 0 & 0 & 0 & 0 \\ 0 & 0 & 0 & 0 \\ 0 & 0 & 1.25 & -1.25 \\ 0 & 0 & -1.25 & 1.25 \end{array}\right] \end{array} \begin{array}{c} ⑥ \\ ② \\ ⑤ \\ ① \end{array}$$

$$K_{BC} = E \times 10^{-3} \begin{array}{c} \begin{array}{cccc} ② & ④ & ① & ③ \end{array} \\ \left[\begin{array}{cccc} 0 & 0 & 0 & 0 \\ 0 & 0 & 0 & 0 \\ 0 & 0 & 1 & -1 \\ 0 & 0 & -1 & 1 \end{array}\right] \end{array} \begin{array}{c} ② \\ ④ \\ ① \\ ③ \end{array}$$

$$K_{CD} = E \times 10^{-3} \begin{array}{c} \begin{array}{cccc} ⑧ & ④ & ⑦ & ③ \end{array} \\ \left[\begin{array}{cccc} 0 & 0 & 0 & 0 \\ 0 & 0 & 0 & 0 \\ 0 & 0 & 1.25 & -1.25 \\ 0 & 0 & -1.25 & 1.25 \end{array}\right] \end{array} \begin{array}{c} ⑧ \\ ④ \\ ⑦ \\ ③ \end{array}$$

$$K_{BD} = E\times10^{-3}\begin{array}{c}\begin{matrix}\textcircled{2} & \textcircled{8} & \textcircled{1} & \textcircled{7}\end{matrix}\\ \begin{bmatrix}0 & 0 & 0 & 0\\ 0 & 0 & 0 & 0\\ 0 & 0 & 1.061 & -1.061\\ 0 & 0 & -1.061 & 1.061\end{bmatrix}\end{array}\begin{matrix}\textcircled{2}\\ \textcircled{8}\\ \textcircled{1}\\ \textcircled{7}\end{matrix}$$

$$K_{AC} = E\times10^{-3}\begin{array}{c}\begin{matrix}\textcircled{6} & \textcircled{4} & \textcircled{5} & \textcircled{3}\end{matrix}\\ \begin{bmatrix}0 & 0 & 0 & 0\\ 0 & 0 & 0 & 0\\ 0 & 0 & 1.061 & -1.061\\ 0 & 0 & -1.061 & 1.061\end{bmatrix}\end{array}\begin{matrix}\textcircled{6}\\ \textcircled{4}\\ \textcircled{5}\\ \textcircled{3}\end{matrix}$$

4.17.2.3 Calculation of Transformation Matrix

$$[T]_{AB} = \begin{bmatrix}0 & 0 & -1 & 0\\ 0 & 0 & 0 & -1\\ 1 & 0 & 0 & 0\\ 0 & 1 & 0 & 0\end{bmatrix}, [T]_{BC} = \begin{bmatrix}1 & 0 & 0 & 0\\ 0 & 1 & 0 & 0\\ 0 & 0 & 1 & 0\\ 0 & 0 & 0 & 1\end{bmatrix}, [T]_{CD} = \begin{bmatrix}0 & 0 & -1 & 0\\ 0 & 0 & 0 & -1\\ 1 & 0 & 0 & 0\\ 0 & 1 & 0 & 0\end{bmatrix}$$

$$[T]_{BD} = \begin{bmatrix}0.7071 & 0 & 0.7071 & 0\\ 0 & 0.7071 & 0 & 0.7071\\ -0.7071 & 0 & 0.7071 & 0\\ 0 & -0.7071 & 0 & 0.7071\end{bmatrix},$$

$$[T]_{AC} = \begin{bmatrix}0.7071 & 0 & -0.7071 & 0\\ 0 & 0.7071 & 0 & -0.7071\\ 0.7071 & 0 & 0.7071 & 0\\ 0 & 0.7071 & 0 & 0.7071\end{bmatrix}$$

4.17.2.4 Calculation of Global Stiffness Matrix

The global stiffness matrix of all the members is calculated by using the following relationship:

$$\left[\bar{K}_T\right]_i = \left[T_T\right]_i^T \left[K_T\right]_i \left[T_T\right]_i$$

$$\bar{K}_{AB} = E\begin{array}{c}\begin{matrix}\textcircled{6} & \textcircled{2} & \textcircled{5} & \textcircled{1}\end{matrix}\\ \begin{bmatrix}0.0013 & -0.0013 & 0 & 0\\ -0.0013 & 0.0013 & 0 & 0\\ 0 & 0 & 0 & 0\\ 0 & 0 & 0 & 0\end{bmatrix}\end{array}\begin{matrix}\textcircled{6}\\ \textcircled{2}\\ \textcircled{5}\\ \textcircled{1}\end{matrix}$$

$$\overline{K}_{BC} = E\begin{array}{c} \begin{matrix} \textcircled{2} & \textcircled{4} & \textcircled{1} & \textcircled{3} \end{matrix} \\ \begin{bmatrix} 0 & 0 & 0 & 0 \\ 0 & 0 & 0 & 0 \\ 0 & 0 & 1 & -1 \\ 0 & 0 & -1 & 1 \end{bmatrix} \end{array} \begin{matrix} \textcircled{2} \\ \textcircled{4} \\ \textcircled{1} \\ \textcircled{3} \end{matrix}$$

$$\overline{K}_{CD} = E\begin{array}{c} \begin{matrix} \textcircled{8} & \textcircled{4} & \textcircled{7} & \textcircled{3} \end{matrix} \\ \begin{bmatrix} 0.0013 & -0.0013 & 0 & 0 \\ -0.0013 & 0.0013 & 0 & 0 \\ 0 & 0 & 0 & 0 \\ 0 & 0 & 0 & 0 \end{bmatrix} \end{array} \begin{matrix} \textcircled{8} \\ \textcircled{4} \\ \textcircled{7} \\ \textcircled{3} \end{matrix}$$

$$\overline{K}_{BD} = E\times 10^{-3}\begin{array}{c} \begin{matrix} \textcircled{2} & \textcircled{8} & \textcircled{1} & \textcircled{7} \end{matrix} \\ \begin{bmatrix} 0.5304 & -0.5304 & -0.5304 & 0.5304 \\ -0.5304 & 0.5304 & 0.5304 & -0.5304 \\ -0.5304 & 0.5304 & 0.5304 & -0.5304 \\ 0.5304 & -0.5304 & -0.5304 & 0.5304 \end{bmatrix} \end{array} \begin{matrix} \textcircled{2} \\ \textcircled{8} \\ \textcircled{1} \\ \textcircled{7} \end{matrix}$$

$$\overline{K}_{AC} = E\times 10^{-3}\begin{array}{c} \begin{matrix} \textcircled{6} & \textcircled{4} & \textcircled{5} & \textcircled{3} \end{matrix} \\ \begin{bmatrix} 0.5304 & -0.5304 & 0.5304 & -0.5304 \\ -0.5304 & 0.5304 & -0.5304 & 0.5304 \\ 0.5304 & -0.5304 & 0.5304 & -0.5304 \\ -0.5304 & 0.5304 & -0.5304 & 0.5304 \end{bmatrix} \end{array} \begin{matrix} \textcircled{6} \\ \textcircled{4} \\ \textcircled{5} \\ \textcircled{3} \end{matrix}$$

The total stiffness matrix is obtained by assembling the global stiffness matrices of all members with respect to the global labels. From the total stiffness matrix, the stiffness matrix for the unrestrained degrees of freedom can be partitioned.

$$[K_{uu}] = E\begin{array}{c} \begin{matrix} \textcircled{1} & \textcircled{2} & \textcircled{3} & \textcircled{4} \end{matrix} \\ \begin{bmatrix} 0.0015 & -0.005 & -0.0010 & 0 \\ -0.005 & 0.0018 & 0 & 0 \\ -0.0010 & 0 & 0.0015 & 0.0005 \\ 0 & 0 & 0.0005 & 0.0018 \end{bmatrix} \end{array} \begin{matrix} \textcircled{1} \\ \textcircled{2} \\ \textcircled{3} \\ \textcircled{4} \end{matrix}$$

$$[K_{uu}]^{-1} = \frac{1}{E}\times 10^{-3}\begin{bmatrix} 1.5534 & 0.4628 & 1.1319 & -0.3372 \\ 0.4628 & 0.6995 & 0.3372 & -0.1005 \\ 1.1319 & 0.3372 & 1.5534 & -0.4628 \\ -0.3372 & -0.1005 & -0.4628 & 0.6995 \end{bmatrix}$$

4.17.2.5 Estimation of Joint Load Vectors

The joint load vector can be written as follows:

$$[\bar{J}_L] = \begin{Bmatrix} +40 \\ 0 \\ 0 \\ -20 \\ 0 \\ 0 \\ 0 \\ 0 \end{Bmatrix} \begin{matrix} (1) \\ (2) \\ (3) \\ (4) \\ (5) \\ (6) \\ (7) \\ (8) \end{matrix}$$

The joint load vector in unrestrained degrees of freedom can be partitioned from the aforementioned matrix as follows:

$$[\bar{J}_{Lu}] = \begin{Bmatrix} +40 \\ 0 \\ 0 \\ -20 \end{Bmatrix} \begin{matrix} (1) \\ (2) \\ (3) \\ (4) \end{matrix}$$

$$\{\bar{\delta}\} = [\bar{K}_{uu}]^{-1} \{\bar{J}_{Lu}\}$$

$$\{\bar{\delta}\}_u = \frac{1\times 10^{-4}}{E} \begin{Bmatrix} 6.881 \\ 2.0521 \\ 5.4532 \\ -2.7479 \end{Bmatrix}$$

4.17.2.6 Calculation of Member Forces

$$\bar{M}_{AB} = \begin{Bmatrix} \bar{V}_6 \\ \bar{V}_2 \\ \bar{H}_5 \\ \bar{H}_1 \end{Bmatrix} = \begin{Bmatrix} -25.6509 \\ 25.6509 \\ 0 \\ 0 \end{Bmatrix}, \bar{M}_{BC} = \begin{Bmatrix} \bar{V}_2 \\ \bar{V}_4 \\ \bar{H}_1 \\ \bar{H}_5 \end{Bmatrix} = \begin{Bmatrix} 0 \\ 0 \\ 14.3491 \\ -14.3491 \end{Bmatrix}, \bar{M}_{CD} = \begin{Bmatrix} \bar{V}_8 \\ \bar{V}_4 \\ \bar{H}_7 \\ \bar{H}_3 \end{Bmatrix} = \begin{Bmatrix} 34.3491 \\ -34.3491 \\ 0 \\ 0 \end{Bmatrix}$$

$$\bar{M}_{BD} = \begin{Bmatrix} \bar{V}_2 \\ \bar{V}_8 \\ \bar{H}_1 \\ \bar{H}_7 \end{Bmatrix} = \begin{Bmatrix} -25.6509 \\ 25.6509 \\ 26.6509 \\ -25.6509 \end{Bmatrix}, \bar{M}_{AC} = \begin{Bmatrix} \bar{V}_6 \\ \bar{V}_4 \\ \bar{H}_5 \\ \bar{H}_3 \end{Bmatrix} = \begin{Bmatrix} -14.3491 \\ 14.3491 \\ -14.3491 \\ 14.3491 \end{Bmatrix}$$

The member and end forces are shown in Figure 1.23 and Figure 1.24.

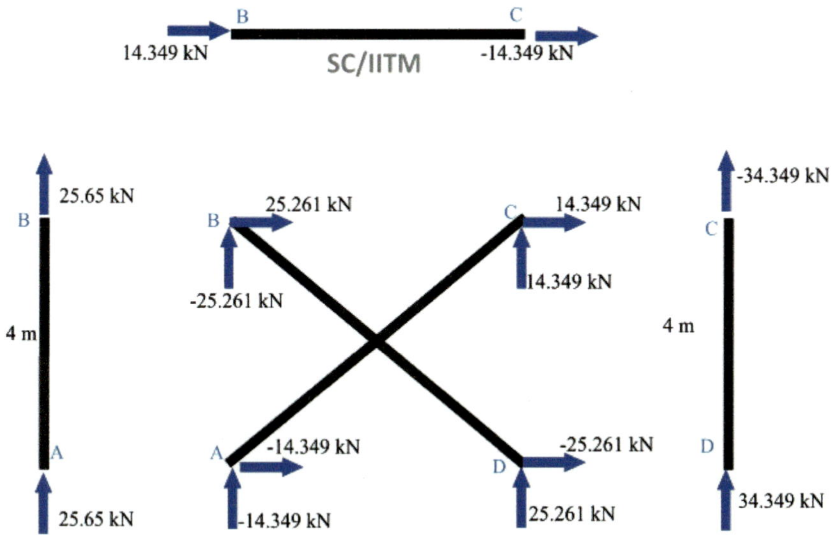

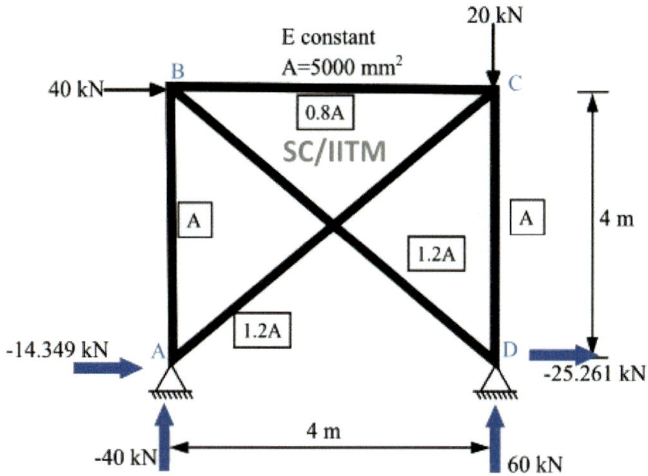

MATLAB PROGRAM

```
%% stiffness matrix method for planar truss
% Input
clc;
clear;
n = 5; % number of members
L = [4 4 4 5.646 5.646]; % length in m
A = [5E-3 4E-3 5E-3 6E-3 6E-3]; % Area in m2
theta = [90 0 90 -45 45]; % angle in degrees
```

```
uu = 4; % Number of unrestrained degrees-of-freedom
ur = 4; % Number of restrained degrees-of-freedom
uul = [1 2 3 4]; % global labels of unrestained dof
url = [5 6 7 8]; % global labels of restained dof
l1 = [6 2 5 1]; % Global labels for member 1
l2 = [2 4 1 3]; % Global labels for member 2
l3 = [8 4 7 3]; % Global labels for member 3
l4 = [2 8 1 7]; % Global labels for member 4
l5 = [6 4 5 3]; % Global labels for member 5
l = [l1; l2; l3; l4; l5];
dof = uu + ur; % Degrees-of-freedom
Ktotal = zeros (dof);
Tt1 = zeros (4); % Transformation matrix for member 1
Tt2 = zeros (4); % Transformation matrix for member 2
Tt3 = zeros (4); % Transformation matrix for member 3
Tt4 = zeros (4); % Transformation matrix for member 4
Tt5 = zeros (4); % Transformation matrix for member 5
fem1 = [0; 0; 0; 0]; % Local Fixed end forces of member 1
fem2 = [0; 0; 0; 0]; % Local Fixed end forces of member 2
fem3 = [0; 0; 0; 0]; % Local Fixed end forces of member 3
fem4 = [0; 0; 0; 0]; % Local Fixed end forces of member 4
fem5 = [0; 0; 0; 0]; % Local Fixed end forces of member 5
%% rotation coefficients for each member
rc = A./L;
cx = cosd(theta);
cy = sind(theta);
%% stiffness matrix 4 by 4
for i = 1:n
  Knew = zeros (dof);
  k1 = [0; 0; 0; 0];
  k2 = [0; 0; 0; 0];
  k3 = [0; 0; rc(i); -rc(i)];
  k4 = -k3;
  K = [k1 k2 k3 k4];
  fprintf ('Member Number =');
  disp (i);
  fprintf ('Local Stiffness matrix of member, [K] = \n');
  disp (K);
  T1 = [cx(i); 0; cy(i); 0];
  T2 = [0; cx(i); 0; cy(i)];
  T3 = [-cy(i); 0; cx(i); 0];
  T4 = [0; -cy(i); 0; cx(i)];
  T = [T1 T2 T3 T4];
  fprintf ('Tranformation matrix of member, [T] = \n');
  disp (T);
  Ttr = T';
```

```
  fprintf ('Tranformation matrix Transpose, [T] = \n');
  disp (Ttr);
  Kg = Ttr*K*T;
  fprintf ('Global Matrix, [K global] = \n');
  disp (Kg);
  for p = 1:4
    for q = 1:4
      Knew((l(i,p)),(l(i,q))) = Kg(p,q);
    end
  end
  Ktotal = Ktotal + Knew;
  if i == 1
    Tt1 = T;
    Kg1 = Kg;
    fembar1 = Tt1'*fem1;
  elseif i == 2
    Tt2 = T;
    Kg2 = Kg;
    fembar2 = Tt2'*fem2;
  elseif i == 3
    Tt3 = T;
    Kg3 = Kg;
    fembar3 = Tt3'*fem3;
  elseif i == 4
    Tt4 = T;
    Kg4 = Kg;
    fembar4 = Tt4'*fem4;
  else
    Tt5 = T;
    Kg5 = Kg;
    fembar5 = Tt5'*fem5;
  end
end
fprintf ('Stiffness Matrix of complete structure,
[Ktotal] = \n');
disp (Ktotal);
Kunr = zeros(uu);
for x = 1:uu
  for y = 1:uu
    Kunr(x,y) = Ktotal(x,y);
  end
end
fprintf ('Unrestrained Stiffness sub-matrix, [Kuu] =
\n');
disp (Kunr);
KuuInv = inv(Kunr);
```

```
fprintf ('Inverse of Unrestrained Stiffness sub-matrix,
[KuuInverse] = \n');
disp (KuuInv);
%% Creation of joint load vector
jl = [40; 0; 0; -20; 0; 0; 0; 0]; % values given in kN
jlu = [40; 0; 0; -20]; % load vector in unrestrained dof
delu = KuuInv*jlu;
fprintf ('Joint Load vector, [Jl] = \n');
disp (jl');
fprintf ('Unrestrained displacements, [DelU] = \n');
disp (delu');
delr = zeros(ur,1);
del = zeros (dof,1);
del = [delu; delr];
deli = zeros (4,1);
for i = 1:n
  for p = 1:4
    deli(p,1) = del((l(i,p)),1) ;
  end
  if i == 1
    delbar1 = deli;
    mbar1 = (Kg1 * delbar1)+fembar1;
    fprintf ('Member Number =');
    disp (i);
    fprintf ('Global displacement matrix [DeltaBar] =
    \n');
    disp (delbar1');
    fprintf ('Global End moment matrix [MBar] = \n');
    disp (mbar1');
  elseif i == 2
    delbar2 = deli;
    mbar2 = (Kg2 * delbar2)+fembar2;
    fprintf ('Member Number =');
    disp (i);
    fprintf ('Global displacement matrix [DeltaBar] =
    \n');
    disp (delbar2');
    fprintf ('Global End moment matrix [MBar] = \n');
    disp (mbar2');
  elseif i == 3
    delbar3 = deli;
    mbar3 = (Kg3 * delbar3)+fembar3;
    fprintf ('Member Number =');
    disp (i);
    fprintf ('Global displacement matrix [DeltaBar] =
    \n');
```

```
    disp (delbar3');
    fprintf ('Global End moment matrix [MBar] = \n');
    disp (mbar3');
  elseif i == 4
    delbar4 = deli;
    mbar4 = (Kg4 * delbar4)+fembar4;
    fprintf ('Member Number =');
    disp (i);
    fprintf ('Global displacement matrix [DeltaBar] =
    \n');
    disp (delbar4');
    fprintf ('Global End moment matrix [MBar] = \n');
    disp (mbar4');
  else
    delbar5 = deli;
    mbar5 = (Kg5 * delbar5)+fembar5;
    fprintf ('Member Number =');
    disp (i);
    fprintf ('Global displacement matrix [DeltaBar] = \n');
    disp (delbar5');
    fprintf ('Global End moment matrix [MBar] = \n');
    disp (mbar5');
  end
end
%% check
mbar = [mbar1'; mbar2'; mbar3'; mbar4'; mbar5'];
jf = zeros(dof,1);
for a = 1:n
  for b = 1:4 % size of k matrix
    d = l(a,b);
    jfnew = zeros(dof,1);
    jfnew(d,1) = mbar(a,b);
    jf = jf+jfnew;
  end
end
fprintf ('Joint forces = \n');
disp (jf');
```

MATLAB OUTPUT

```
Member Number = 1
Local Stiffness matrix of member, [K] =
        0        0        0        0
        0        0        0        0
        0        0   0.0013  -0.0013
        0        0  -0.0013   0.0013
```

```
Transformation matrix of member, [T] =
        0         0        -1         0
        0         0         0        -1
        1         0         0         0
        0         1         0         0
Transformation matrix Transpose, [T] =
        0         0         1         0
        0         0         0         1
       -1         0         0         0
        0        -1         0         0
Global Matrix, [K global] =
   0.0013   -0.0013         0         0
  -0.0013    0.0013         0         0
        0         0         0         0
        0         0         0         0
Member Number = 2
Local Stiffness matrix of member, [K] =
  1.0e-03 *
        0         0         0         0
        0         0         0         0
        0         0    1.0000   -1.0000
        0         0   -1.0000    1.0000
Transformation matrix of member, [T] =
        1         0         0         0
        0         1         0         0
        0         0         1         0
        0         0         0         1
Transformation matrix Transpose, [T] =
        1         0         0         0
        0         1         0         0
        0         0         1         0
        0         0         0         1
Global Matrix, [K global] =
  1.0e-03 *
        0         0         0         0
        0         0         0         0
        0         0    1.0000   -1.0000
        0         0   -1.0000    1.0000
Member Number = 3
Local Stiffness matrix of member, [K] =
        0         0         0         0
        0         0         0         0
        0         0    0.0013   -0.0013
        0         0   -0.0013    0.0013
Transformation matrix of member, [T] =
        0         0        -1         0
        0         0         0        -1
```

```
        1         0         0         0
        0         1         0         0
Transformation matrix Transpose, [T] =
        0         0         1         0
        0         0         0         1
       -1         0         0         0
        0        -1         0         0
Global Matrix, [K global] =
   0.0013   -0.0013         0         0
  -0.0013    0.0013         0         0
        0         0         0         0
        0         0         0         0
Member Number = 4
Local Stiffness matrix of member, [K] =
        0         0         0         0
        0         0         0         0
        0         0    0.0011   -0.0011
        0         0   -0.0011    0.0011
Transformation matrix of member, [T] =
   0.7071         0    0.7071         0
        0    0.7071         0    0.7071
  -0.7071         0    0.7071         0
        0   -0.7071         0    0.7071
Transformation matrix Transpose, [T] =
   0.7071         0   -0.7071         0
        0    0.7071         0   -0.7071
   0.7071         0    0.7071         0
        0    0.7071         0    0.7071
Global Matrix, [K global] =
  1.0e-03 *
   0.5313   -0.5313         0    0.5313    0.5313
  -0.5313    0.5313    0.5313   -0.5313
  -0.5313    0.5313    0.5313   -0.5313
   0.5313   -0.5313         0    0.5313    0.5313
Member Number = 5
Local Stiffness matrix of member, [K] =
        0         0         0         0
        0         0         0         0
        0         0    0.0011   -0.0011
        0         0   -0.0011    0.0011
Transformation matrix of member, [T] =
   0.7071         0   -0.7071         0
        0    0.7071         0   -0.7071
   0.7071         0    0.7071         0
        0    0.7071         0    0.7071
Transformation matrix Transpose, [T] =
   0.7071         0    0.7071         0
        0    0.7071         0    0.7071
```

```
  -0.7071         0   0.7071         0
        0   -0.7071        0    0.7071
Global Matrix, [K global] =
  1.0e-03 *
   0.5313  -0.5313   0.5313  -0.5313
  -0.5313   0.5313  -0.5313   0.5313
   0.5313  -0.5313   0.5313  -0.5313
  -0.5313   0.5313  -0.5313   0.5313
Stiffness Matrix of complete structure, [Ktotal] =
   0.0015  -0.0005  -0.0010        0        0        0
  -0.0005   0.0005
  -0.0005   0.0018        0        0        0  -0.0013
   0.0005  -0.0005
  -0.0010        0   0.0015   0.0005  -0.0005  -0.0010
        0   0.0015   0.0005        0        0
        0        0   0.0005   0.0018  -0.0005  -0.0005
        0  -0.0013
        0        0  -0.0005  -0.0005   0.0005   0.0005
        0        0
        0  -0.0013  -0.0005  -0.0005   0.0005   0.0018
        0        0
  -0.0005   0.0005        0        0        0        0
   0.0005  -0.0005
   0.0005  -0.0005        0  -0.0013        0        0
  -0.0005   0.0018
Unrestrained Stiffness sub-matrix, [Kuu] =
   0.0015  -0.0005  -0.0010        0
  -0.0005   0.0018        0        0
  -0.0010        0   0.0015   0.0005
        0        0   0.0005   0.0018
Inverse of Unrestrained Stiffness sub-matrix,
[KuuInverse] =
  1.0e + 03 *
   1.5517   0.4629   1.1303  -0.3371
   0.4629   0.6994   0.3371  -0.1006
   1.1303   0.3371   1.5517  -0.4629
  -0.3371  -0.1006  -0.4629   0.6994
Joint Load vector, [Jl] =
       40        0        0      -20        0        0
        0        0
Unrestrained displacements, [DelU] =
  1.0e+04 *
   6.8812   2.0525   5.4468  -2.7475
Member Number = 1
Global displacement matrix [DeltaBar] =
  1.0e+04 *
        0   2.0525        0   6.8812
```

```
Global End moment matrix [MBar] =
  -25.6568   25.6568         0         0
Member Number = 2
Global displacement matrix [DeltaBar] =
   1.0e+04 *
    2.0525   -2.7475    6.8812    5.4468
Global End moment matrix [MBar] =
         0         0   14.3432  -14.3432
Member Number = 3
Global displacement matrix [DeltaBar] =
   1.0e+04 *
         0   -2.7475         0    5.4468
Global End moment matrix [MBar] =
   34.3432  -34.3432         0         0
Member Number = 4
Global displacement matrix [DeltaBar] =
   1.0e+04 *
    2.0525         0    6.8812         0
Global End moment matrix [MBar] =
  -25.6568   25.6568   25.6568  -25.6568
Member Number = 5
Global displacement matrix [DeltaBar] =
   1.0e+04 *
         0   -2.7475         0    5.4468
Global End moment matrix [MBar] =
  -14.3432   14.3432  -14.3432   14.3432
Joint forces =
   40.0000    0.0000   -0.0000  -20.0000  -14.3432
  -40.0000  -25.6568   60.0000
```

4.17.3 Example Problem 2: Planar Truss

Analyze the planar truss system shown in Figure using the stiffness method. E is constant for all the members.

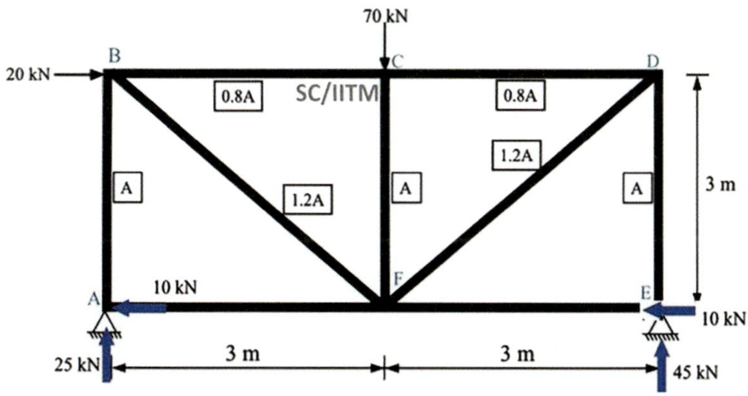

4.17.4 Solution

4.17.4.1 Marking Restrained and Unrestrained Degrees of Freedom

Unrestrained degrees of freedom = 4 $(\delta_1, \delta_2, \delta_3, \delta_4, \delta_5, \delta_6, \delta_7, \delta_8)$ (marked in green color)

Restrained degrees of freedom = 4 $(\delta_9, \delta_{10}, \delta_{11}, \delta_{12})$ (marked in red color)

The unrestrained and restrained degrees of freedom are marked in the truss, as shown in Figure 1.26.

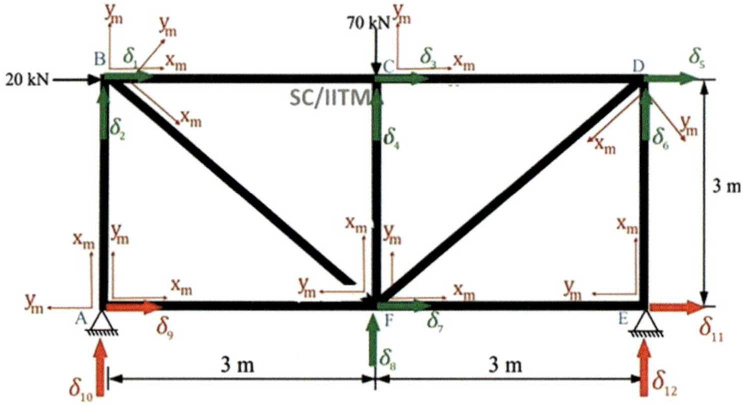

Member	Ends j	k	Length (m)	θ (Degrees)	C_x	C_y	Global Labels
AB	A	B	3	90	0	1	(10, 2, 9, 1)
BC	B	C	3	0	1	0	(2, 4, 1, 3)
CD	C	D	3	0	1	0	(4, 6, 3, 5)
DE	E	D	3	90	0	1	(12, 6, 11, 5)
AF	A	F	3	0	1	0	(10, 8, 9, 7)
FE	F	E	3	0	1	0	(8, 12, 7, 11)
BF	B	F	4.242	−45	0.707	−0.707	(2, 8, 1, 7)
CF	F	C	3	+90	0	1	(8, 4, 7, 3)
DF	D	F	4.242	−135	−0.707	0.707	(6, 8, 5, 7)

4.17.4.2 Calculation of Local Stiffness Matrix

The local stiffness matrix for the truss element is given by:

$$[K_T]_i = \begin{bmatrix} 0 & 0 & 0 & 0 \\ 0 & 0 & 0 & 0 \\ 0 & 0 & \frac{AE}{l} & -\frac{AE}{l} \\ 0 & 0 & -\frac{AE}{l} & \frac{AE}{l} \end{bmatrix}$$

$$K_{AB} = E\times10^{-3}\begin{bmatrix}0&0&0&0\\0&0&0&0\\0&0&1.7&-1.7\\0&0&-1.7&1.7\end{bmatrix},\ K_{BC} = E\times10^{-3}\begin{bmatrix}0&0&0&0\\0&0&0&0\\0&0&1.3&-1.3\\0&0&-1.3&1.3\end{bmatrix}$$

$$K_{CD} = E\times10^{-3}\begin{bmatrix}0&0&0&0\\0&0&0&0\\0&0&1.3&-1.3\\0&0&-1.3&1.3\end{bmatrix},\ K_{DE} = E\times10^{-3}\begin{bmatrix}0&0&0&0\\0&0&0&0\\0&0&1.7&-1.7\\0&0&-1.7&1.7\end{bmatrix}$$

$$K_{AF} = E\times10^{-3}\begin{bmatrix}0&0&0&0\\0&0&0&0\\0&0&1.3&-1.3\\0&0&-1.3&1.3\end{bmatrix},\ K_{FE} = E\times10^{-3}\begin{bmatrix}0&0&0&0\\0&0&0&0\\0&0&1.3&-1.3\\0&0&-1.3&1.3\end{bmatrix}$$

$$K_{BF} = E\times10^{-3}\begin{bmatrix}0&0&0&0\\0&0&0&0\\0&0&1.4&-1.4\\0&0&-1.4&1.4\end{bmatrix},\ K_{CF} = E\times10^{-3}\begin{bmatrix}0&0&0&0\\0&0&0&0\\0&0&1.7&-1.7\\0&0&-1.7&1.7\end{bmatrix}$$

$$K_{DF} = E\times10^{-3}\begin{bmatrix}0&0&0&0\\0&0&0&0\\0&0&1.4&-1.4\\0&0&-1.4&1.4\end{bmatrix}$$

4.17.4.3 Calculation of Transformation Matrix

$$[T]_{AB} = \begin{bmatrix}0&0&-1&0\\0&0&0&-1\\1&0&0&0\\0&1&0&0\end{bmatrix},\ [T]_{BC} = \begin{bmatrix}1&0&0&0\\0&1&0&0\\0&0&1&0\\0&0&0&1\end{bmatrix},\ [T]_{CD} = \begin{bmatrix}1&0&0&0\\0&1&0&0\\0&0&1&0\\0&0&0&1\end{bmatrix}$$

$$[T]_{DE} = \begin{bmatrix}0&0&-1&0\\0&0&0&-1\\1&0&0&0\\0&1&0&0\end{bmatrix},\ [T]_{AF} = \begin{bmatrix}1&0&0&0\\0&1&0&0\\0&0&1&0\\0&0&0&1\end{bmatrix},\ [T]_{FE} = \begin{bmatrix}1&0&0&0\\0&1&0&0\\0&0&1&0\\0&0&0&1\end{bmatrix}$$

$$[T]_{BF} = \begin{bmatrix}0.7071&0&0.7071&0\\0&0.7071&0&0.7071\\-0.7071&0&0.7071&0\\0&-0.7071&0&0.7071\end{bmatrix},\ [T]_{CF} = \begin{bmatrix}0&0&-1&0\\0&0&0&-1\\1&0&0&0\\0&1&0&0\end{bmatrix}$$

$$[T]_{BF} = \begin{bmatrix}0.7071&0&0.7071&0\\0&0.7071&0&0.7071\\-0.7071&0&0.7071&0\\0&-0.7071&0&0.7071\end{bmatrix}$$

4.17.4.4 Calculation of Global Stiffness Matrix

The global stiffness matrix of all the members is calculated by using the following relationship:

$$\left[\overline{K}_T\right]_i = \left[T_T\right]_i^T \left[K_T\right]_i \left[T_T\right]_i$$

$$\overline{K}_{AB} = E\begin{array}{c}\begin{array}{cccc} \textcircled{10} & \textcircled{2} & \textcircled{9} & \textcircled{1}\end{array}\\ \begin{bmatrix} 0.0017 & -0.0017 & 0 & 0 \\ -0.0017 & 0.0017 & 0 & 0 \\ 0 & 0 & 0 & 0 \\ 0 & 0 & 0 & 0 \end{bmatrix}\end{array}\begin{array}{c}\textcircled{10}\\ \textcircled{2}\\ \textcircled{9}\\ \textcircled{1}\end{array} \qquad \overline{K}_{BC} = E\begin{array}{c}\begin{array}{cccc} \textcircled{2} & \textcircled{4} & \textcircled{1} & \textcircled{3}\end{array}\\ \begin{bmatrix} 0 & 0 & 0 & 0 \\ 0 & 0 & 0 & 0 \\ 0 & 0 & 0.0013 & -0.0013 \\ 0 & 0 & -0.0013 & 0.0013 \end{bmatrix}\end{array}\begin{array}{c}\textcircled{2}\\ \textcircled{4}\\ \textcircled{1}\\ \textcircled{3}\end{array}$$

$$\overline{K}_{CD} = E\begin{array}{c}\begin{array}{cccc} \textcircled{4} & \textcircled{6} & \textcircled{3} & \textcircled{5}\end{array}\\ \begin{bmatrix} 0 & 0 & 0 & 0 \\ 0 & 0 & 0 & 0 \\ 0 & 0 & 0.0013 & -0.0013 \\ 0 & 0 & -0.0013 & 0.0013 \end{bmatrix}\end{array}\begin{array}{c}\textcircled{4}\\ \textcircled{6}\\ \textcircled{3}\\ \textcircled{5}\end{array} \qquad \overline{K}_{DE} = E\begin{array}{c}\begin{array}{cccc} \textcircled{12} & \textcircled{6} & \textcircled{11} & \textcircled{5}\end{array}\\ \begin{bmatrix} 0.0017 & -0.0017 & 0 & 0 \\ -0.0017 & 0.0017 & 0 & 0 \\ 0 & 0 & 0 & 0 \\ 0 & 0 & 0 & 0 \end{bmatrix}\end{array}\begin{array}{c}\textcircled{12}\\ \textcircled{6}\\ \textcircled{11}\\ \textcircled{5}\end{array}$$

$$\overline{K}_{AF} = E\begin{array}{c}\begin{array}{cccc} \textcircled{10} & \textcircled{8} & \textcircled{9} & \textcircled{7}\end{array}\\ \begin{bmatrix} 0 & 0 & 0 & 0 \\ 0 & 0 & 0 & 0 \\ 0 & 0 & 0.0013 & -0.0013 \\ 0 & 0 & -0.0013 & 0.0013 \end{bmatrix}\end{array}\begin{array}{c}\textcircled{10}\\ \textcircled{8}\\ \textcircled{9}\\ \textcircled{7}\end{array} \qquad \overline{K}_{FE} = E\begin{array}{c}\begin{array}{cccc} \textcircled{8} & \textcircled{12} & \textcircled{7} & \textcircled{11}\end{array}\\ \begin{bmatrix} 0 & 0 & 0 & 0 \\ 0 & 0 & 0 & 0 \\ 0 & 0 & 0.0013 & -0.0013 \\ 0 & 0 & -0.0013 & 0.0013 \end{bmatrix}\end{array}\begin{array}{c}\textcircled{8}\\ \textcircled{12}\\ \textcircled{7}\\ \textcircled{11}\end{array}$$

$$\overline{K}_{BF} = E\times 10^{-3}\begin{array}{c}\begin{array}{cccc} \textcircled{2} & \textcircled{8} & \textcircled{1} & \textcircled{7}\end{array}\\ \begin{bmatrix} 0.7072 & -0.7072 & -0.7072 & 0.7072 \\ -0.7072 & 0.7072 & 0.7072 & -0.7072 \\ -0.7072 & 0.7072 & 0.7072 & -0.7072 \\ 0.7072 & -0.7072 & -0.7072 & 0.7072 \end{bmatrix}\end{array}\begin{array}{c}\textcircled{2}\\ \textcircled{8}\\ \textcircled{1}\\ \textcircled{7}\end{array}$$

$$\overline{K}_{CF} = E\begin{array}{c}\begin{array}{cccc} \textcircled{8} & \textcircled{4} & \textcircled{7} & \textcircled{3}\end{array}\\ \begin{bmatrix} 0.0017 & -0.0017 & 0 & 0 \\ -0.0017 & 0.0017 & 0 & 0 \\ 0 & 0 & 0 & 0 \\ 0 & 0 & 0 & 0 \end{bmatrix}\end{array}\begin{array}{c}\textcircled{8}\\ \textcircled{4}\\ \textcircled{7}\\ \textcircled{3}\end{array}$$

$$\overline{K}_{DF} = E\times 10^{-3}\begin{array}{c}\begin{array}{cccc} \textcircled{6} & \textcircled{8} & \textcircled{5} & \textcircled{7}\end{array}\\ \begin{bmatrix} 0.7072 & -0.7072 & 0.7072 & -0.7072 \\ -0.7072 & 0.7072 & -0.7072 & 0.7072 \\ 0.7072 & -0.7072 & 0.7072 & -0.7072 \\ -0.7072 & 0.7072 & -0.7072 & 0.7072 \end{bmatrix}\end{array}\begin{array}{c}\textcircled{6}\\ \textcircled{8}\\ \textcircled{5}\\ \textcircled{7}\end{array}$$

The total stiffness matrix is obtained by assembling the global stiffness matrices of all members with respect to the global labels.

From the total stiffness matrix, the stiffness matrix for the unrestrained degrees of freedom can be partitioned.

$$
\begin{array}{c} \\ [K_{uu}] = E\times 10^{-3} \end{array}
\begin{array}{c}
\begin{array}{cccccccc} (1) & (2) & (3) & (4) & (5) & (6) & (7) & (8) \end{array} \\
\begin{bmatrix}
2 & -0.7 & -1.3 & 0 & 0 & 0 & -0.7 & 0.7 \\
-0.7 & 2.4 & 0 & 0 & 0 & 0 & 0.7 & -0.7 \\
-1.3 & 0 & 2.7 & 0 & -1.3 & 0 & 0 & 0 \\
0 & 0 & 0 & 1.7 & 0 & 0 & 0 & -1.7 \\
0 & 0 & -1.3 & 0 & 1.2 & 0.7 & -0.7 & -0.7 \\
0 & 0 & 0 & 0 & 0.7 & 2.4 & -0.7 & -0.7 \\
-0.7 & 0.7 & 0 & 0 & -0.7 & -0.7 & 4.1 & 0 \\
0.7 & -0.7 & 0 & -1.7 & -0.7 & -0.7 & 0 & 3.1
\end{bmatrix}
\end{array}
\begin{array}{c} \\ (1) \\ (2) \\ (3) \\ (4) \\ (5) \\ (6) \\ (7) \\ (8) \end{array}
$$

$$
[K_{uu}]^{-1} = \frac{1}{E}\times 10^{-3}
\begin{bmatrix}
1.757 & 0.3 & 1.382 & -0.375 & 1.007 & -0.3 & 0.375 & -0.375 \\
0.3 & 0.6 & 0.3 & 0.3 & 1.3 & 0 & 0 & 0.3 \\
1.382 & 0.3 & 1.7570 & 0 & 1.382 & -0.3 & 0.375 & 0 \\
-0.375 & 0.3 & 0 & 1.982 & 0.375 & 0.3 & 0 & 1.382 \\
1.007 & 0.3 & 1.382 & 0.375 & 1.757 & -0.3 & 0.375 & 0.375 \\
-0.3 & 0 & -0.3 & 0.3 & -0.3 & 0.6 & 0 & 0.3 \\
0.375 & 0 & 0.375 & 0 & 0.375 & 0 & 0.375 & 0 \\
-0.375 & 0.3 & 0 & 1.382 & 0.375 & 0.3 & 0 & 1.382
\end{bmatrix}
$$

4.17.4.5 Estimation of Joint Load Vectors

The joint load vector can be written as follows:

$$
[\bar{J}_L] = \begin{Bmatrix} 20 \\ 0 \\ 0 \\ -70 \\ 0 \\ 0 \\ 0 \\ 0 \\ 0 \\ 0 \\ 0 \\ 0 \end{Bmatrix}
\begin{array}{c} (1) \\ (2) \\ (3) \\ (4) \\ (5) \\ (6) \\ (7) \\ (8) \\ (9) \\ (10) \\ (11) \\ (12) \end{array}
$$

The joint load vector in unrestrained degrees of freedom can be partitioned from the preceding matrix as follows:

$$\left[\bar{J}_{Lu}\right]=\begin{Bmatrix}20\\0\\0\\-70\\0\\0\\0\\0\end{Bmatrix}$$

$$\{\bar{\delta}\}=\left[\bar{K}_{uu}\right]^{-1}\{\bar{J}_{Lu}\}$$

$$\{\bar{\delta}\}_U=\frac{1\times10^{-5}}{E}\begin{Bmatrix}0.6139\\-0.15\\0.2764\\-1.4624\\-0.0611\\-0.27\\0.075\\-1.0424\end{Bmatrix}$$

4.17.4.6 Calculation of Member Forces

$$\bar{M}_{AB}=\begin{Bmatrix}\bar{V}_{10}\\\bar{V}_2\\\bar{H}_9\\\bar{H}_1\end{Bmatrix}=\begin{Bmatrix}25\\-25\\0\\0\end{Bmatrix},\ \bar{M}_{BC}=\begin{Bmatrix}\bar{V}_2\\\bar{V}_4\\\bar{H}_1\\\bar{H}_3\end{Bmatrix}=\begin{Bmatrix}0\\0\\45\\-45\end{Bmatrix},\ \bar{M}_{CD}=\begin{Bmatrix}\bar{V}_4\\\bar{V}_6\\\bar{H}_3\\\bar{H}_5\end{Bmatrix}=\begin{Bmatrix}0\\0\\45\\-45\end{Bmatrix}$$

$$\bar{M}_{DE}=\begin{Bmatrix}\bar{V}_{12}\\\bar{V}_6\\\bar{H}_{11}\\\bar{H}_5\end{Bmatrix}=\begin{Bmatrix}45\\-45\\0\\0\end{Bmatrix},\ \bar{M}_{AF}=\begin{Bmatrix}\bar{V}_{10}\\\bar{V}_8\\\bar{H}_9\\\bar{H}_7\end{Bmatrix}=\begin{Bmatrix}0\\0\\-10\\10\end{Bmatrix},\ \bar{M}_{FE}=\begin{Bmatrix}\bar{V}_8\\\bar{V}_{12}\\\bar{H}_7\\\bar{H}_{11}\end{Bmatrix}=\begin{Bmatrix}0\\0\\10\\-10\end{Bmatrix}$$

$$\bar{M}_{BF}=\begin{Bmatrix}\bar{V}_2\\\bar{V}_8\\\bar{H}_1\\\bar{H}_7\end{Bmatrix}=\begin{Bmatrix}25\\-25\\-25\\25\end{Bmatrix},\ \bar{M}_{CF}=\begin{Bmatrix}\bar{V}_8\\\bar{V}_4\\\bar{H}_7\\\bar{H}_3\end{Bmatrix}=\begin{Bmatrix}70\\-70\\0\\0\end{Bmatrix},\ \bar{M}_{DF}=\begin{Bmatrix}\bar{V}_6\\\bar{V}_8\\\bar{H}_5\\\bar{H}_7\end{Bmatrix}=\begin{Bmatrix}45\\-45\\45\\-45\end{Bmatrix}$$

The member end forces and final end forces are given in Figure 1.27 and Figure 1.28, respectively.

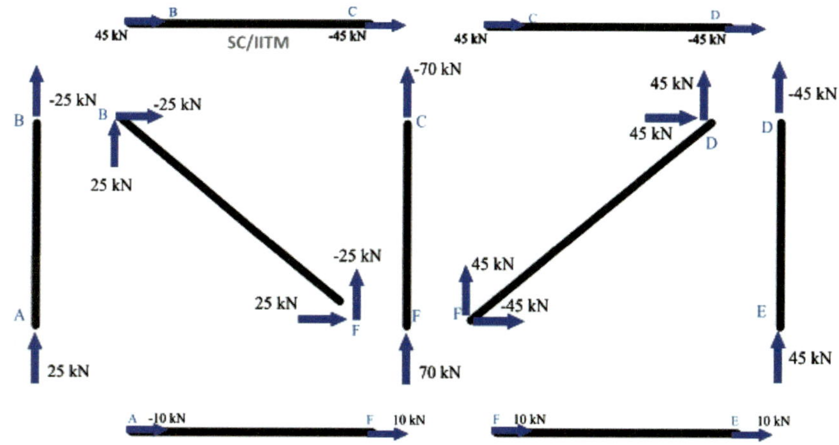

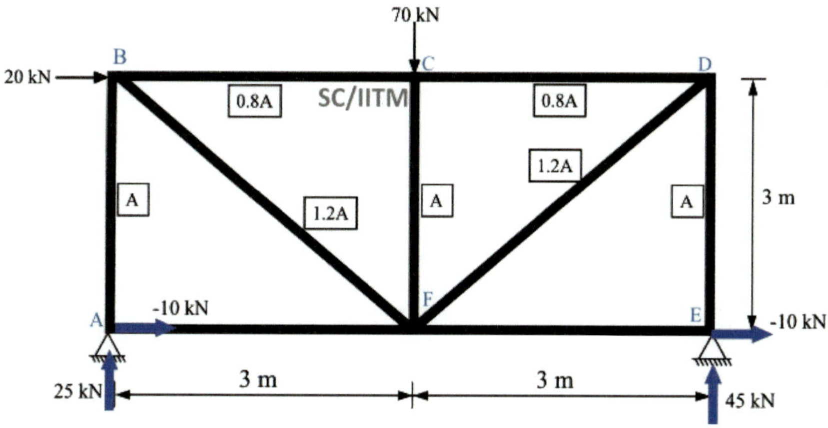

MATLAB PROGRAM

```
%% stiffness matrix method for planar truss
% Input
clc;
clear;
n = 9; % number of members
L = [3 3 3 3 3 3 4.242 3 4.242]; % length in m
A = [5E-3 4E-3 4E-3 5E-3 4E-3 4E-3 6E-3 5E-3 6E-3]; %
Area in m2
theta = [90 0 0 90 0 0 -45 90 -135]; % angle in degrees
uu = 8; % Number of unrestrained degrees-of-freedom
```

```
ur = 4; % Number of restrained degrees-of-freedom
uul = [1 2 3 4 5 6 7 8]; % global labels of unrestained dof
url = [9 10 11 12]; % global labels of restained dof
l1 = [10 2 9 1]; % Global labels for member 1
l2 = [2 4 1 3]; % Global labels for member 2
l3 = [4 6 3 5]; % Global labels for member 3
l4 = [12 6 11 5]; % Global labels for member 4
l5 = [10 8 9 7]; % Global labels for member 5
l6 = [8 12 7 11]; % Global labels for member 6
l7 = [2 8 1 7]; % Global labels for member 7
l8 = [8 4 7 3]; % Global labels for member 8
l9 = [6 8 5 7]; % Global labels for member 9
l = [l1; l2; l3; l4; l5; l6; l7; l8; l9];
dof = uu + ur; % Degrees-of-freedom
Ktotal = zeros (dof);
Tt1 = zeros (4); % Transformation matrix for member 1
Tt2 = zeros (4); % Transformation matrix for member 2
Tt3 = zeros (4); % Transformation matrix for member 3
Tt4 = zeros (4); % Transformation matrix for member 4
Tt5 = zeros (4); % Transformation matrix for member 5
Tt6 = zeros (4); % Transformation matrix for member 6
Tt7 = zeros (4); % Transformation matrix for member 7
Tt8 = zeros (4); % Transformation matrix for member 8
Tt9 = zeros (4); % Transformation matrix for member 9
fem1 = [0; 0; 0; 0]; % Local Fixed end forces of member 1
fem2 = [0; 0; 0; 0]; % Local Fixed end forces of member 2
fem3 = [0; 0; 0; 0]; % Local Fixed end forces of member 3
fem4 = [0; 0; 0; 0]; % Local Fixed end forces of member 4
fem5 = [0; 0; 0; 0]; % Local Fixed end forces of member 5
fem6 = [0; 0; 0; 0]; % Local Fixed end forces of member 6
fem7 = [0; 0; 0; 0]; % Local Fixed end forces of member 7
fem8 = [0; 0; 0; 0]; % Local Fixed end forces of member 8
fem9 = [0; 0; 0; 0]; % Local Fixed end forces of member 9
%% rotation coefficients for each member
rc = A./L;
cx = cosd(theta);
cy = sind(theta);
%% stiffness matrix 4 by 4
for i = 1:n
  Knew = zeros (dof);
  k1 = [0; 0; 0; 0];
  k2 = [0; 0; 0; 0];
  k3 = [0; 0; rc(i); -rc(i)];
  k4 = -k3;
  K = [k1 k2 k3 k4];
  fprintf ('Member Number =');
```

```
disp (i);
fprintf ('Local Stiffness matrix of member, [K] = \n');
disp (K);
T1 = [cx(i); 0; cy(i); 0];
T2 = [0; cx(i); 0; cy(i)];
T3 = [-cy(i); 0; cx(i); 0];
T4 = [0; -cy(i); 0; cx(i)];
T = [T1 T2 T3 T4];
fprintf ('Transformation matrix of member, [T] = \n');
disp (T);
Ttr = T';
fprintf ('Transformation matrix Transpose, [T] = \n');
disp (Ttr);
Kg = Ttr*K*T;
fprintf ('Global Matrix, [K global] = \n');
disp (Kg);
for p = 1:4
  for q = 1:4
    Knew((l(i,p)),(l(i,q))) = Kg(p,q);
  end
end
Ktotal = Ktotal + Knew;
if i == 1
  Tt1 = T;
  Kg1 = Kg;
  fembar1 = Tt1'*fem1;
elseif i == 2
  Tt2 = T;
  Kg2 = Kg;
  fembar2 = Tt2'*fem2;
elseif i == 3
  Tt3 = T;
  Kg3 = Kg;
  fembar3 = Tt3'*fem3;
elseif i == 4
  Tt4 = T;
  Kg4 = Kg;
  fembar4 = Tt4'*fem4;
elseif i == 5
  Tt5 = T;
  Kg5 = Kg;
  fembar5 = Tt5'*fem5;
elseif i == 6
  Tt6 = T;
  Kg6 = Kg;
  fembar6 = Tt6'*fem6;
```

```
  elseif i == 7
    Tt7 = T;
    Kg7 = Kg;
    fembar7 = Tt7'*fem7;
  elseif i == 8
    Tt8 = T;
    Kg8 = Kg;
    fembar8 = Tt8'*fem8;
  else
    Tt9 = T;
    Kg9 = Kg;
    fembar9 = Tt9'*fem9;
  end
end
fprintf ('Stiffness Matrix of complete structure,
[Ktotal] = \n');
disp (Ktotal);
Kunr = zeros(uu);
for x = 1:uu
  for y = 1:uu
    Kunr(x,y) = Ktotal(x,y);
  end
end
fprintf ('Unrestrained Stiffness sub-matrix, [Kuu] =
\n');
disp (Kunr);
KuuInv = inv(Kunr);
fprintf ('Inverse of Unrestrained Stiffness sub-matrix,
[KuuInverse] = \n');
disp (KuuInv);
%% Creation of joint load vector
jl = [20; 0; 0; -70; 0; 0; 0; 0; 0; 0; 0; 0]; % values
given in kN
jlu = [20; 0; 0; -70; 0; 0; 0; 0]; % load vector in
unrestrained dof
delu = KuuInv*jlu;
fprintf ('Joint Load vector, [Jl] = \n');
disp (jl');
fprintf ('Unrestrained displacements, [DelU] = \n');
disp (delu');
delr = zeros(ur,1);
del = zeros (dof,1);
del = [delu; delr];
deli = zeros (4,1);
for i = 1:n
  for p = 1:4
```

```
    deli(p,1) = del((l(i,p)),1) ;
  end
  if i == 1
    delbar1 = deli;
    mbar1 = (Kg1 * delbar1)+fembar1;
    fprintf ('Member Number =');
    disp (i);
    fprintf ('Global displacement matrix [DeltaBar] =
    \n');
    disp (delbar1');
    fprintf ('Global End moment matrix [MBar] = \n');
    disp (mbar1');
  elseif i == 2
    delbar2 = deli;
    mbar2 = (Kg2 * delbar2)+fembar2;
    fprintf ('Member Number =');
    disp (i);
    fprintf ('Global displacement matrix [DeltaBar] =
    \n');
    disp (delbar2');
    fprintf ('Global End moment matrix [MBar] = \n');
    disp (mbar2');
  elseif i == 3
    delbar3 = deli;
    mbar3 = (Kg3 * delbar3)+fembar3;
    fprintf ('Member Number =');
    disp (i);
    fprintf ('Global displacement matrix [DeltaBar] =
    \n');
    disp (delbar3');
    fprintf ('Global End moment matrix [MBar] = \n');
    disp (mbar3');
  elseif i == 4
    delbar4 = deli;
    mbar4 = (Kg4 * delbar4)+fembar4;
    fprintf ('Member Number =');
    disp (i);
    fprintf ('Global displacement matrix [DeltaBar] =
    \n');
    disp (delbar4');
    fprintf ('Global End moment matrix [MBar] = \n');
    disp (mbar4');
  elseif i == 5
    delbar5 = deli;
    mbar5 = (Kg5 * delbar5)+fembar5;
    fprintf ('Member Number =');
```

```
    disp (i);
    fprintf ('Global displacement matrix [DeltaBar] =
    \n');
    disp (delbar5');
    fprintf ('Global End moment matrix [MBar] = \n');
    disp (mbar5');
  elseif i == 6
    delbar6 = deli;
    mbar6 = (Kg6 * delbar6)+fembar6;
    fprintf ('Member Number =');
    disp (i);
    fprintf ('Global displacement matrix [DeltaBar] = \n');
    disp (delbar6');
    fprintf ('Global End moment matrix [MBar] = \n');
    disp (mbar6');
  elseif i == 7
    delbar7 = deli;
    mbar7 = (Kg7 * delbar7)+fembar7;
    fprintf ('Member Number =');
    disp (i);
    fprintf ('Global displacement matrix [DeltaBar] =
    \n');
    disp (delbar7');
    fprintf ('Global End moment matrix [MBar] = \n');
    disp (mbar7');
  elseif i == 8
    delbar8 = deli;
    mbar8 = (Kg8 * delbar8)+fembar8;
    fprintf ('Member Number =');
    disp (i);
    fprintf ('Global displacement matrix [DeltaBar] = \n');
    disp (delbar8');
    fprintf ('Global End moment matrix [MBar] = \n');
    disp (mbar8');
  else
    delbar9 = deli;
    mbar9 = (Kg9 * delbar9)+fembar9;
    fprintf ('Member Number =');
    disp (i);
    fprintf ('Global displacement matrix [DeltaBar] =
    \n');
    disp (delbar9');
    fprintf ('Global End moment matrix [MBar] = \n');
    disp (mbar9');
  end
end
```

```
%% check
mbar = [mbar1'; mbar2'; mbar3'; mbar4'; mbar5'; mbar6';
mbar7'; mbar8'; mbar9'];
jf = zeros(dof,1);
for a = 1:n
  for b = 1:4 % size of k matrix
    d = l(a,b);
    jfnew = zeros(dof,1);
    jfnew(d,1) = mbar(a,b);
    jf = jf+jfnew;
  end
end
fprintf ('Joint forces = \n');
disp (jf');
```

MATLAB OUTPUT

```
Member Number = 1
Local Stiffness matrix of member, [K] =
        0         0         0         0
        0         0         0         0
        0         0    0.0017   -0.0017
        0         0   -0.0017    0.0017
Transformation matrix of member, [T] =
        0         0        -1         0
        0         0         0        -1
        1         0         0         0
        0         1         0         0
Transformation matrix Transpose, [T] =
        0         0         1         0
        0         0         0         1
       -1         0         0         0
        0        -1         0         0
Global Matrix, [K global] =
   0.0017   -0.0017         0         0
  -0.0017    0.0017         0         0
        0         0         0         0
        0         0         0         0
Member Number = 2
Local Stiffness matrix of member, [K] =
        0         0         0         0
        0         0         0         0
        0         0    0.0013   -0.0013
        0         0   -0.0013    0.0013
Transformation matrix of member, [T] =
        1         0         0         0
        0         1         0         0
```

```
        0         0         1         0
        0         0         0         1
Transformation matrix Transpose, [T] =
        1         0         0         0
        0         1         0         0
        0         0         1         0
        0         0         0         1
Global Matrix, [K global] =
        0         0         0         0
        0         0         0         0
        0         0    0.0013   -0.0013
        0         0   -0.0013    0.0013
Member Number = 3
Local Stiffness matrix of member, [K] =
        0         0         0         0
        0         0         0         0
        0         0    0.0013   -0.0013
        0         0   -0.0013    0.0013
Transformation matrix of member, [T] =
        1         0         0         0
        0         1         0         0
        0         0         1         0
        0         0         0         1
Transformation matrix Transpose, [T] =
        1         0         0         0
        0         1         0         0
        0         0         1         0
        0         0         0         1
Global Matrix, [K global] =
        0         0         0         0
        0         0         0         0
        0         0    0.0013   -0.0013
        0         0   -0.0013    0.0013
Member Number = 4
Local Stiffness matrix of member, [K] =
        0         0         0         0
        0         0         0         0
        0         0    0.0017   -0.0017
        0         0   -0.0017    0.0017
Transformation matrix of member, [T] =
        0         0        -1         0
        0         0         0        -1
        1         0         0         0
        0         1         0         0
Transformation matrix Transpose, [T] =
        0         0         1         0
        0         0         0         1
```

```
      -1        0        0        0
       0       -1        0        0
Global Matrix, [K global] =
   0.0017  -0.0017        0        0
  -0.0017   0.0017        0        0
       0        0        0        0
       0        0        0        0
Member Number = 5
Local Stiffness matrix of member, [K] =
       0        0        0        0
       0        0        0        0
       0        0   0.0013  -0.0013
       0        0  -0.0013   0.0013
Transformation matrix of member, [T] =
       1        0        0        0
       0        1        0        0
       0        0        1        0
       0        0        0        1
Transformation matrix Transpose, [T] =
       1        0        0        0
       0        1        0        0
       0        0        1        0
       0        0        0        1
Global Matrix, [K global] =
       0        0        0        0
       0        0        0        0
       0        0   0.0013  -0.0013
       0        0  -0.0013   0.0013
Member Number = 6
Local Stiffness matrix of member, [K] =
       0        0        0        0
       0        0        0        0
       0        0   0.0013  -0.0013
       0        0  -0.0013   0.0013
Transformation matrix of member, [T] =
       1        0        0        0
       0        1        0        0
       0        0        1        0
       0        0        0        1
Transformation matrix Transpose, [T] =
       1        0        0        0
       0        1        0        0
       0        0        1        0
       0        0        0        1
Global Matrix, [K global] =
       0        0        0        0
       0        0        0        0
```

```
        0         0    0.0013   -0.0013
        0         0   -0.0013    0.0013
Member Number = 7
Local Stiffness matrix of member, [K] =
        0         0         0         0
        0         0         0         0
        0         0    0.0014   -0.0014
        0         0   -0.0014    0.0014
Transformation matrix of member, [T] =
   0.7071         0    0.7071         0
        0    0.7071         0    0.7071
  -0.7071         0    0.7071         0
        0   -0.7071         0    0.7071
Transformation matrix Transpose, [T] =
   0.7071         0   -0.7071         0
        0    0.7071         0   -0.7071
   0.7071         0    0.7071         0
        0    0.7071         0    0.7071
Global Matrix, [K global] =
  1.0e-03 *
   0.7072   -0.7072   -0.7072    0.7072
  -0.7072    0.7072    0.7072   -0.7072
  -0.7072    0.7072    0.7072   -0.7072
   0.7072   -0.7072   -0.7072    0.7072
Member Number = 8
Local Stiffness matrix of member, [K] =
        0         0         0         0
        0         0         0         0
        0         0    0.0017   -0.0017
        0         0   -0.0017    0.0017
Transformation matrix of member, [T] =
        0         0        -1         0
        0         0         0        -1
        1         0         0         0
        0         1         0         0
Transformation matrix Transpose, [T] =
        0         0         1         0
        0         0         0         1
       -1         0         0         0
        0        -1         0         0
Global Matrix, [K global] =
   0.0017   -0.0017         0         0
  -0.0017    0.0017         0         0
        0         0         0         0
        0         0         0         0
```

```
Member Number = 9
Local Stiffness matrix of member, [K] =
        0         0         0         0
        0         0         0         0
        0         0    0.0014   -0.0014
        0         0   -0.0014    0.0014
Transformation matrix of member, [T] =
  -0.7071         0    0.7071         0
        0   -0.7071         0    0.7071
  -0.7071         0   -0.7071         0
        0   -0.7071         0   -0.7071
Transformation matrix Transpose, [T] =
  -0.7071         0   -0.7071         0
        0   -0.7071         0   -0.7071
   0.7071         0   -0.7071         0
        0    0.7071         0   -0.7071
Global Matrix, [K global] =
  1.0e-03 *
   0.7072   -0.7072    0.7072   -0.7072
  -0.7072    0.7072   -0.7072    0.7072
   0.7072   -0.7072    0.7072   -0.7072
  -0.7072    0.7072   -0.7072    0.7072
Stiffness Matrix of complete structure, [Ktotal] =
Columns 1 through 9
   0.0020   -0.0007   -0.0013         0         0         0
  -0.0007    0.0007         0
  -0.0007    0.0024         0         0         0         0
   0.0007   -0.0007         0
  -0.0013         0    0.0027         0   -0.0013         0
        0         0         0
        0         0         0    0.0017         0         0
        0   -0.0017         0
        0         0   -0.0013         0    0.0020    0.0007
  -0.0007   -0.0007         0
        0         0         0         0    0.0007    0.0024
  -0.0007   -0.0007         0
  -0.0007    0.0007         0         0   -0.0007   -0.0007
   0.0041         0   -0.0013
   0.0007   -0.0007         0   -0.0017   -0.0007   -0.0007
        0    0.0031         0
        0         0         0         0         0         0
  -0.0013         0    0.0013
        0   -0.0017         0         0         0         0
        0         0         0
        0         0         0         0         0         0
  -0.0013         0         0
```

```
        0         0         0         0         0   -0.0017
        0         0         0
Columns 10 through 12
        0         0         0
  -0.0017         0         0
        0         0         0
        0         0         0
        0         0         0
        0         0   -0.0017
        0   -0.0013         0
        0         0         0
        0         0         0
   0.0017         0         0
        0    0.0013         0
        0         0    0.0017
Unrestrained Stiffness sub-matrix, [Kuu] =
   0.0020   -0.0007   -0.0013         0         0         0
  -0.0007    0.0007
  -0.0007    0.0024         0         0         0         0
   0.0007   -0.0007
  -0.0013         0    0.0027         0   -0.0013         0
        0         0
        0         0         0    0.0017         0         0
        0   -0.0017
        0         0   -0.0013         0    0.0020    0.0007
  -0.0007   -0.0007
        0         0         0         0    0.0007    0.0024
  -0.0007   -0.0007
  -0.0007    0.0007         0         0   -0.0007   -0.0007
   0.0041         0
   0.0007   -0.0007         0   -0.0017   -0.0007   -0.0007
        0    0.0031
Inverse of Unrestrained Stiffness sub-matrix,
[KuuInverse] =
  1.0e+03 *
   1.7570    0.3000    1.3820   -0.3750    1.0070   -0.3000
   0.3750   -0.3750
   0.3000    0.6000    0.3000    0.3000    0.3000         0
   0.0000    0.3000
   1.3820    0.3000    1.7570   -0.0000    1.3820   -0.3000
   0.3750   -0.0000
  -0.3750    0.3000   -0.0000    1.9820    0.3750    0.3000
  -0.0000    1.3820
   1.0070    0.3000    1.3820    0.3750    1.7570   -0.3000
   0.3750    0.3750
  -0.3000         0   -0.3000    0.3000   -0.3000    0.6000
  -0.0000    0.3000
```

```
   0.3750   0.0000   0.3750  -0.0000   0.3750  -0.0000
   0.3750  -0.0000
  -0.3750   0.3000  -0.0000   1.3820   0.3750   0.3000
  -0.0000   1.3820
Joint Load vector, [Jl] =
      20        0        0      -70        0        0
       0        0        0        0        0        0
Unrestrained displacements, [DelU] =
  1.0e+05 *
   0.6139  -0.1500   0.2764  -1.4624  -0.0611  -0.2700
   0.0750  -1.0424
Member Number = 1
Global displacement matrix [DeltaBar] =
   1.0e+04 *
        0  -1.5000        0   6.1390
Global End moment matrix [MBar] =
   25.0000  -25.0000        0         0
Member Number = 2
Global displacement matrix [DeltaBar] =
   1.0e+05 *
   -0.1500   -1.4624   0.6139   0.2764
Global End moment matrix [MBar] =
         0         0   45.0000  -45.0000
Member Number = 3
Global displacement matrix [DeltaBar] =
   1.0e+05 *
   -1.4624   -0.2700   0.2764   -0.0611
Global End moment matrix [MBar] =
         0         0   45.0000  -45.0000
Member Number = 4
Global displacement matrix [DeltaBar] =
   1.0e+04 *
         0   -2.7000        0   -0.6110
Global End moment matrix [MBar] =
   45.0000  -45.0000        0         0
Member Number = 5
Global displacement matrix [DeltaBar] =
   1.0e+05 *
         0   -1.0424        0    0.0750
Global End moment matrix [MBar] =
         0         0  -10.0000   10.0000
Member Number = 6
Global displacement matrix [DeltaBar] =
   1.0e+05 *
   -1.0424         0    0.0750         0
Global End moment matrix [MBar] =
         0         0   10.0000  -10.0000
```

```
Member Number = 7
Global displacement matrix [DeltaBar] =
   1.0e+05 *
   -0.1500   -1.0424    0.6139    0.0750
Global End moment matrix [MBar] =
   25.0000  -25.0000  -25.0000   25.0000
Member Number = 8
Global displacement matrix [DeltaBar] =
   1.0e+05 *
   -1.0424   -1.4624    0.0750    0.2764
Global End moment matrix [MBar] =
        70       -70         0         0
Member Number = 9
Global displacement matrix [DeltaBar] =
   1.0e+05 *
   -0.2700   -1.0424   -0.0611    0.0750
Global End moment matrix [MBar] =
   45.0000  -45.0000   45.0000  -45.0000
Joint forces =
   20.0000    0.0000    0.0000  -70.0000         0
    0.0000   -0.0000   -0.0000  -10.0000   25.0000
  -10.0000   45.0000
```

4.18 SHEAR FORCE AND BENDING MOMENT DIAGRAMS

EXAMPLE 1

Analyze the continuous beam shown in the following, and assess the order of static determinacy:

Let us neglect axial reactions.

The number of unknowns is $(V_A, M_A, V_B, V_C, V_D = 5)$.
The number of equations are 2 $(\sum F_Y = 0,\ \sum M = 0)$.

As we neglected the axial deformation in the analysis, the sum of axial forces cannot be included in the static equations of equilibrium. Hence, $\sum F_X = 0$ cannot be used.

The structure is statically indeterminate by order (5 – 2) = 3, which is the degree of static indeterminacy.

EXAMPLE 2

Analyze the simply supported beam shown in the following, and assess the order of static determinacy:

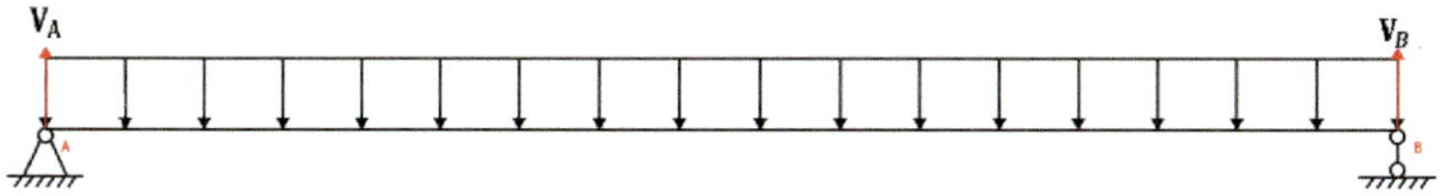

Number of reactions $= 2\ (V_A, V_B)$

Number of equations $= 2\ (\sum F_Y =,\ 0 \sum M = 0)$

As we neglected the axial deformation in the analysis, the sum of axial forces cannot be included in the static equations of equilibrium. Hence, $\sum F_X = 0$ cannot be used.

Hence, the beam is statically determinate.

EXAMPLE 3

Analyze the cantilever that follows, and assess the order of static determinacy:

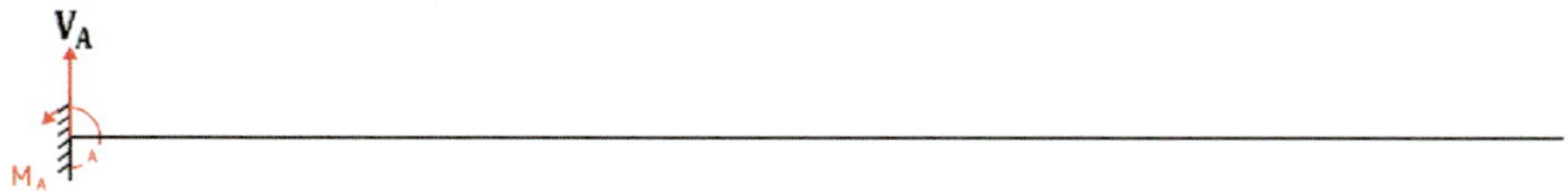

Number of reactions $= 2\ (V_A, M_A)$

Number of equations $= 2\ (\sum F_Y =,\ 0 \sum M = 0)$

As we neglected the axial deformation in the analysis, the sum of axial forces cannot be included in the static equations of equilibrium. Hence, $\sum F_X = 0$ cannot be used.

Hence, the cantilever beam is statically determinate.

EXAMPLE 4

Analyze the beam with an overhanging span, shown in what follows, and assess its order of static indeterminacy:

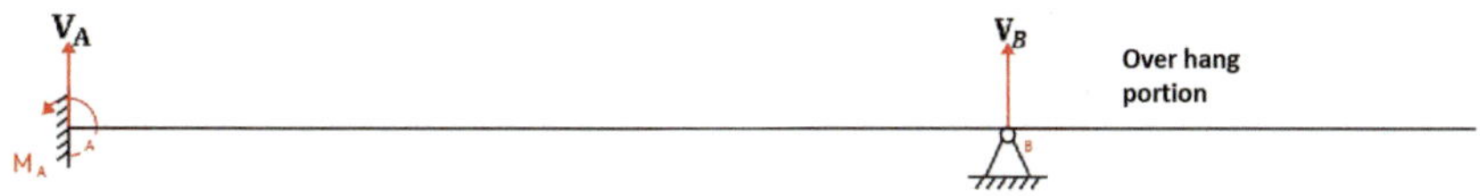

The number of unknowns are $(V_A, M_A, V_B = 3)$.

The number of equations are $2(\sum F_Y =,\ 0 \sum M = 0)$.

As we neglected the axial deformation in the analysis, the sum of axial forces cannot be included in the static equations of equilibrium. Hence, $\sum F_X = 0$ cannot be used.

The structure is statically indeterminate by the order (3 − 2) = 1.

4.18.1 Shear Force Diagram (SFD)

A *shear force diagram* is a schematic representation of net force acting at any section in the beam. It is the algebraic sum of all the forces at any section.

Sign convention: At any section, either sum up all forces to the left of the section; any force acting upward is considered positive.

4.18.2 Bending Moment Diagram

To calculate the moment at any section, we need to first cut a section. Sections are cut on the beam at any place of our choice. The moment will be the algebraic sum of moments of all forces at that chosen section, considering sagging as positive and hogging moment as negative.

EXAMPLE 1

Analyze the beam shown in the following, and draw the SFD and the BMD.

This is a simply supported beam *AB* with a span of 5 m. The beam is statically determinate. Unknown reactions are therefore (V_A, V_E).

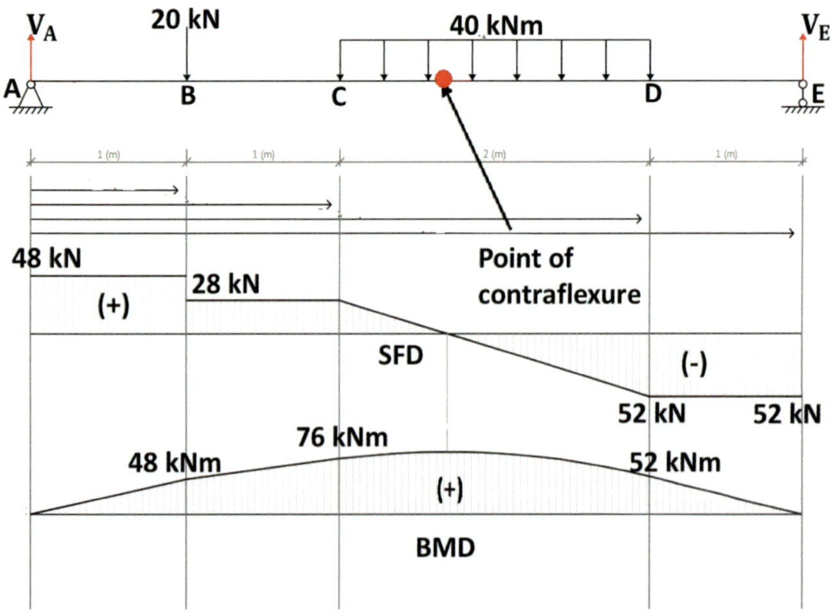

SOLUTION

$$\sum F_Y = 0,$$

$$V_A + V_E - 20 - (40 \times 2) = 0.$$

$$V_A + V_E = 100$$

$\sum M_A = 0$. Taking clockwise moments as positive, we get:

$$20 \times (1) + 40 \times 2 \times (3) - V_E \times 5 = 0$$

$$V_E = 52 \text{ kN}$$

From (1):

$$V_A = 48 \text{ kN}$$

4.18.2.1 SFD

To calculate the shear force at any section, we need to first cut a section. Sections are cut on the beam just to the left and just to the right of point loads.

As three is a reaction force acting at A (which is treated as a point load), let us cut a section just left of A, and look to the left to sum up all the vertical forces.

Shear force just left of A = 0.

Similarly, cut a section, just to the right of A, and look to its left and sum up all vertical forces.

Shear force at the right of A = +48 kN (upward).

Shear force just left of B = +48 kN (upward).

Shear force at C $= +48 - 20 = +28$ kN.

Shear force at D $= +48 - 20 - (40 \times 2) = -52$ kN (downward).

Shear force just left of E $= -52$ kN.

Shear force just right of E $= -52 + 52 = 0$.

Bending moment at A $= 0$.

Bending moment at B $= +(48 \times 1) = +48$ kNm. (It is positive due to the fact that upward force at support A will induce sagging moment about the section at *B*.)

Bending moment at C $= +(48 \times 2) - (20 \times 1) = +76$ kNm.

Bending moment at D $= +(48 \times 4) - (20 \times 3) - (40 \times 2 \times 1) = +52$ kNm.

Bending moment at E $= +(48 \times 5) - (20 \times 4) - (40 \times 2 \times (1 + 1)) = 0$.

BMD in CD will be curved due to UDL.

The maximum bending moment is 76 kNm, which is now considered as the design moment

EXAMPLE 2

A beam 250 × 500 mm is simply supported over a span of 6 m. Permissible stress in the material is 8 N/mm^2. Take E as 2.1×10^5 N/mm^2.

1. Find the moment of resistance of the beam.
2. Find the safe UDL it can carry without exceeding permissible stress.
3. Find the safe UDL it can carry without exceeding permissible deflection of Span/325.

SOLUTION

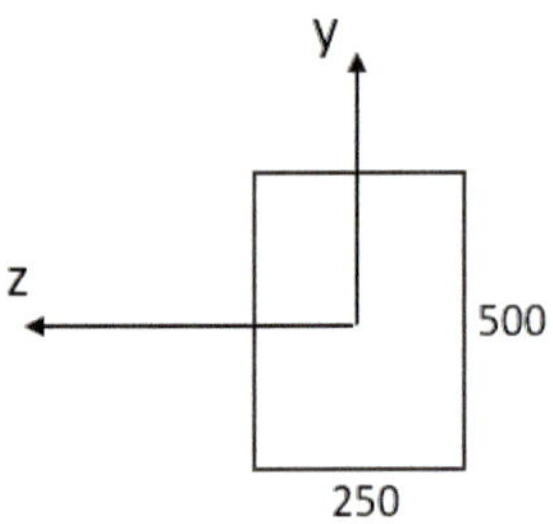

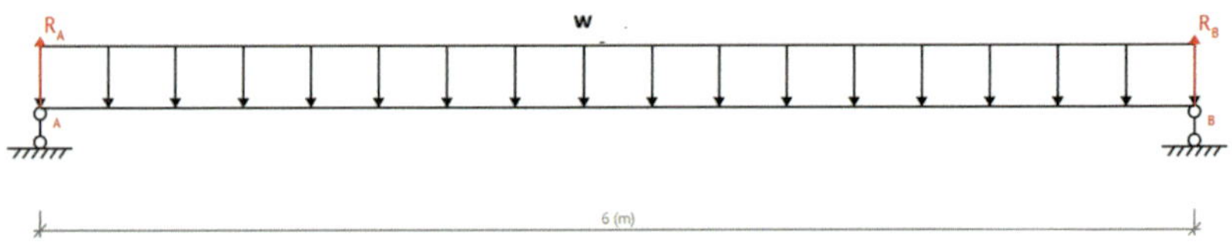

$$\frac{M}{I} = \frac{\sigma}{y}$$

$$I_{zz} = \frac{bd^3}{12} = \frac{250\times500^3}{12} = 2.604\times10^9 \text{ mm}^4$$

$$y_{max} = d/2 = 500/2 = 250 \text{ mm.}$$

$M = \dfrac{8\times2.604\times10^9}{250} = 83.328\times106 = 83.328$ kNm. This is the moment of resistance (MoR) of the beam, which is also noted as its maximum capacity. For simply supported beams, the maximum bending moment is given as:

$$M_{max} = \frac{wl^2}{8}$$

Equating this to the MoR, we get:

$$83.328 = \frac{w\times6^2}{8}$$

$$w = 18.52 \text{ kN/m}$$

A UDL of 18.52 kN/m (including self-weight) is the safe load the beam can carry without exceeding the permissible stress.

Type of Beam	Maximum Bending Moment	Maximum Deflection and Slope
Simply supported with UDL 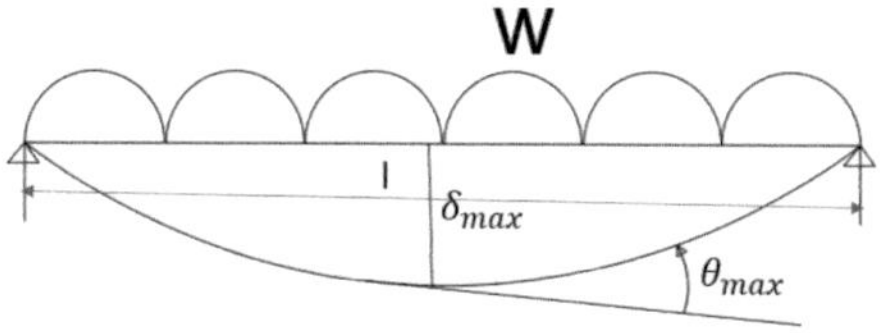	$M_{max} = \frac{wl^2}{8}$	$\delta_{max} = \frac{5wl^4}{384EI}$ $\theta_{max} = \frac{wl^3}{24EI}$
Simply supported with a central point load 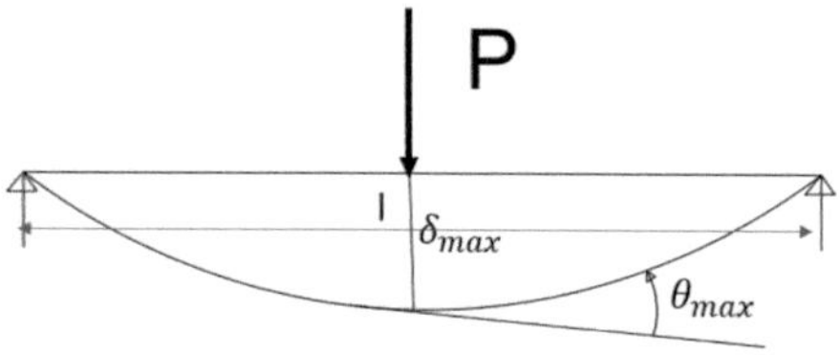	$M_{max} = \frac{Pl}{4}$	$\delta_{max} = \frac{Pl^3}{48EI}$ $\theta_{max} = \frac{Pl^2}{16EI}$
Cantilever with a point load at its free end 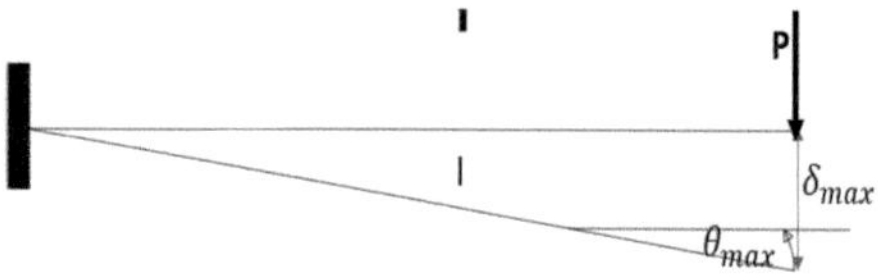	$M_{max} = Pl$	$\delta_{max} = \frac{Pl^3}{3EI}$ $\theta_{max} = \frac{Pl^2}{2EI}$
Cantilever under UDL 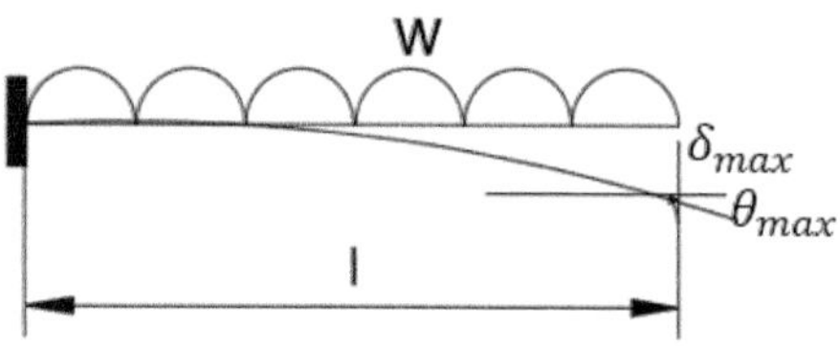	$M_{max} = \frac{wl^2}{2}$	$\delta_{max} = \frac{wl^4}{8EI}$ $\theta_{max} = \frac{wl^3}{6EI}$
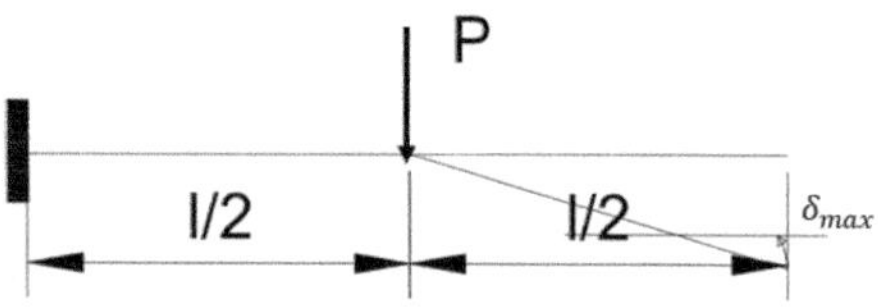	$M_{max} = \frac{pl}{2}$	$\delta_{max} = \frac{5pl^3}{48EI}$ $\theta_{max} = \frac{pl^2}{8EI}$

The maximum deflection of the beam will occur at its mid-span and is given by:

$$\delta_{max} = \frac{5wl^4}{384EI}$$

Let w kN/m be the load causing this maximum deflection. Now, equating this to the maximum permissible deflection of Span/325, we get:

Hence:

$$y = 110/2 = 55 \text{ mm}$$

Solving, we get the permissible UDL on the beam, based on the deflection criteria, as 34.19 kN/m.

EXAMPLE 3

A beam 300 × 600 mm is simply supported over a span of 6 m. Permissible stress in the material is 6 N/mm². Take E as 2.1×10^5 N/mm².

1. Find the moment of resistance of the beam.
2. Find the safe UDL it can carry without exceeding permissible stress.
3. Find the safe UDL it can carry without exceeding permissible deflection of Span/325.

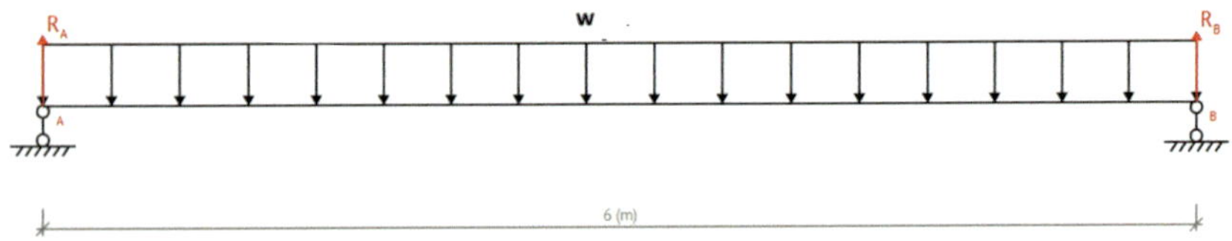

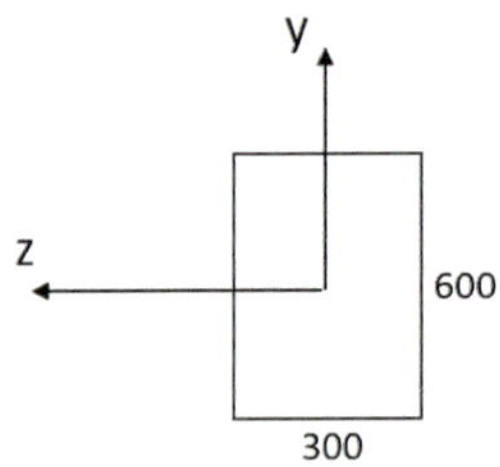

$$\frac{M}{I} = \frac{\sigma}{y}$$

$$I_{zz} = \frac{bd^3}{12} = \frac{300\times600^3}{12} = 5.4\times109 \text{ mm}^4$$

$$y_{max} = d/2 = 600/2 = 300 \text{ mm.}$$

$$M = \frac{6\times5.4\times10^9}{300} = 108 \text{ kNm} = \text{MoR}$$

Equating MoR to the maximum bending moment of a simply supported beam, we get:

$$108 = \frac{w\times6^2}{8}$$

$$w = 24 \text{ kNm}$$

A UDL of 24 kNm (including self-weight) is the safe load the beam can carry without exceeding the permissible stress.

$$\delta_{max} = \frac{5wl^4}{384EI}$$

Let w be kN/m.

Hence:

$$\delta_{max} = \frac{5wl^4}{384EI} = \frac{5(w \times 6) \times 10^3 \times 6000^3}{1.2 \times 10^4 \times 2.604 \times 10^9}$$

$$\text{Span}/325 = 6000/325 = \frac{5(w \times 6) \times 10^3 \times 6000^3}{1.2 \times 10^4 \times 5.4 \times 10^9}$$

$$\text{w} = 70.89 \text{ kN/m (based on deflection)}$$

EXAMPLE 4

Determine the maximum bending moment the beam can carry. Also draw the SFD and BMD.

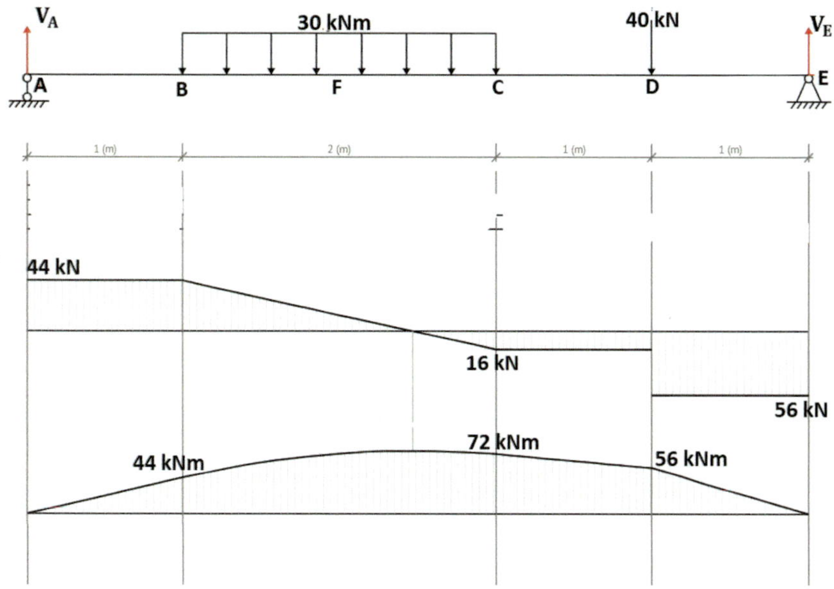

Unknown reactions are V_A and V_E.

$$\sum F_Y = 0,$$

$$V_A + V_E - 40 - (30 \times 2) = 0.$$

$$V_A + V_E = 100 \tag{1}$$

$$\sum M_A = 0,$$
$$40 \times (4) + 30 \times 2 \times (2) - V_E \times 5 = 0$$
$$V_E = 56 \text{ kN}$$

From (1):

$$V_A = 41 \text{ kN}$$

4.18.2.2 SFD

Shear force just right of $A = 44$ kN

Shear force at $B = 44$ kN

Shear force at $C = 44 - 30 \times 2 = -16$ kN

Shear force left of $D = -16$ kN

Shear force right of $D = -16 - 40 = -56$ kN

Shear force just left of $E = -56$ kN

Shear force just right of $E = 0$

4.18.2.3 BMD

Sagging = positive.

Hogging = negative.

Bending moment at $A = 0$.

Bending moment at $B = 44 \times 1 = 44$ kNm.

Bending moment at $C = 44 \times 3 - 30 \times 2 \times 1 = 72$ kNm.

Bending moment at $F = 44 \times 2 - 30 \times 1 \times 0.5 = 73$ kNm.

Bending moment at $D = 44 \times 4 - 30 \times 2 \times 2 = 56$ kNm.

Bending moment at $E = 44 \times 5 - 30 \times 2 \times 3 - 40 \times 1 = 0$.

Now it is necessary to find the maximum bending moment and the point where it acts on the beam.

It is known that M_{max} will be in BC.

Let us cut a section xx, x from B $(0 < x < 2)$.

$$\begin{aligned} Mx &= +44\,(x + 1) - (30x^2)/2 \\ &= 44x + 44 - 1.5x^2 \end{aligned}$$

For maximum $Mx = \dfrac{dM_x}{dx} = 0$:

$$\begin{aligned} 0 &= 44 - 30x \\ x &= 1.47 \\ M_{max} &= 44\,(x + 1) - (30x^2)/2 \\ &= 44\,(1.47 + 1) - (30 \times 1.47^2)/2 \\ &= 76.27 \text{ kNm} \end{aligned}$$

EXAMPLE 5

Find the concentrated load at the mid-span of a beam of 5 m span. The section is a hollow rectangle of size 85 × 110 mm, 10 mm thick. Take the safe bending stress = 100 MPa.

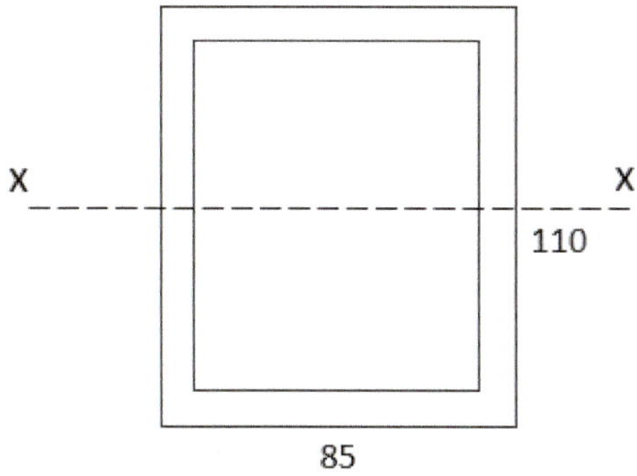

SOLUTION

Moment of inertia about the *xx*-axis.

$$I_{xx} = \left(\frac{bd^3}{12}\right)_{Outer} - \left(\frac{bd^3}{12}\right)_{Inner} = \frac{85 \times 110^3}{12} - \frac{65 \times 90^3}{12} = 5.48 \times 10^6 \text{ mm}^4$$

$$y = 110/2 = 55 \text{ mm}$$

$$\text{Section modulus } Z = \frac{I_{xx}}{y} = \frac{5.48 \times 10^6}{(55)} = 99.64 \times 10^3.$$

$$\frac{M}{I} = \frac{\sigma}{y}$$

$$M = \left(\frac{I}{y}\right)\sigma$$

$$M = \sigma Z$$

Moment of resistance = $100 \times 99.64 \times 10^3 = 9.96$ kNm.

The maximum bending moment of a simply supported beam under a centrally concentrated load is given as:

$$M = \frac{Pl}{4}$$

Now, equating the maximum *BM* to this value, one can compute thus:

$$P = \frac{4 \times 9.96}{5}$$

$$P = 8.96 \text{ kN}$$

EXAMPLE 6

Calculate the maximum stress induced in a cast-iron pipe with an external diameter of 40 mm, internal diameter of 20 mm, and length of 4 m. The pipe is simply supported at both ends and carries a point load of 8 kN at the center.

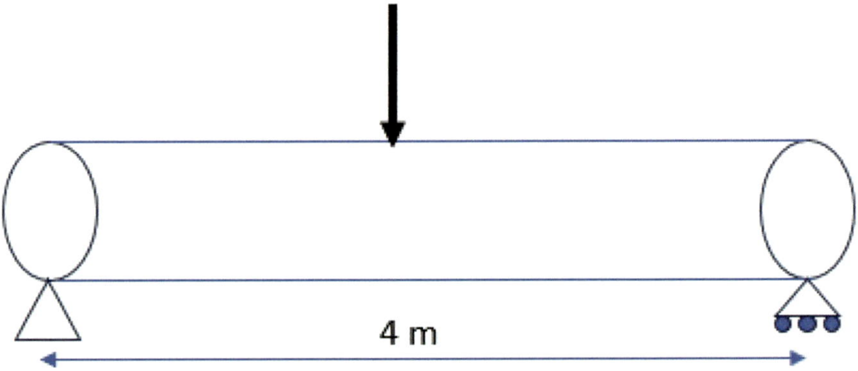

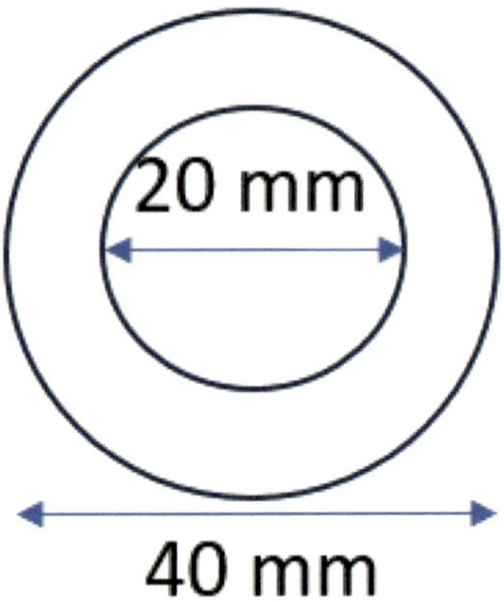

SOLUTION

$$\frac{M}{I} = \frac{\sigma}{y}$$

$$M = \frac{pl}{4} = \frac{(8 \times 4)}{4} = 8\ kNm$$

$$I_{center} = \frac{\pi}{64}\left(2.56 \times 10^6 - 1.60 \times 10^5\right) = 117.81 \times 10^3\ \text{mm}^4$$

$$y_{max} = \frac{40}{2} = 20\,mm$$

$$= 592500\ \text{mm}^3$$

Table of Deflection for Standard Load Cases

Beam Type	Slope	Deflection at Any Section x	Maximum Deflection
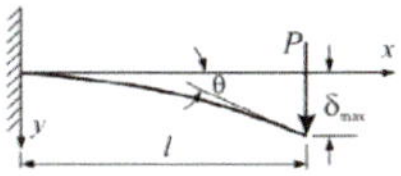	$\theta=\dfrac{Pl^2}{2EI}$	$y=\dfrac{Px^2}{6EI}(3l-x)$	$\delta_{max}=\dfrac{Pl^3}{3EI}$
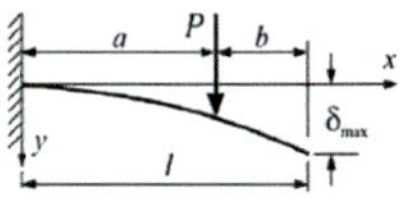	$\theta=\dfrac{Pa^2}{2EI}$	$y=\dfrac{Px^2}{6EI}(3a-x)$ for $0<x<a$; $y=\dfrac{Pa^2}{6EI}(3x-a)$ for $a<x<l$	$\delta_{max}=\dfrac{Pa^2}{6EI}(3l-a)$
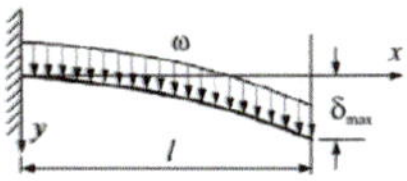	$\theta=\dfrac{\omega l^3}{6EI}$	$y=\dfrac{\omega x^2}{24EI}(x^2+6l^2-4lx)$	$\delta_{max}=\dfrac{\omega l^4}{8EI}$
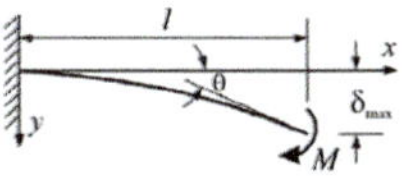	$\theta=\dfrac{Ml}{EI}$	$y=\dfrac{Mx^2}{2EI}$	$\delta_{max}=\dfrac{Ml^2}{2EI}$
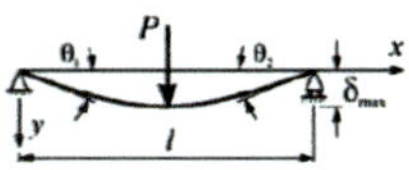	$\theta_1=\theta_2=\dfrac{Pl^2}{16EI}$	$y=\dfrac{Px}{12EI}\left(\dfrac{3l^2}{4}-x^2\right)$ for $0<x<\dfrac{1}{2}$	$\delta_{max}=\dfrac{Pl^3}{48EI}$
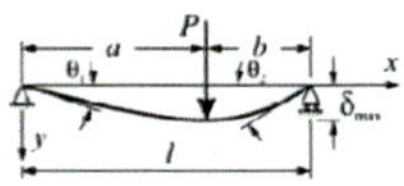	$\theta_1=\dfrac{Pb(l^2-b^2)}{6lEI}$; $\theta_2=\dfrac{Pab(2l-b)}{6lEI}$	$y=\dfrac{Pbx}{6lEI}(l^2-x^2-b^2)$ for $0<x<a$; $y=\dfrac{Pb}{6lEI}\left[\dfrac{l}{b}(x-a)^3+(l^2-b^2)x-x^3\right]$ for $a<xl$	$\delta_{max}=\dfrac{Pb(l^2-b^2)^{3/2}}{9\sqrt{3}lEI}$ at $x=\sqrt{(l^2-b^2)/3}$; $\delta_{max}=\dfrac{Pb}{48EI}(3l^2-4b^2)$ at the center, if $a>b$
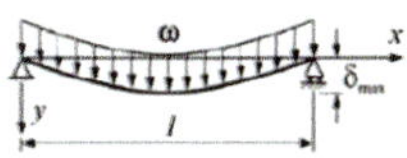	$\theta_1=\theta_2=\dfrac{\omega l^3}{24EI}$	$y=\dfrac{\omega x^2}{24EI}(l^3-2lx^2+x^3)$	$\delta_{max}=\dfrac{5\omega l^4}{384EI}$

4.19 MACAULAY'S METHOD

It is also called the double integration method. It is used to compute the maximum deflection in a beam based on the analytical method.

EXAMPLE 1

Find the maximum deflection of a simply supported beam, span 6 m. Point loads (30, 30) are acting at distances (2, 3 m) from the left support A. Take E as 2×10^5 N/mm^2, and the cross section of the beam as $(100\times200$ mm$)$.

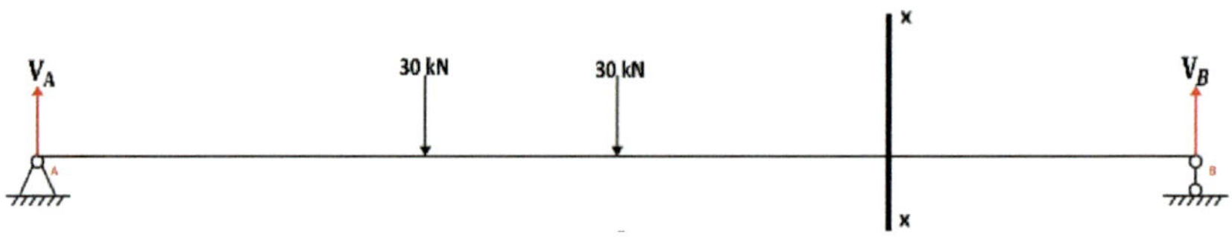

SOLUTION

The MoI of the section about the bending axis is given as:

$$I_{ZZ} = \frac{100(200)^3}{12} = 66.67 \times 10^6\ mm^4$$

Unknown reactions are V_A and V_B. Applying the condition $\sum M_A = 0$, we get as follows.
Sign convention used: Clockwise moments are considered as positive.

$$30\ (2) + 30\ (3) - V_B\ (6) = 0$$
$$V_B = 25\ KN$$

Now, applying the condition $\sum F_Y = 0$, and considering upward forces (and reactions) as positive, we get:

$$V_A + V_B = 30 + 30$$
$$V_A = 35\ kN$$

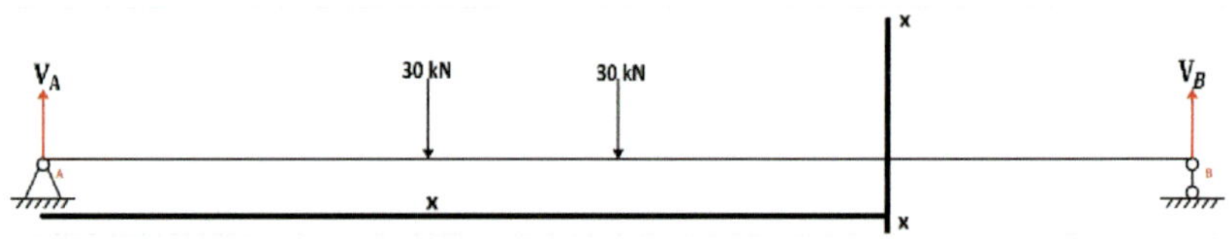

Cut a section *xx* at *x* from *A*, as shown. Taking moment about the *xx* section, considering clockwise moments as positive, and equating $\sum M_x = 0$, we get:
35 (x) – 30 (x – 2) – 30 (x – 3) = M_x

$$EI\frac{d^2y}{dx^2} = M$$

$$EI\frac{d^2y}{dx^2} = [35\ (x) - 30\ (x-2) - 30(x-3)]$$

Integrate once we get:

$$EI\frac{dy}{dx} = \left[\frac{35x^2}{2} - \frac{30(x-2)^2}{2} - \frac{30(x-3)^2}{2}\right] + C_1$$

Integrating again:

$$EIy = \left[\frac{35x^3}{6} + C_1 x + C_2 - \frac{15(x-2)^3}{3} - \frac{15(x-3)^3}{3}\right]$$

Boundary conditions:

First condition: At x = 0, y = 0.
Second condition: At x = 6, y = 0.

Substituting the first condition and neglecting the negative terms within the parenthesis, we get:

$$0 = 0 + 0 + C_2$$

(Note that the last two terms are not included, as they give negative values, since the parentheses terms will return negative values.)

This makes one the constants, C_2 as zero. Hence, the equation is updated as thus:

$$EIy = \left[\frac{35x^3}{6} + C_1 x - \frac{15(x-2)^3}{3} - \frac{15(x-3)^3}{3}\right]$$

Applying the second condition, we get:

$$0 = \left[\frac{35(6^3)}{6} + C_1(6) - \frac{15(6-2)^3}{3} - \frac{15(6-3)^3}{3}\right]$$

$$C_1 = -134.17$$

$$EI\frac{dy}{dx} = \left[\frac{35x^2}{2} - 15(x-2)^2 - 15(x-3)^2\right] - 134.17$$

$$EIy = \left[\frac{35x^3}{6} - 134.17x - \frac{15(x-2)^3}{3} - \frac{15(x-3)^3}{3}\right]$$

$$EIy = \left[\frac{35x^3}{6} - 134.17x - 5(x-2)^3 - 5(x-3)^3\right]$$

Let us now observe the units of the LHS and RHS of the preceding equation.

Units of LHS are $\frac{N}{mm^2} mm^4 \, mm = N\, mm^3$.

Units of RHS are kN m^3.

Hence, it should be converted to Nmm^3 to make it compatible.

Hence, the equation is now updated as follows:

$$EIy = \left[\frac{35x^3}{6} - 134.17x - 5(x-2)^3 - 5(x-3)^3\right]10^{12}$$

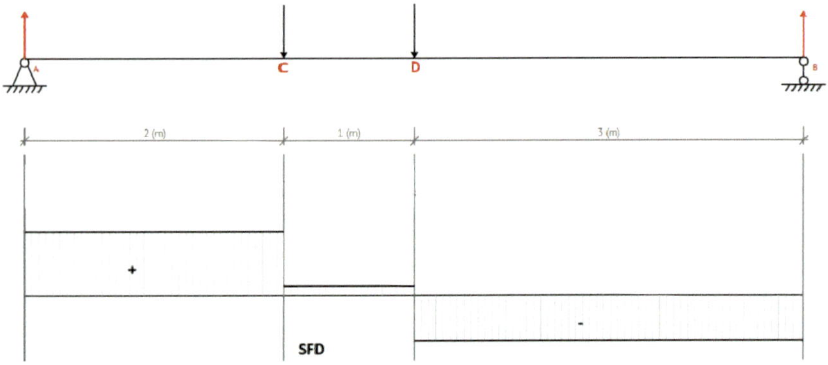

Let us try to find the deflection at x = 2 m to get deflection at *C*.

$$yc = \frac{1}{EI}\left[\frac{35x^3}{6} - 134.17x - 5(x-2)^3 - 5(x-3)^3\right]10^{12}$$

$$yc = \frac{1}{EI}\left[\frac{35(2)^3}{6} - 134.17(2)\right]10^{12}$$

Note that the negative terms of the equation should not be considered in the calculation.

$$= \frac{1}{(2x10^5)(66.67x10^6)}\left[\frac{35(2)^3}{6} - 134.17(2)\right]10^{12} = -16.63mm$$

A negative sign indicates that the deflection is measured along the negative Y-axis, downward.

Let us try to find the deflection at x = 3 m to get deflection at *D*.

$$YD = \frac{1}{EI}\left[\frac{35x^3}{6} - 134.17x - 5(x-2)^3 - 5(x-3)^3\right]10^{12}$$

$$YD = \frac{1}{EI}\left[\frac{35(3)^3}{6} - 134.17(3) - 5(3-2)^3\right]10^{12}$$

Note that the negative terms of the equation should not be considered in the calculation.

$$= \frac{1}{(2\times10^5)(66.67\times10^6)}\left[\frac{35(3)^3}{6} - 134.17(3) - 5\right]10^{12} = -18.75\ mm$$

A negative sign indicates that the deflection is measured along the negative Y-axis, downward.

To find the maximum deflection, the following condition should be satisfied:

$$\frac{dy}{dx} = 0$$

The SFD drawn earlier shows that at *D*, shear force changes its sign. Then, at this section, the following condition should be satisfied:

$$\frac{dy}{dx} = 0$$

$$EI\frac{dy}{dx} = \left[\frac{35x^2}{2} - 15(x-2)^2 - 15(x-3)^2\right] - 134.17 = 0$$

Neglecting the last term since at D (x = 3), the bending moment is maximum.

$$0 = \left[17.5\,x^2 - 15(x^2 + 4 - 4x)\right] - 134.17$$

Solving, we get x = 2.88 m.

$$y_{max} = \text{y at 2.88 m.}$$

The maximum deflection is not occurring at *D*, but very close to section D.

$$EIy_{max} = \left[\frac{35(2.88)^3}{6} - 134.17(2.88) - 5(2.88-2)^3 - 5(2.88-3)^3\right]10^{12}$$

Neglecting the negative term in the parenthesis (the last term), we get:

$$y_{max} = \frac{1}{EI}\left[\frac{35(2.88)^3}{6} - 134.17(2.88) - 5(2.88-2)^3\right]10^{12}$$

$$y_{max} = \frac{1}{(2\times10^5)(66.67\times10^6)}\left[\frac{35(2.88)^3}{6} - 134.17(2.88) - 5(2.88-2)^3\right]10^{12}$$

$$= -18.78\ mm$$

EXAMPLE 2

Find the deflection of the cantilever beam using the double integration method.

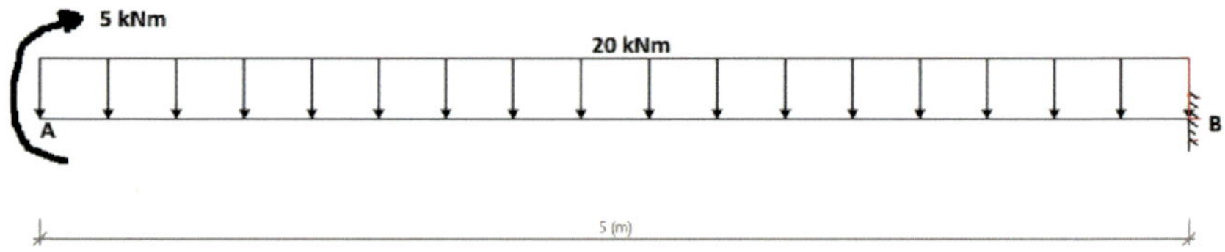

Consider a cantilever beam under the consideration of UDL and end moment as shown.

Let us determine the deflection and slope at the free end (*A*).

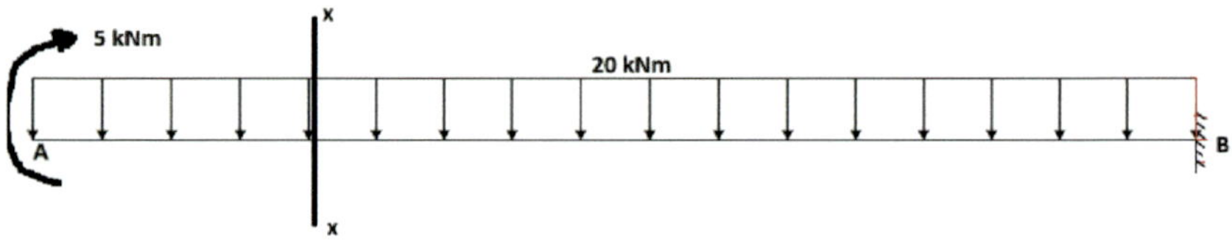

Consider a section *xx*, *x* distance from *A*.

$$\text{Moment at xx} = M_x = 5 - \frac{20x^2}{2} \text{ (considering sagging positive)}$$

Substituting the preceding in the equation of simple bending, we get:

$$EI\frac{d^2y}{dx^2} = M$$

$$EI\frac{d^2y}{dx^2} = 5 - \frac{20x^2}{2}$$

Integrating once, we get slope at *A*:

$$EI\frac{dy}{dx} = 5x - \frac{20x^3}{6} + C_1$$

Applying the boundary condition at the fixed end as at x = 5 m, $\frac{dy}{dx} = 0$, we get:

$$0 = 5(5) - \frac{20(5)^3}{6} + C_1$$

$$C_1 = 391.67$$

Hence, the updated equation is given by:

$$EI\frac{dy}{dx} = 5x - \frac{20x^3}{6} + 391.67$$

To obtain deflection, integrate the preceding equation once again.

$$EIy = \frac{5x^2}{2} - \frac{20x^4}{24} + 391.67x + C_2$$

By applying the second boundary condition at x = 5 m, y = 0 (at the fixed end, deflection is zero), we get:

$$0 = \frac{5(5)^2}{2} - \frac{20(5)^4}{24} + 391.67(5) + C_2$$
$$C_2 = -1500.02$$

Hence, the updated equation is given as:

$$EIy = \frac{5x^2}{2} - \frac{20x^4}{24} + 391.67x - 1500.02$$

Deflection:

$$y = \frac{1}{EI}\left[\frac{5x^2}{2} - \frac{20x^4}{24} + 391.67x - 1500.02\right]$$

Slope:

$$\frac{dy}{dx} = \frac{1}{EI}\left[5x - \frac{20x^3}{6} + 391.67\right]$$

Using the previous two equations, one can find deflection and slope at *A*.
At A, x = 0.

$$\text{Hence, slope} = \left(\frac{dy}{dx}\right)_A = \frac{1}{EI}(391.67).$$

$$\text{Deflection: } (y)_A = \frac{1}{EI}(-1500.01).$$

A negative indicates that the deflection is in the downward direction.

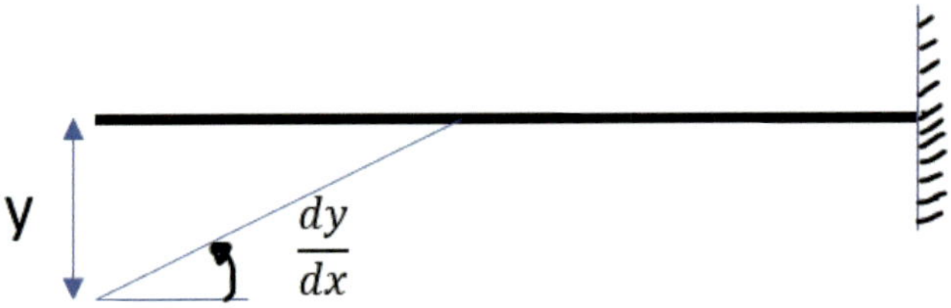

4.20 SHEAR STRESS DISTRIBUTION

Shear failure in structural members is a serious concern. The shear force causes the body to fail along the plane, as shown in the figure. It bifurcates the beam in segments and thus reduces the load-carrying capacity of the beam. Shear stress depends on the area of cross section available at that section. Let us consider a simply supported beam under a centrally concentrated load, as shown:

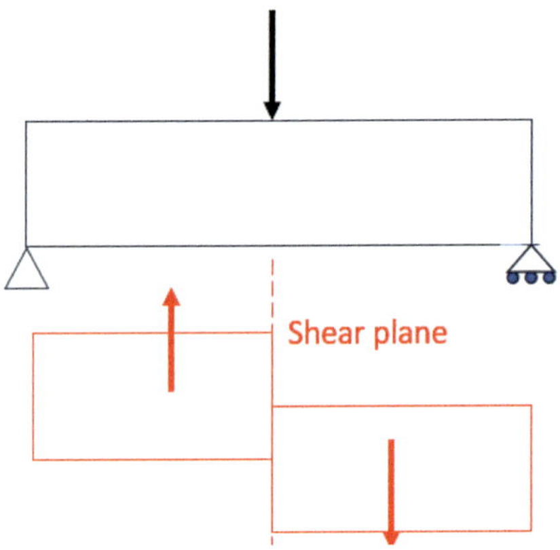

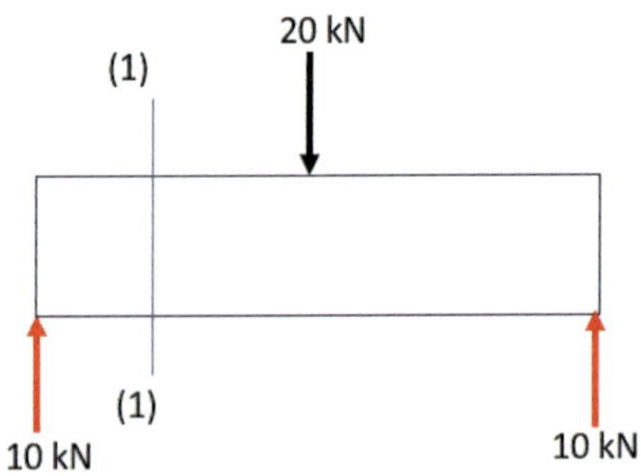

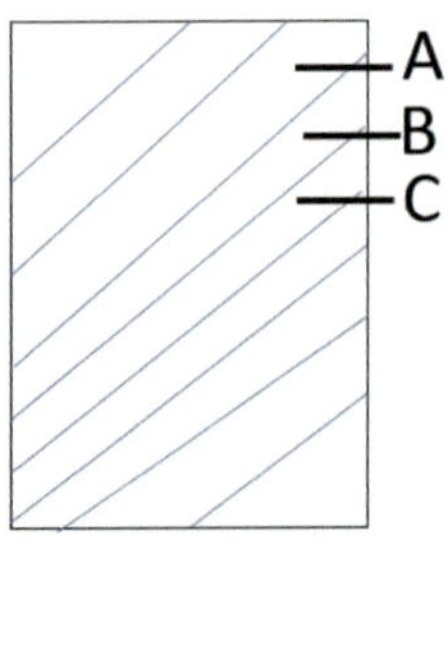

Consider different fibers (A, B, C) in the cross section. Shear stress is not uniform throughout the cross section but varies with the depth of the section. Let us consider a cross section and mark at B-B. Calculate the longitudinal shear stress.

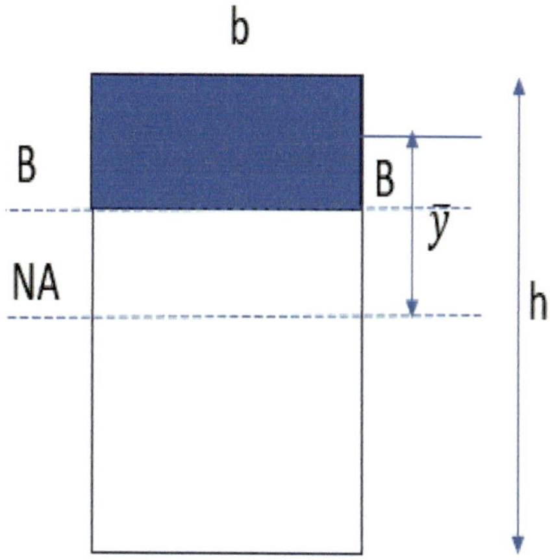

The distance of CG of the area above BB is marked as $\overline{y}$.
$\overline{y}$ is the distance of the CG of the hatched area from the neutral axis.

$$\tau_B = Shear\,stress\,at\,B = \frac{QV}{Ib}\,\frac{N}{mm^2}$$

Where:

Q = first moment of area (mm^3) of the upper cross section of the beam above the B-B plane
V = vertical shear force (N)
I = moment of inertia of the beam (mm^4)
b = width of the beam (mm)

EXAMPLE

Find the shear stress at the beam's neutral axis under a shear force of 10 kN.

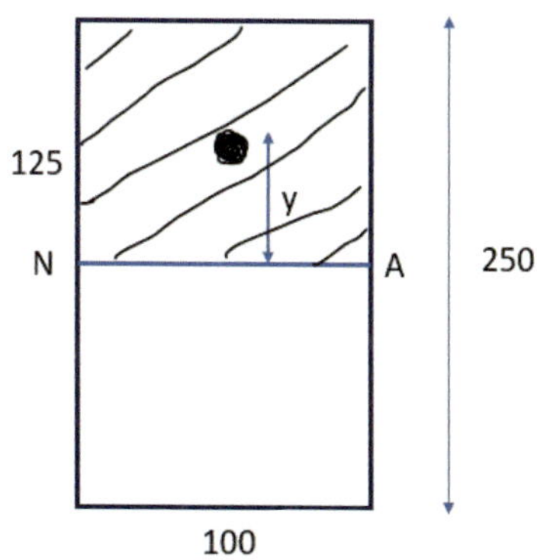

$$\bar{y} = \left(\frac{125}{2}\right) = 62.5\ mm$$

$$Q = A\ \bar{y} = 100 \times 125 \times 62.5 = 781250\ \text{mm}^3$$

$$V = 10\ \text{kN.}$$

$$I = \frac{100 \times 250^3}{12} = 1.3 \times 10^8\,\text{mm}^4$$

$$b = 100$$

$$\tau_A = \frac{(10 \times 10^3) \times (781250)}{(1.3 \times 10^8) \times 100} = 0.6 \frac{N}{mm^2}$$

EXAMPLE 3

For the I section shown in what follows, determine the shear stress at the sections (A-A).

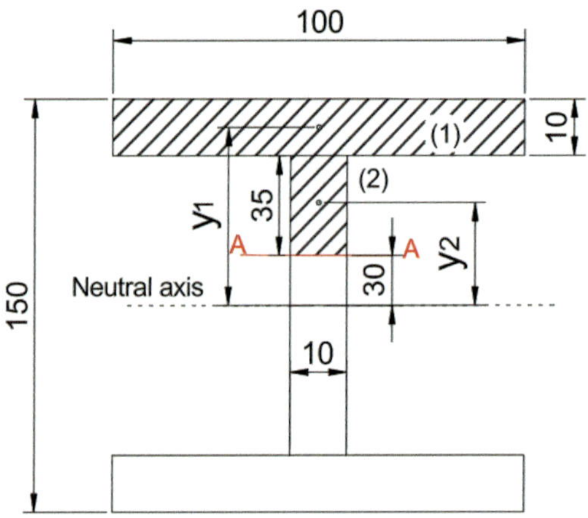

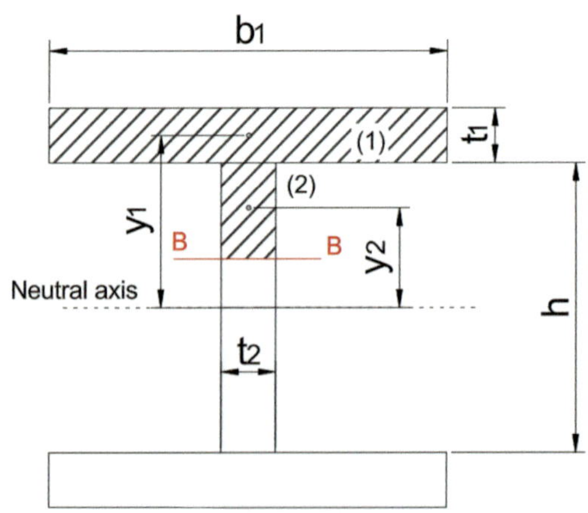

Part (1) Q = $A_1\ y_1 + A_2\ y_2$.

(y_1, y_2) distance of centroid of areas (A_1, A_2) from the neutral axis.

To find τ_{AA}:

$$A_1 = 100 \times 10 = 1000\ mm^2$$

$$y_1 = 70\ mm$$

$$y_2 = \left(\frac{35}{2} + 60\right) = 47.5\ mm$$

$$A_2 = 10 \times 35 = 350\ mm^2$$

$$Q = A_1 y_1 + A_2 y_2$$

$$= (1000 \times 70) + (350 \times 47.5) = 86625\ mm^3$$

$$\mathrm{V} = 25\ \mathrm{kN}$$

$$\mathrm{I} = 1.16 \times 10^7\ mm^4$$

where b is the breadth of the section at AA = 10 mm.

$$\tau_{AA} = \frac{\left(25 \times 10^3\right)(86625)}{\left(1.16 \times 10^7\right)} = 18.67 \frac{N}{mm^2}$$

4.20.1 Shear Stress Variation

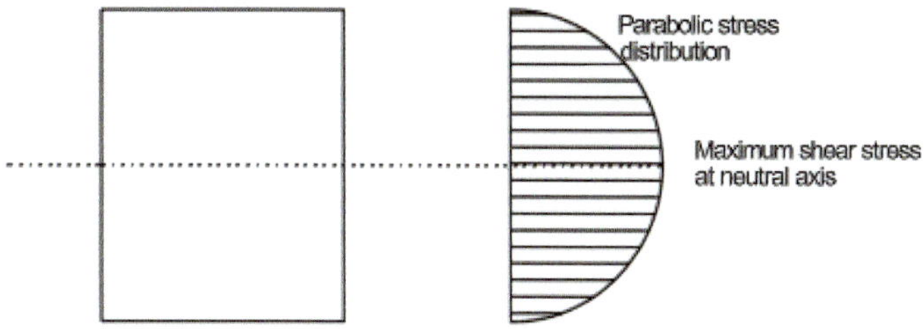

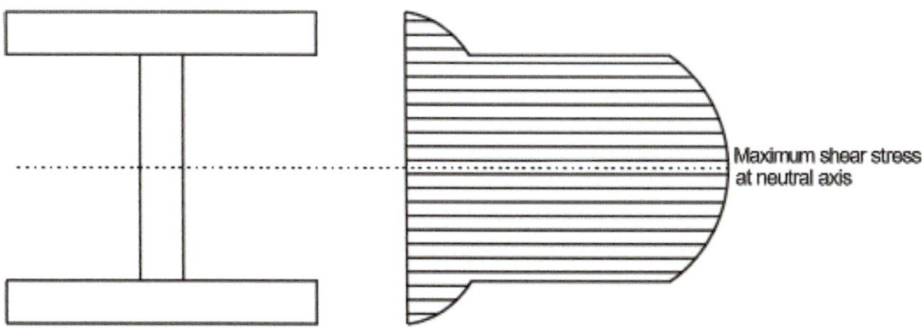

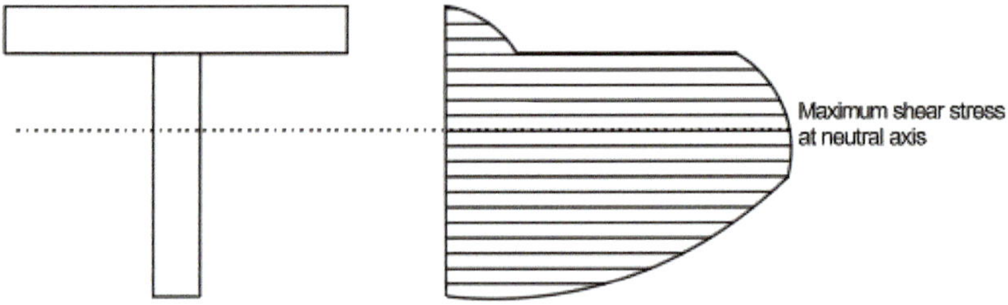

The figure shows the typical shear stress variations along the depth of the cross section of different shapes of the cross section. Let us consider the following example to understand the shear stress variations.

EXAMPLE

A steel I beam has a flange 150 × 20, web 300 × 10, and shear of 50 kN. Determine the shear stress distribution for the beam. Draw the shear stress distribution diagram.

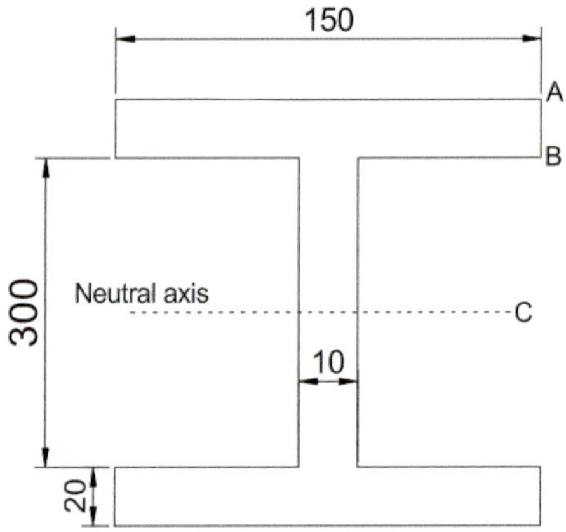

$$\tau = Shear\ stress = \frac{QV}{Ib} \frac{N}{mm^2}$$

$$\tau_A = 0 \text{ because } Q = Ay = 0$$

To find τ_B:

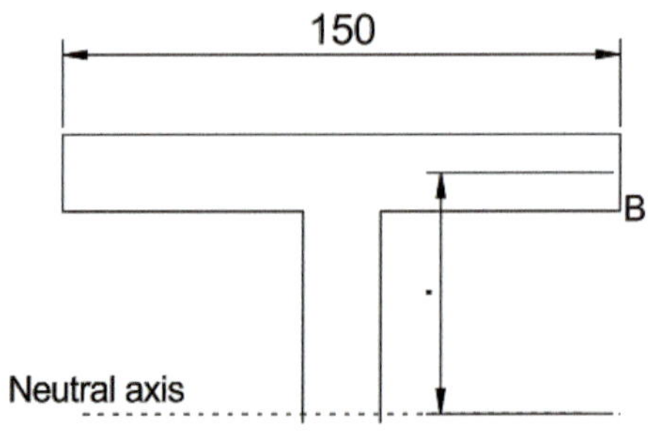

$$A = 150 \times 20 \text{ mm}^2$$

$$\bar{y} = \left(150 + \frac{20}{2}\right) = 160\,mm$$

$$Q = A\bar{y} = (150 \times 20) \times 160 = 48 \times 10^4\,mm^3$$

Moment of inertia:

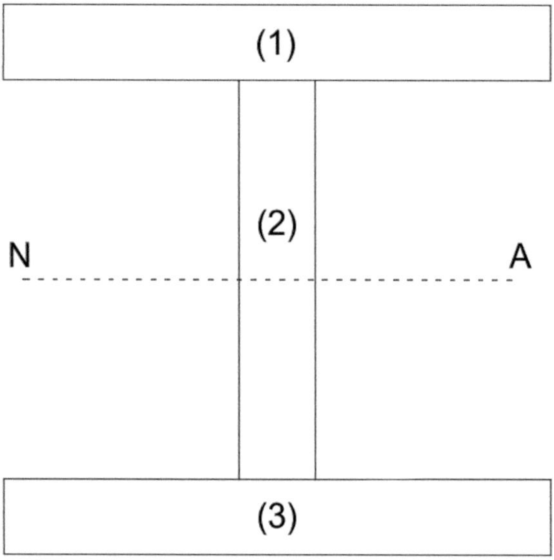

$$I_{(1-1)} = \frac{bd^3}{12} + Ak^2 \text{ (parallel axis theorem)}$$

$$= \frac{150 \times 20^3}{12} + (150 \times 20)160^2$$

$$= 76.9 \times 10^6\,mm^4$$

$$I_{(2-2)} = \frac{10 \times 300^3}{12}$$

$$= 76.9 \times 10^6\,mm^4$$

Total I:

$$I_{\text{Total}} = \left(I_{(1-1)} \times 2\right) + I_{(2-2)}$$

$$= \left(76.9 \times 10^6 \times 2\right) + 22.5 \times 10^6$$

$$= 176.3 \times 10^6\,mm^4$$

where b is breadth at BB = 150 mm.

$$\tau_B = \frac{(50\times10^3)\times(48\times10^4)}{(176.3\times10^6)150} = 0.908\frac{N}{mm^2}$$

To find $\tau_{NA} = \frac{QV}{Ib}$:

$$V = 50 \text{ kN}$$

$$Q = \sum A\overline{y}$$

$$b \text{ at NA} = 10 \text{ mm.}$$

$$\begin{aligned} Q &= A_1\overline{y_1} + A_2\overline{y_2} \\ &= (150\times20)\,160 + (150\times10)\,(150/2) \\ &= 592500 \text{ mm}^3 \end{aligned}$$

$$\tau_B = \frac{(50\times10^3)\times(592500)}{(176.3\times10^6)10} = 16.80\frac{N}{mm^2}$$

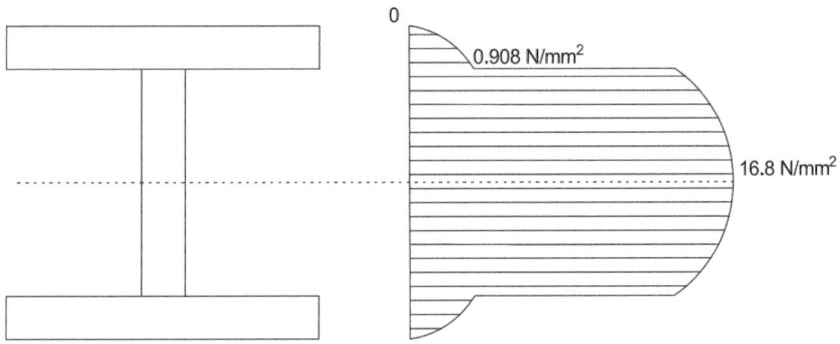

4.21 STABILITY AND EULER'S BUCKLING LOAD FOR COLUMNS

The stability of structural systems in general and steel structures in particular is extremely important when structures tend to resist the loads by the geometry (shells, folded plates, etc.). While the structural geometry of a form-dominant design is posed to a challenge under the given loads and boundary conditions, it needs to satisfy the static equilibrium, compatibility, and force–displacement relationships to assess it to be stable. An unstable condition is often referred to as a failed state. Hence, stability demands an understanding of failure.

4.21.1 Conditions of Stability

Stability is affected significantly under compressive forces. Further, geometric stability is important as it may challenge the functional requirements of the structural member. Therefore, *stability* refers to a stable state of equilibrium and is defined as the ability of any structural system to remain (or continue to remain) in its geometric form, which is capable of performing the intended function even if the geometric position is disturbed by external forces. But structures don't need to remain (or continue to remain) in their original geometric form to be classified as stable. They may continue to remain stable even under the deformed geometric position if they continue to perform their intended function successfully.

Following are the three criteria of checking stability: (1) Euler's static criterion, (2) Lyapunov's dynamic criterion, and (3) the potential energy stability criterion. *Euler's static criterion* is applicable under the non-trivial equilibrium state. It evaluates the stability of a structural system by examining the optimum geometric configuration of the system other than the original (initially straight) configuration, at which the structural system can still disburse the applied load $\left(P_{cr}\right)$. Under the given boundary conditions and initially perfect straight geometry, structures are examined. Euler's criterion evaluates whether the structure is capable of carrying the load (maybe in lesser magnitude, which is P_{cr}, where $P>P_{cr}$) instead of remaining in a state where it is unable to carry any load at all. It is interesting to note that the load-carrying capacity of the structural member is reduced from P to P_{cr}, but the important fact is that the structural member can carry at least P_{cr} even at a changed geometric form, which is quite weak from that of the initial form. This value of load with lesser magnitude in comparison to that of the originally intended load (*P*) is termed critical load or buckling load. It can be easily seen from the standard literature that P_{cr} is easily computed from the boundary conditions, cross-sectional properties of the member, and slenderness ratio. However, instability occurs when two or more adjacent equilibrium positions correspond to different mode shapes.

Lyapunov's condition examines the stability of the structural system under dynamic excitations. If a member is subjected to a continuously varying disturbing force, it is necessary to examine whether the condition of equilibrium under the dynamic forces is satisfied. As explained in the literature (Chandrasekaran, 2015b, 2016a, 2016b, 2017), stability can be influenced by both varying amplitudes of the exciting force and its period of excitation. While the former can influence the load-carrying capacity of the member, which is the strength-dependent criterion, the latter can result in an unconditional response at the near-resonance state of vibration. Hence, Lyapunov's assessment of stability is focused on the dynamic response behavior of the structural system instead of purely assessing its load-carrying capacity, as in the case of Euler's criterion.

Lyapunov's condition is more significant for structures that are designed to perform their intended function under varying external forces. The most challenging part in the context of stability is that the structure shall also be assuming a different geometric position with time. Structural systems like ships, offshore complaint platforms, and floating production platforms fall under this category of stability check. Stability calculations of ships focus on estimating the center of gravity, the center of buoyancy, the metacenter of vessels, and their interaction. The saving part of such systems is that they are designed to remain hydrostatically stable at any instant of time. In case of structures that are permitted to undergo large displacements, as in the case of TLPs, then the geometric design ensures proper recentering, which means that the structural system continuously tries to regain its original geometric position with the help of dedicated members present within the system; in case of TLP, it is tethers. A potential energy stability criterion applies to structural systems for which the potential energy of the system ceases to be the minimum, which is more relevant for conservative systems.

4.21.2 Buckling and Instability

The loss of structural stability is termed *instability*, which can be assessed based on (1) material properties, (2) geometric configuration, and (3) the nature and magnitude of loads acting on the structure. Assessment based on material strength highlights where the applied load exceeds the accepted strength level of the material, classifying the structure as unstable. Assessment based on the geometric configuration explores the functional success of the member even under the deformed (displaced) position, while that based on the nature and magnitude of the loads assesses dynamic instability. In general terms, the loss of stability is expressed by a load–deformation relationship. Figure 4.35 shows an unstable situation; even a small disturbance, the ball can roll over, causing instability.

A significant change in the geometric configuration of a structure or any structural component under compression resulting in the loss of the ability to resist the encountered loads is termed *instability*. It is important as this can lead to a catastrophic failure, as the load-carrying capacity of an unstable structural system is close to zero; hence, it should be properly accounted for in the design. Instability is a strength-related limit state. Structural systems can become unstable upon application of excessive loads. Beyond the threshold limits, stresses developed in the member can magnify the deflections, causing failure. Any structural system whose deformation exceeds the permissible limits is termed *unstable*. Therefore, the stability of any structural system depends upon its ability to withstand the encountered load without undergoing excessive deformation. It is important to note that a structure which is found to be stable for a load applied at one section may become unstable under the same magnitude and nature of the load but applied at a different section. Stability can

FIGURE 4.35 Unstable condition.

be simply illustrated with the help of figures shown aside. The ball remains in equilibrium in all three cases, as the specific locations, as shown. However, a slight disturbance imposed on the ball can change its behavior differently in the three cases. As seen in case (a), the ball can return to its original position upon the removal of any disturbing external force and therefore assumes a stable position. It corresponds to a position in the material stress–strain curve, within the elastic limit. In case (b), the ball is considered as neutrally stable, as the ball can continue to move under the influence of disturbing external forces. This state corresponds to a point on the ultimate strength region of the stress–strain curve of the material.

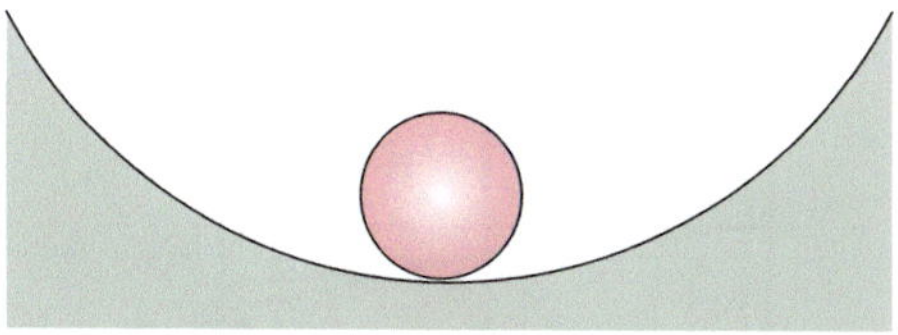

(a) Stable

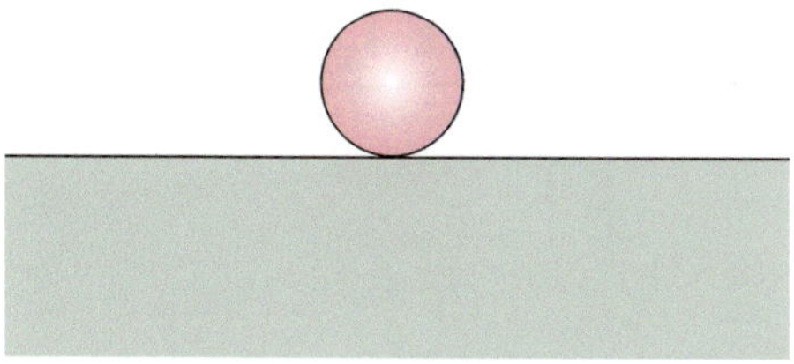

(b) Neutrally stable

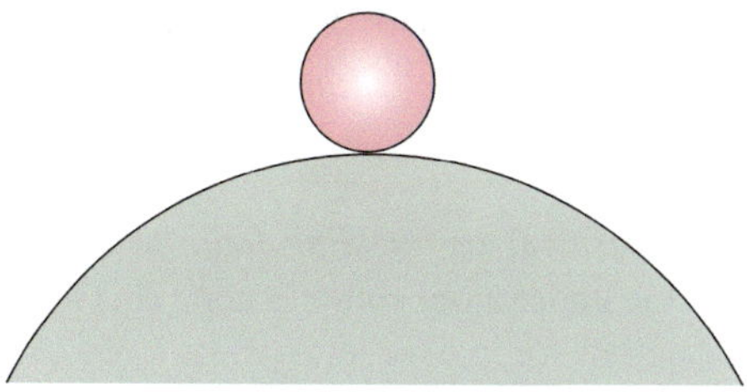

(c) Unstable

However, in case (c), the ball will not be able to return to its original position even under a small disturbance caused by an external force. It is a classical state of instability. It is important to note that loss of stability arising either from a change in geometry or a change in the structural configuration adds a new set of additional loads to the structural system. It may lead to the further classification of instability as flexural buckling, torsional buckling, flexural-torsional buckling, and lateral-torsional buckling (Gambhir, 2004).

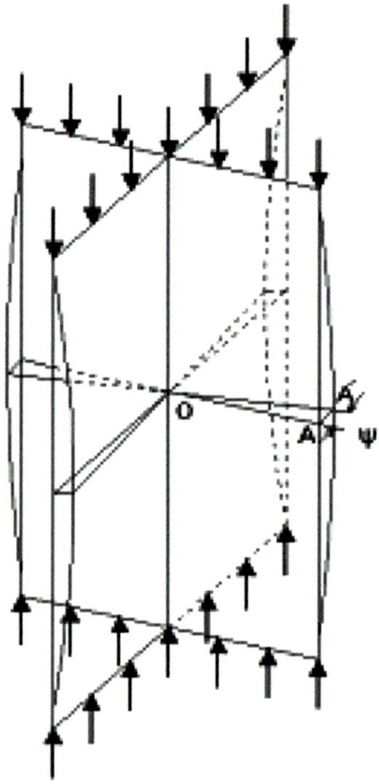

In structural engineering, *stability* refers to a column's ability to resist buckling when subjected to axial compressive forces. Buckling is an instability phenomenon where a column suddenly bends or deflects due to compressive loads, even if the stress level is below the material's yield strength. This instability is influenced by the column's length, end conditions, and cross-sectional properties.

4.21.3 Euler's Buckling Load

Euler's formula gives the critical or *buckling load*, at which a slender, perfectly straight column will buckle under axial compression. This load is given by:

$$P_{cr} = \frac{\pi^2 EI}{(KL)^2}$$

where P_{cr} is the critical buckling load; E is the Young's modulus of the material; I is the moment of inertia of the column's cross section about the bending plane; K is the effective length factor, which depends on the column's end conditions; and L is the actual (unsupported) length of the column. The critical load is the load at which even a small disturbance will cause a significant lateral deflection of the column, leading to buckling.

4.21.4 End Conditions and Effective Lengths

The effective length of a column depends on its end support conditions, which affect its tendency to buckle. Here are the four common cases with their effective lengths and critical load formulas:

1. Both ends pinned (hinged) (Figure 4.36a)
   - Effective length KL = L
   - Buckling load $P_{cr} = \dfrac{\pi^2 EI}{(KL)^2}$
2. One end fixed, other end free (Figure 4.36b)
   - Effective length KL = 2 L
   - Buckling load $P_{cr} = \dfrac{\pi^2 EI}{(4L)^2}$
3. Both ends fixed (Figure 4.36c)
   - Effective length KL = 0.5 L
   - Buckling Load, $P_{cr} = \dfrac{4\pi^2 EI}{(L)^2}$

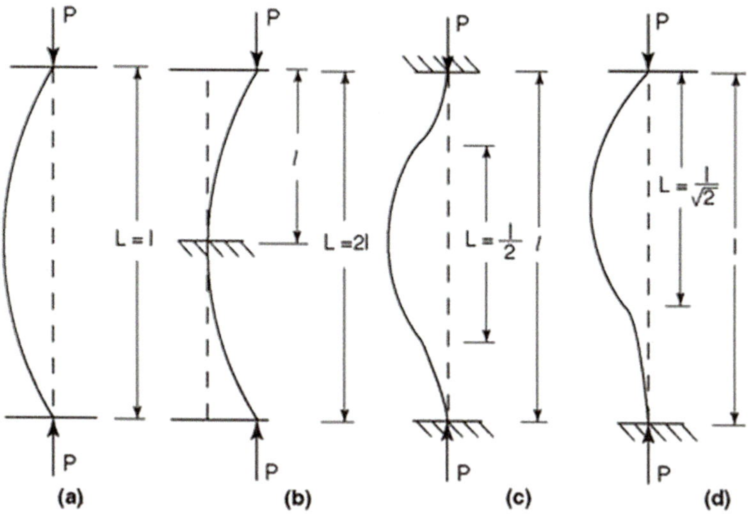

FIGURE 4.36 End conditions of column.

4. One end fixed, other end pinned (Figure 4.36d)

- Effective length KL $= \dfrac{L}{\sqrt{2}}$

- Buckling load $P_{cr} = \dfrac{2\pi^2 EI}{(L)^2}$

EXAMPLE

A structural engineer designs a steel column 6 m long with one end fixed and the other pinned to support an axial load of 500 kN without buckling. The moment of inertia I about the weakest axis is $4 \times 10^{-6}\,\text{m}^4$. Calculate the required cross-sectional properties to ensure the column does not buckle.

SOLUTION

1. *Effective length calculation.* Since the column has one end fixed and the other pinned, the effective length is:

$$\text{KL} = \frac{L}{\sqrt{2}} = \frac{6}{\sqrt{2}} = 4.24\ m$$

2. *Critical load using Euler's formula.* Since 440.5 kN, which is the capacity of the column, is lesser than the applied load (demand) of 500 kN, the current configuration would buckle under the load. To increase the column's load capacity, we need to increase the moment of inertia I by adjusting the cross-sectional design.

$$P_{cr} = \frac{\pi^2 EI}{(KL)^2} = \frac{\pi^2 \times 200 \times 10^9 \times 4 \times 10^{-6}}{(4.24)^2} = 440.5\,kN$$

3. *New moment of inertia calculation for stability.* To prevent buckling, we need Pcr ≥ 500 kN. Rearranging Euler's formula:

$$I = \frac{P_{cr}(KL)^2}{\pi^2 E} = \frac{500 \times 10^3 \times (4.24)^2}{\pi^2 \times 200 \times 10^9} = 4.54 \times 10^{-6}\ m^4$$

The column needs a moment of inertia of at least $4.54 \times 10^{-6}\ m^4$ to safely support the load without buckling. Based on this value, for the known cross-sectional shape of the column, one can determine the dimensions of the column.

4.21.4.1 Exercise Problems

EXERCISE 1

Using the stiffness method of analysis, find the moments and reactions of the continuous beam shown in the figure.

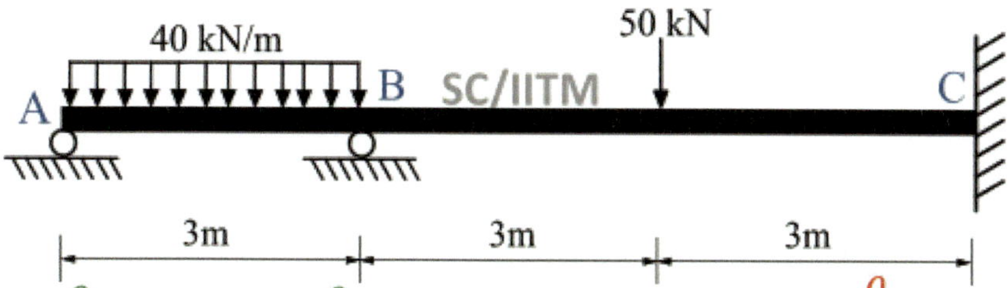

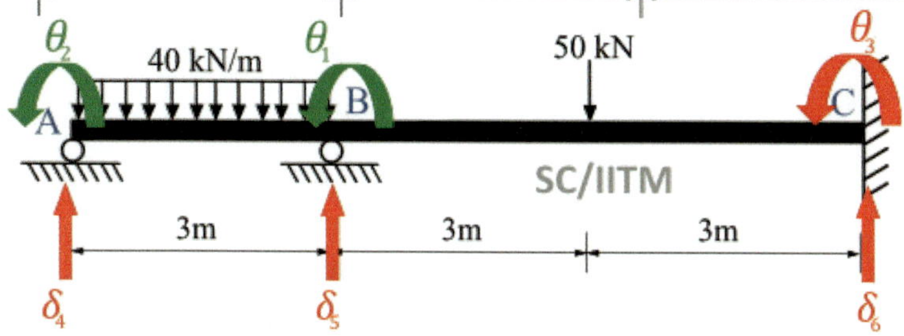

Technical Discussions

No. of members = 2 (member 1 = AB; member 2 = BC)
Length of each member = {3, 6}
No. of unrestrained degrees of freedom = 2 (rotation at A; rotation at B)
No. of restrained degrees of freedom = 4 (rotation at C; displacement at A, B, and C)
Axial deformation in the beam neglected, making $[K]$ as 4×4 only
EI considered to be uniform throughout

Solution

Details as required by the computer program are given in what follow.

Input

n = 2, l = [3 6], I = [1 1]
uu = 2, ur = 4

Global Labels

AB = [2 1 4 5]
BC = [1 3 5 6]

Fixed End Moments

AB = [30–30 60 60]

BC = [37.5–37.5 25 25]

Joint Load

[–7.5–30 37.5–60–85–25]^T

Final Solution

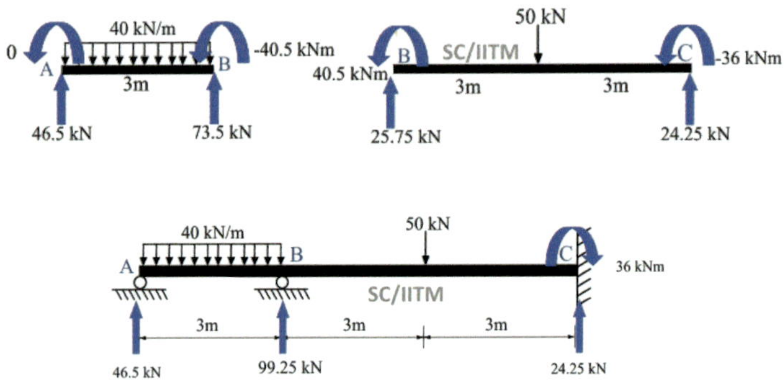

EXERCISE 2

Using the stiffness method of analysis, find the moments and reactions of the continuous beam shown in the figure.

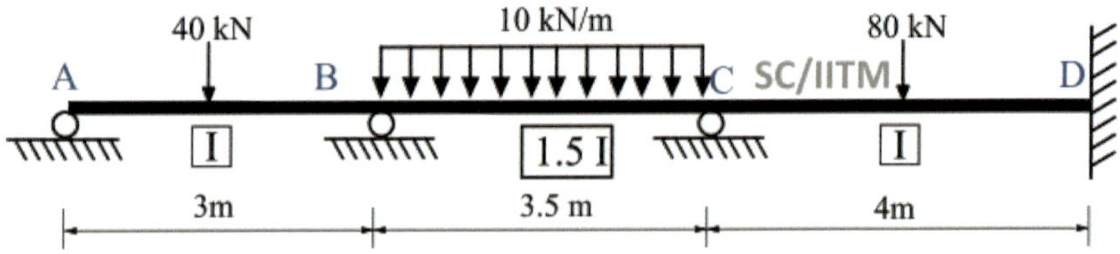

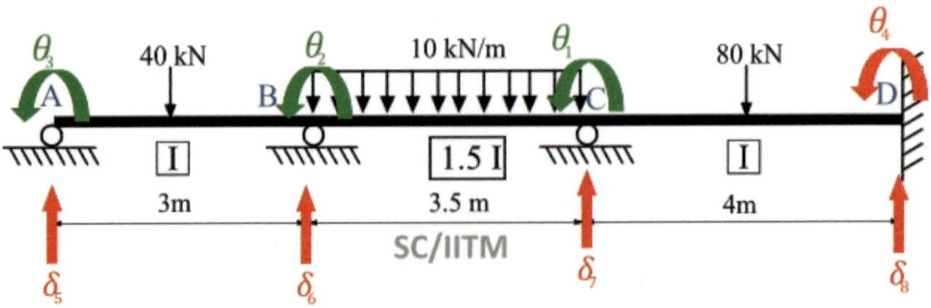

Technical Discussions

No. of members = 3 (member 1 = AB; member 2 = BC; member 3 = CD)
Length of each member = {3, 3.5, 4}
No. of unrestrained degrees of freedom = 3 (rotation at *A*, *B*, and *C*)
No. of restrained degrees of freedom = 5 (rotation at *D*; displacement at *A*, *B*, *C*, and *D*)
Axial deformation in the beam neglected, making [*K*] 4 × 4 only

Solution

Details as required by the computer program are given in what follow.

Input

n = 3, l = [3 3.5 4], I = [1 1.5 1]
uu = 3, ur = 5

Global Labels

AB = [3 2 5 6]
BC = [2 1 6 7]
CD = [1 4 7 8]

Fixed End Moments

AB = [15–15 20 20]
BC = [10.208–10.208 17.5 17.5]
CD = [40–40 40 40]

Joint Load

= [–29.792 4.792–15 40–20–37.5–57.5–40]^T

Final Solution

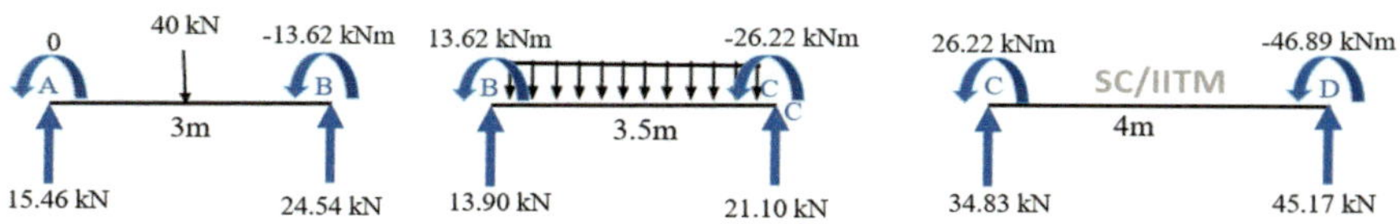

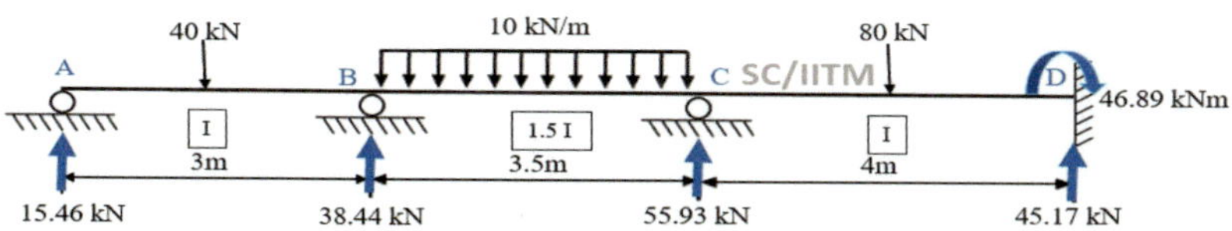

EXERCISE 3

Using the stiffness method of analysis, find the moments and reactions of the continuous beam shown in the figure.

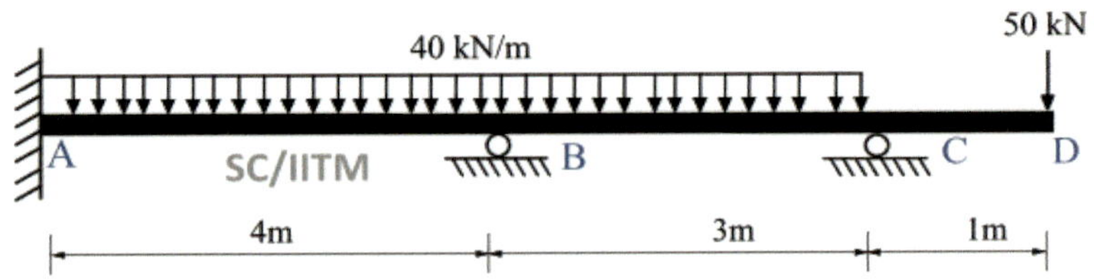

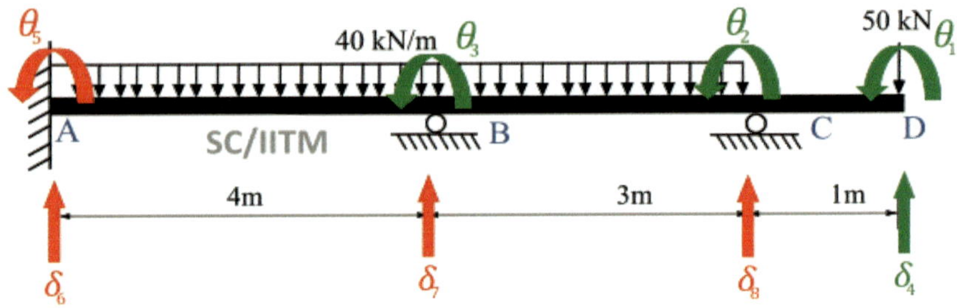

Technical Discussions

No. of members = 3
Length of each member = {4, 3, 1}
No. of unrestrained degrees of freedom = 4 (rotation at B, C, and D; displacement at D)
No. of restrained degrees of freedom = 4 (rotation at A; displacement at A, B, and C)
Axial deformation in the beam neglected, making [*K*] 4 × 4 only
EI considered to be uniform throughout

Solution

Details as required by the computer program are given in what follow.

Input

n = 3, l = [4 3 1], I = [1 1 1]
uu = 4, ur = 4

Global Labels

AB = [5 3 6 7]
BC = [3 2 7 8]
CD = [2 1 8 4]

Fixed End Moments

AB = [53.333–53.333 80 80]
BC = [30–30 60 60]
CD = [0 0 0 50]

Joint Load

= $[0\ 30\ 23.333-50-53.333-80-140-60]^T$

Final Solution

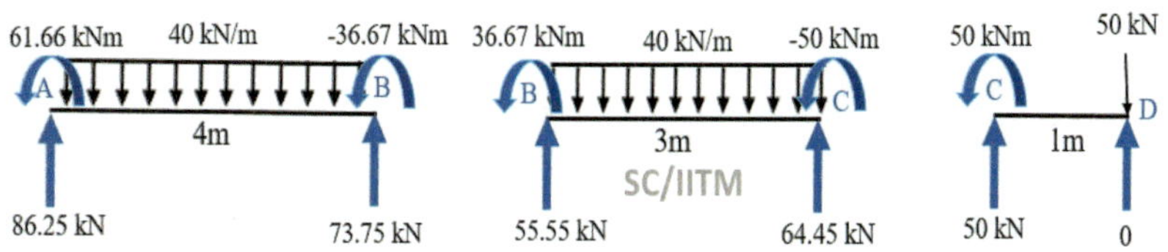

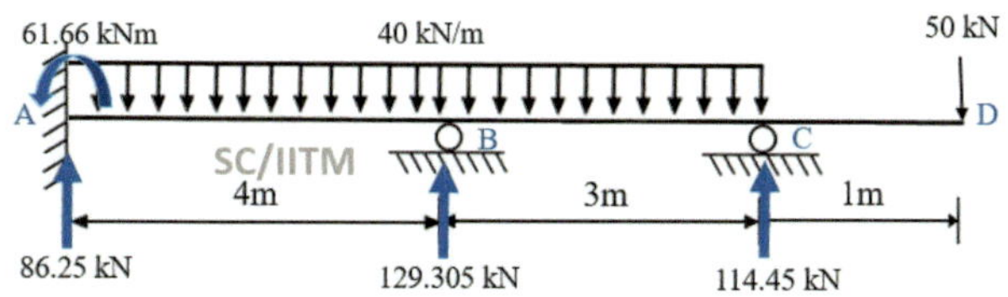

EXERCISE 4

Using the stiffness method of analysis, find the moments and reactions of the continuous beam shown in the figure.

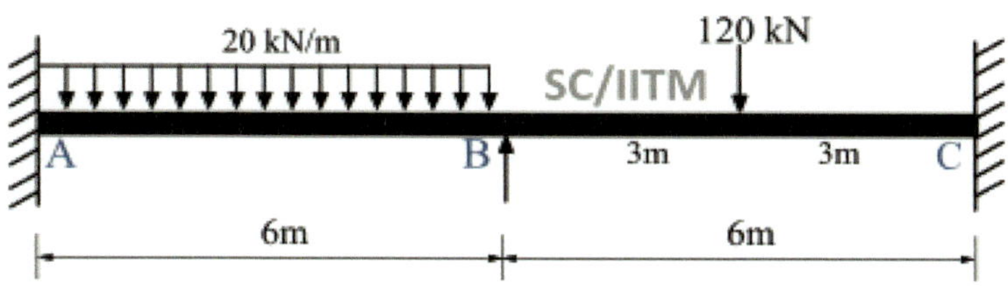

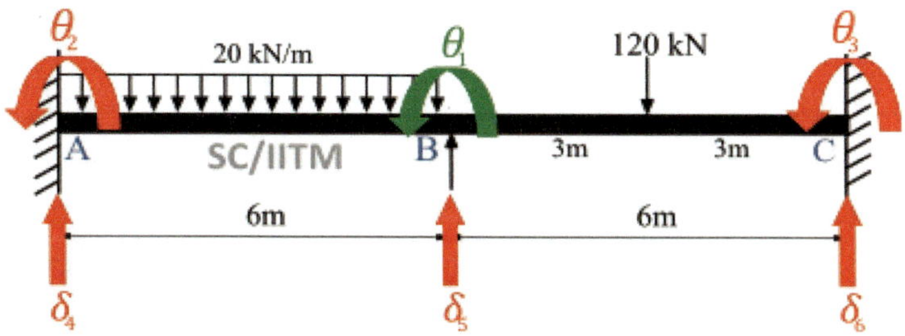

Technical Discussions

No. of members = 2 (member 1 = AB; member 2 = BC)
Length of each member = {6, 6}
No. of unrestrained degrees of freedom = 1 (rotation at *A*)
No. of restrained degrees of freedom = 5 (rotation at A and C; displacement at A, B, and C)
Axial deformation in the beam neglected, making [*K*] 4 × 4 only
EI considered to uniform throughout

Input

n = 2, l = [6 6], I = [1 1]
uu = 1, ur = 5

Global Labels

AB = [2 1 4 5]
BC = [1 3 5 6]

Fixed End Moments

AB = [60–60 60 60]
BC = [90–90 60 60]

Joint Load

= [–30–60 90–60–120–60]^T

Output

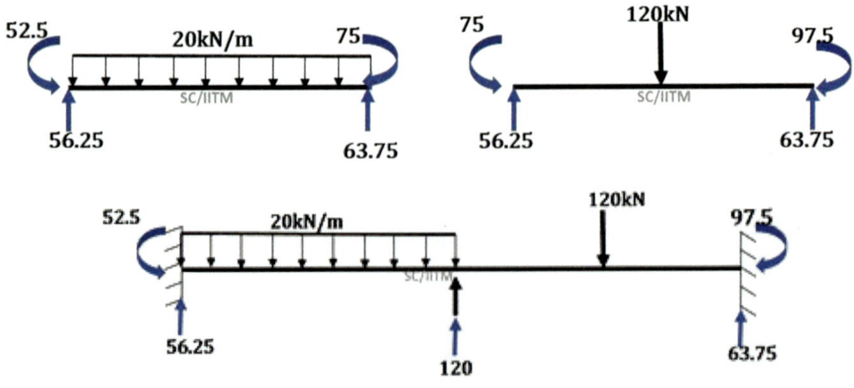

EXERCISE 5

Using the stiffness method of analysis, find the moments and reactions of the frame shown in the figure.

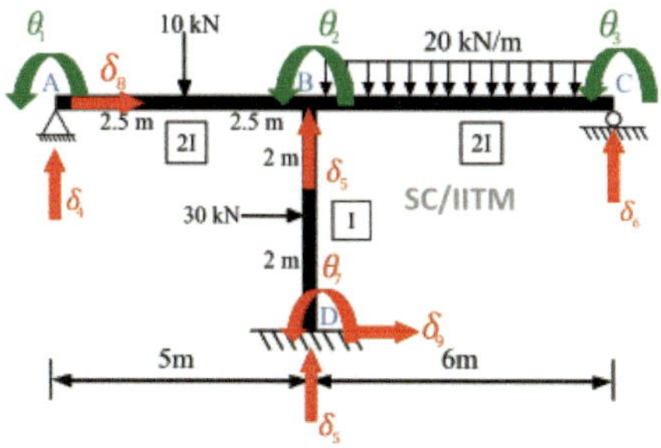

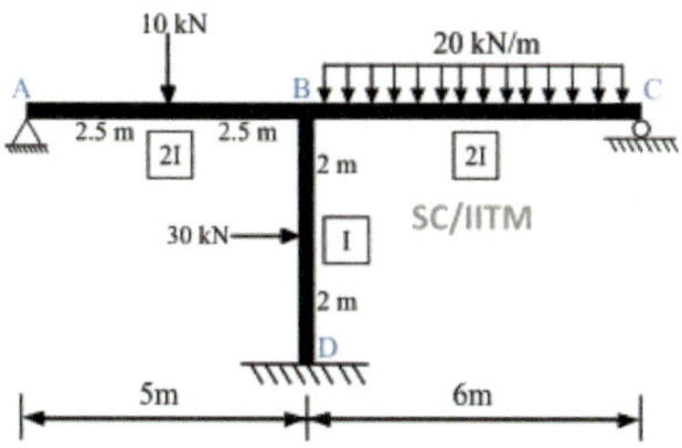

Technical Discussions

No. of members = 3 (member 1 = AB; member 2 = BC; member 3 = BD)
Length of each member = {5, 6, 4}
No. of unrestrained degrees of freedom = 3 (rotation at A, B, C)
No. of restrained degrees of freedom = 6 (rotation at D; displacement at A, B, C, D)
Axial deformation in the beam neglected, making $[K]$ 4×4 only

Solution

Details as required by the computer program are given in what follow.

Input

n = 3, l = [5 6 4], I = [2 2 1]
uu = 3, ur = 6

Global Labels

AB = [1 2 4 5]
BC = [2 3 5 6]
BD = [2 7 8 9]

Fixed End Moments

AB = [6.25–6.25 5 5]
BC = [60–60 60 60]
BD = [–15 15–15–15]

Joint Load

$= [-6.25 -38.75\ 60 -5 -65 -60 -15\ 15\ 15]^T$

Output

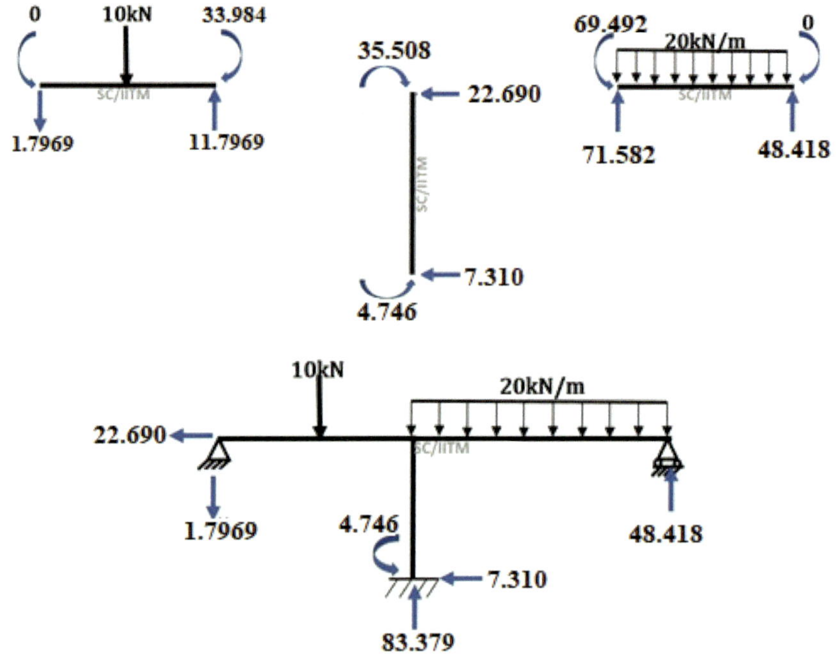

EXERCISE 6

Using the stiffness method of analysis, find the moments and reactions of the frame shown in the figure.

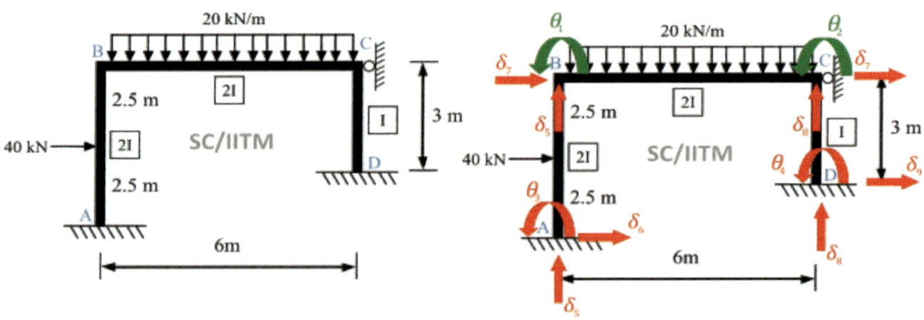

Technical Discussions

No. of members = 3 (member 1 = BA; member 2 = BC; member 3 = CD)
Length of each member = {5, 6, 3}
No. of unrestrained degrees of freedom = 2 (rotation at B, C)
No. of restrained degrees of freedom = 7 (rotation at A, D; displacement at A, B, C, D)
Axial deformation in the beam neglected, making [K] 4 × 4 only
E considered uniform throughout

Solution

Details as required by the computer program are given in what follow.

Input

n = 3, l = [5 6 3], I = [2 2 1]
uu = 2, ur = 7

Global Labels

BA = [1 3 7 6]
BC = [1 2 5 8]
CD = [2 4 7 9]

Fixed End Moments

BA = [–25 25–20–20]
BC = [60–60 60 60]
CD = [0 0 0 0]

Joint Load

= [–35 60–25 0–60 20 20–60 0]^T

Output: Final Solution

1. Member end forces and moments

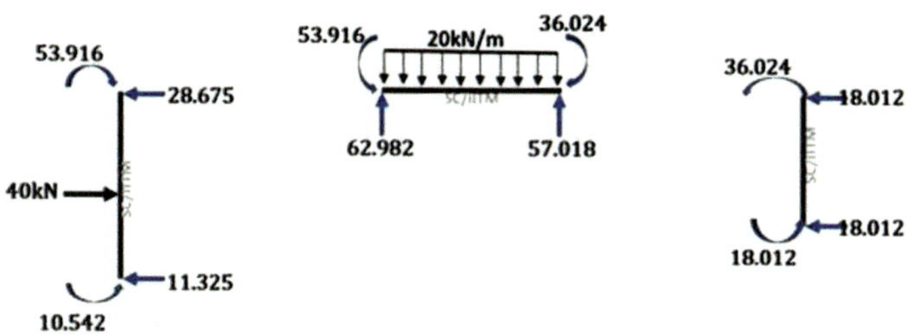

2. Final moments and reactions

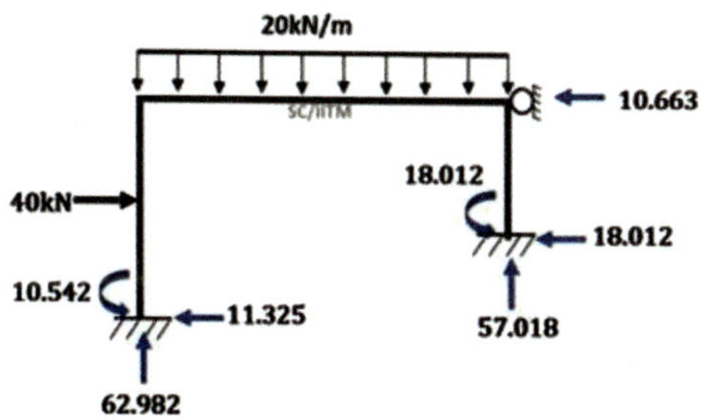

EXERCISE 7

Using the stiffness method of analysis, find the moments and reactions of the frame shown in the figure.

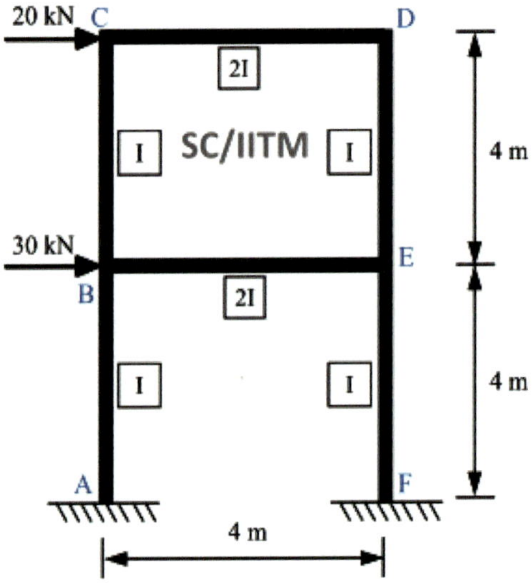

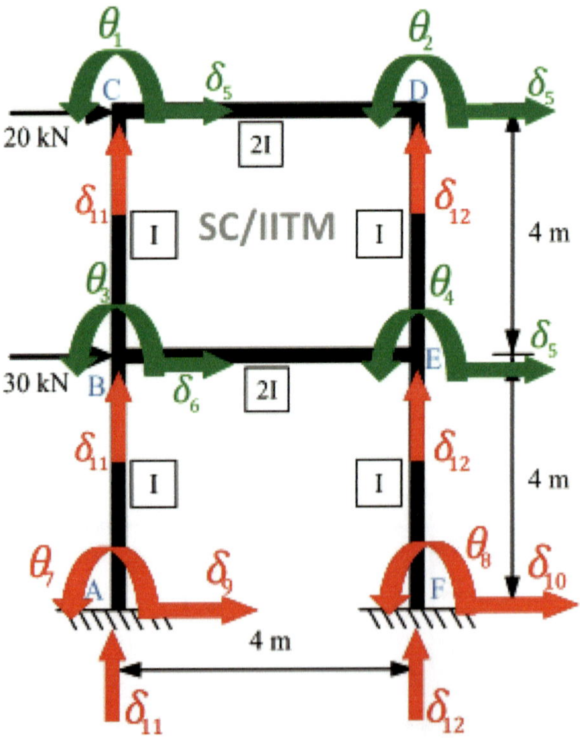

Technical Discussions

No. of members = 6 (member 1 = CD; member 2 = CB; member 3 = DE; member 4 = BE; member 5 = BA; member 6 = EF)
Length of each member = {4, 4, 4, 4, 4, 4}
No. of unrestrained degrees of freedom = 6 (rotation at B, C, D, E; displacement at B, C, D, E)
No. of restrained degrees of freedom = 6 (rotation at A, F; displacement at A, B, C, D, E, F)
Axial deformation in the beam neglected, making $[K]$ 4×4 only

Solution

Details as required by the computer program are given in what follow.

Input

n = 6, l = [4 4 4 4 4 4], I = [2 1 1 2 1 1]
uu = 6, ur = 6

Global Labels

CD = [1 2 11 12]
CB = [1 3 5 6]
DE = [2 4 5 6]
BE = [3 4 11 12]
BA = [3 7 6 9]
EF = [4 8 6 10]

Fixed End Moments

CD = [0 0 0 0]
CB = [0 0 0 0]
DE = [0 0 0 0]
BE = [0 0 0 0]
BA = [0 0 0 0]
EF = [0 0 0 0]

Joint Load

$= [0\ 0\ 0\ 0\ 20\ 30\ 0\ 0\ 0\ 0\ 0\ 0]^T$

Output: Final Solution

1. Member end forces and moments

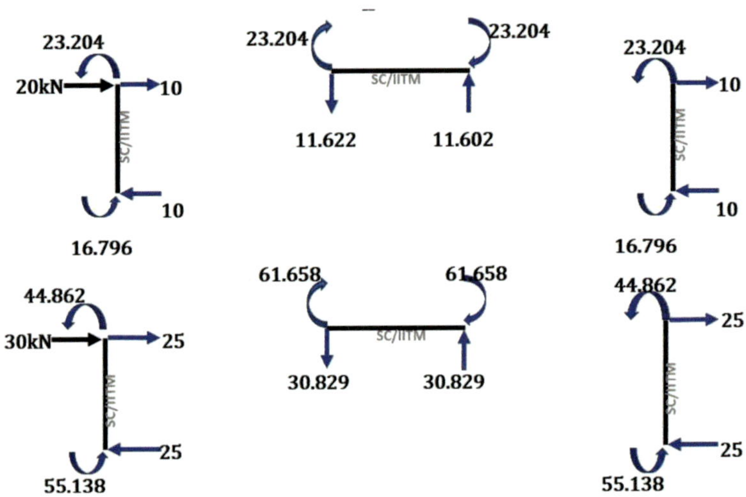

2. Final moments and reactions

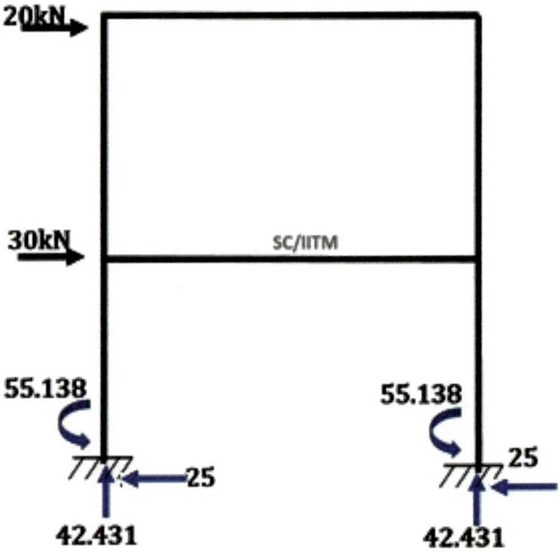

EXERCISE 8

Using the stiffness method of analysis, find the moments and reactions of the continuous beam shown in the figure.

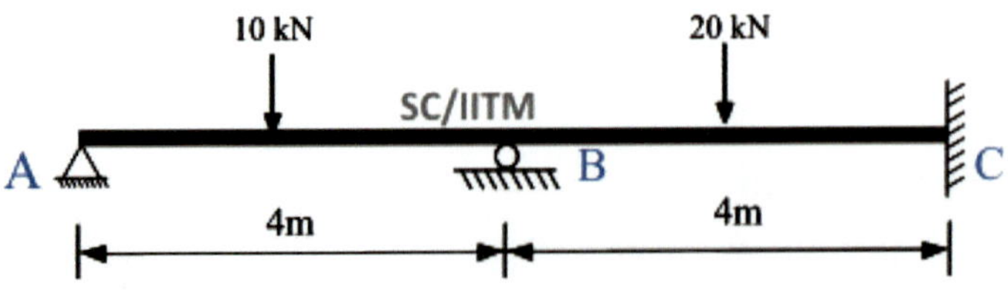

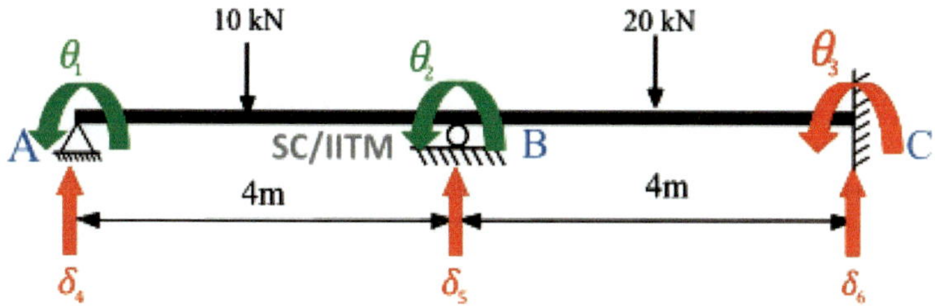

Technical Discussions

No. of members = 2 (member 1 = AB; member 2 = BC)
Length of each member = {4, 4}
No. of unrestrained degrees of freedom = 2 (rotation at A, B)
No. of restrained degrees of freedom = 4 (rotation at C; displacement at A, B, C)
Axial deformation in the beam neglected, making [*K*] 4 × 4 only
EI uniform throughout

Solution

Details as required by the computer program are given in what follow.

Input

n = 2, l = [4 4], I = [1 1]
uu = 2, ur = 4

Global Labels

AB = [1 2 4 5]
BC = [2 3 5 6]

Fixed End Moments

AB = [5–5 5 5]
BC = [10–10 10 10]

Joint Load

= [-5–5 10–5–15–10]^T

Output: Final Solution

1. Member end forces and moments

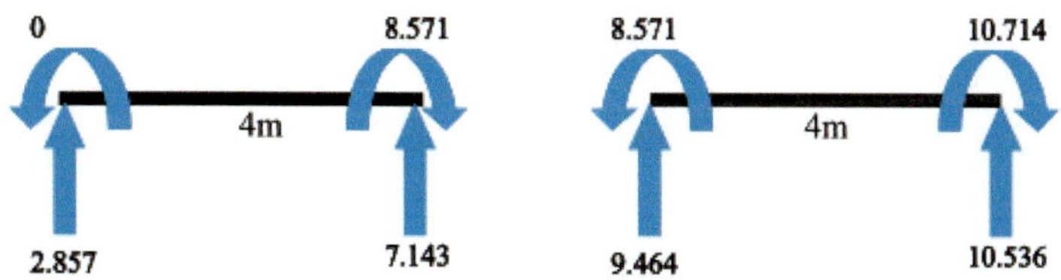

2. Final moments and reactions

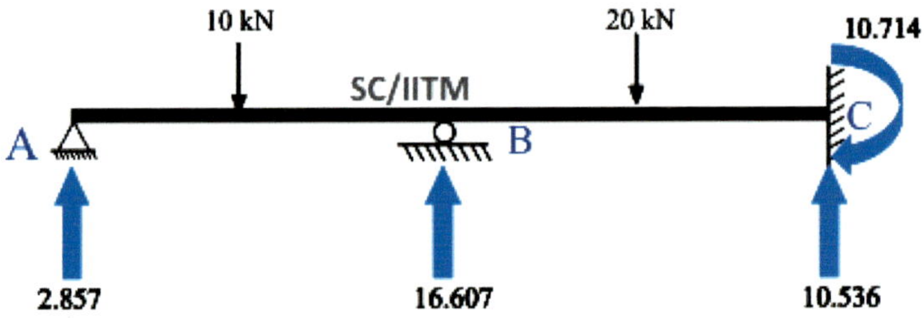

EXERCISE 9

Using the stiffness method of analysis, find the moments and reactions of the continuous beam shown in the figure.

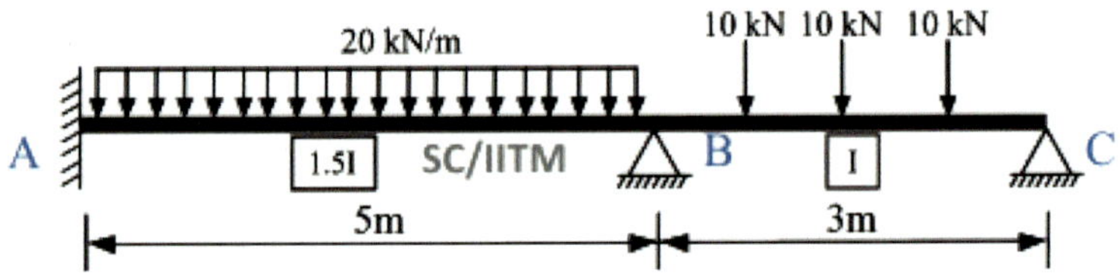

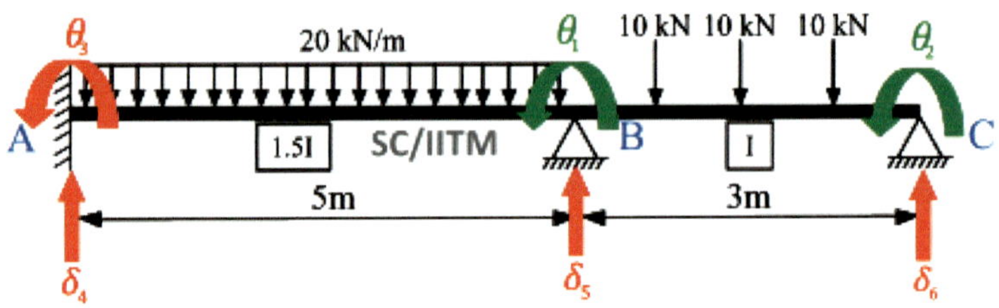

Technical Discussions

No. of members = 2
Member 1 = AB; Member 2 = BC
Length of each member = {5, 3}
No. of unrestrained degrees of freedom = 2 (rotation at B, C)
No. of restrained degrees of freedom = 4 (rotation at A; displacement at A, B, C)
Axial deformation in the beam neglected, making [K] 4 × 4 only

Solution

Details as required by the computer program are given in what follow.

Input

n = 2, l = [5 3], I = [1.5 1]
uu = 2, ur = 4

Global Labels

AB = [3 1 4 5]
BC = [1 2 5 6]

Fixed End Moments

AB = [41.67–41.67 50 50]
BC = [9.375–9.375 15 15]

Joint Load

= [32.295 9.375–41.667–50–65–15]^T

Output

Final Solution

1. ***Member end forces and moments***

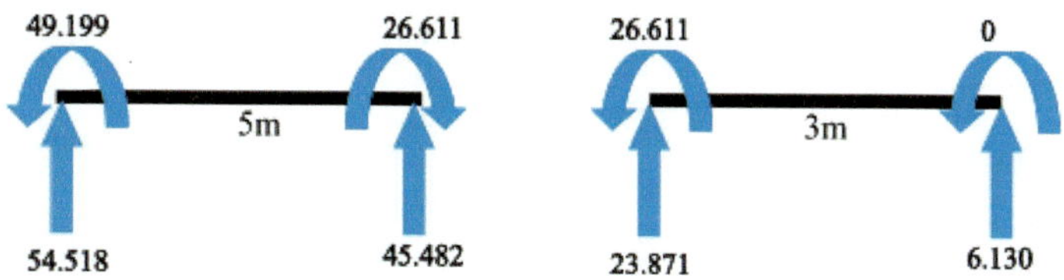

2. Final moments and reactions

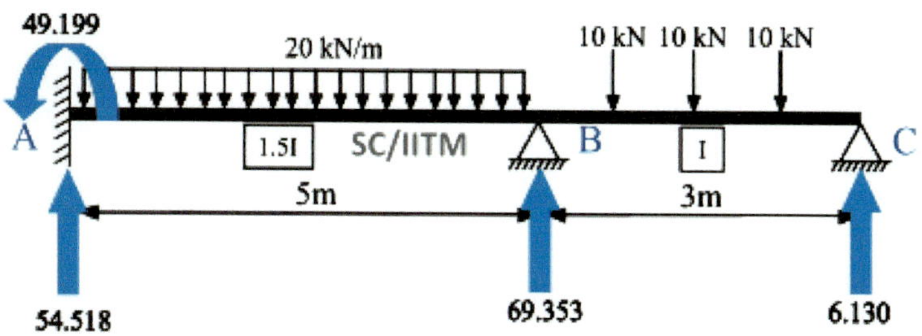

EXERCISE 10

Using the stiffness method of analysis, find the moments and reactions of the continuous beam shown in the figure.

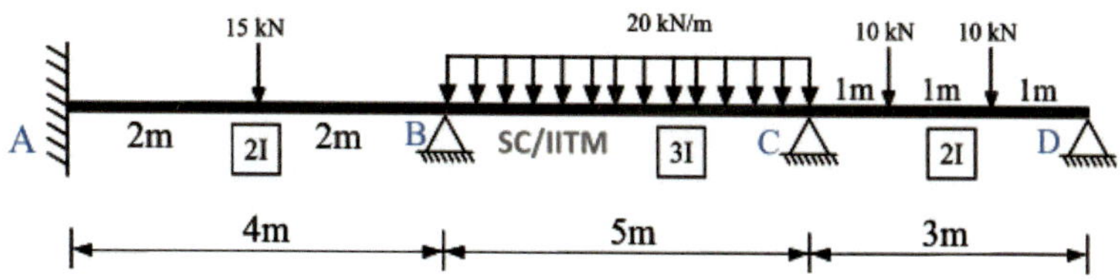

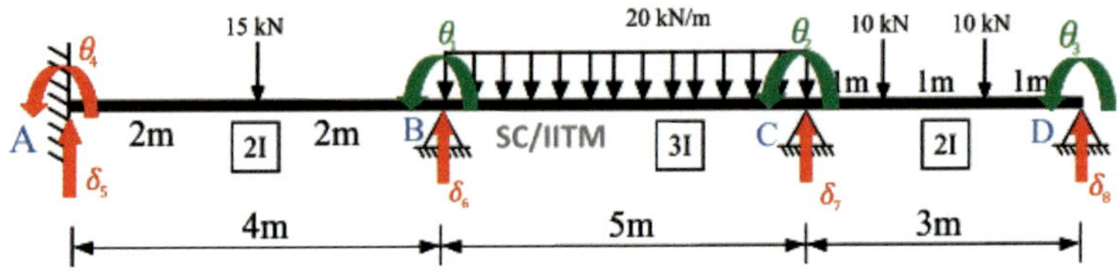

Technical Discussions

No. of members = 3 (member 1 = AB; member 2 = BC; member 3 = CD)
Length of each member = {4, 5, 3}
No. of unrestrained degrees of freedom = 3
No. of restrained degrees of freedom = 5
Axial deformation in the beam neglected, making [*K*] 4 × 4 only

Solution

Details as required by the computer program are given in what follow.

Input

n = 3, l = [4 5 3], I = [2 3 2]
uu = 4, ur = 9

Global Labels

AB = [4 1 5 6]
BC = [1 2 6 7]
CD = [2 3 7 8]

Fixed End Moments

AB = [7.5–7.5 7.5 7.5]
BC = [41.667–41.667 50 50]
CD = [6.667–6.667 10 10]

Joint Load

= [–34.167 35 6.667–7.5–7.5–57.5–60–10]^T

Final Solution

1. Member end forces and moments

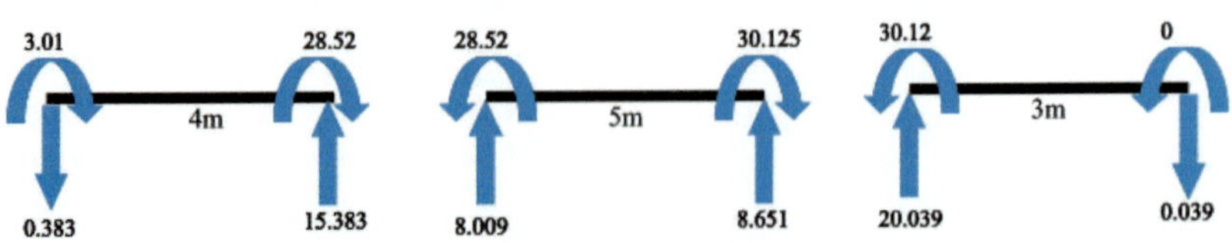

2. Final moments and reactions

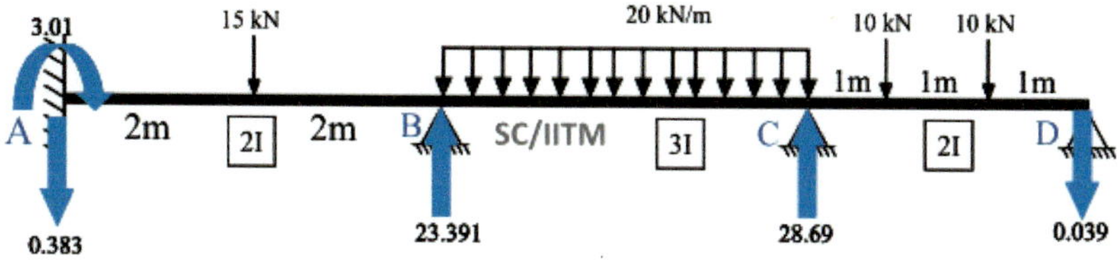

EXERCISE 11

Using the stiffness method of analysis, find the moments and reactions of the continuous beam shown in the figure.

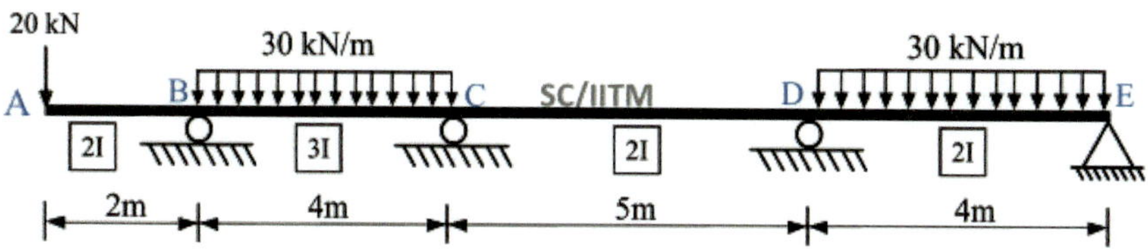

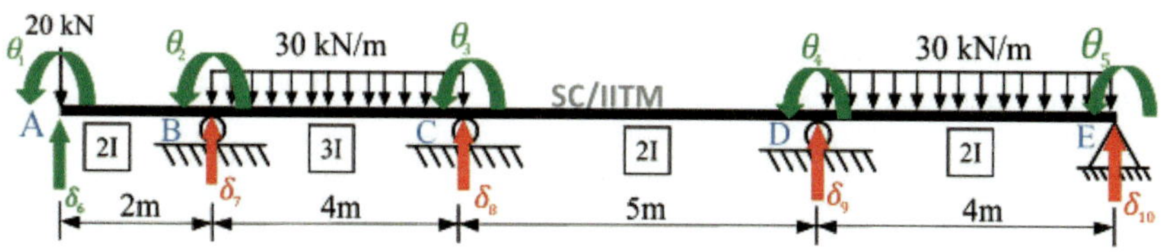

Technical Discussions

No. of members = 4 (member 1 = AB; member 2 = BC; member 3 = CD; member 4 = DE)
Length of each member = {2, 4, 5, 4}
No. of unrestrained degrees of freedom = 6
No. of restrained degrees of freedom = 4
Axial deformation in the beam neglected, making $[K]$ 4×4 only

Solution

Details as required by the computer program are given in what follow.

Input

n = 4, l = [2 4 5 4], I = [2 3 2 2]
uu = 6, ur = 4

Global Labels

AB = [1 2 6 7]
BC = [2 3 7 8]

CD = [3 4 8 9]
DE = [4 5 9 10]

Fixed End Moments

AB = [0 0 20 0]
BC = [26.667–26.667 40 40]
CD = [25–25 20 20]
DE = [40–40 60 60]

Joint Load

= $[0–26.667\ 1.667–15\ 40–20–40–60–80–60]^T$

Final Solution

1. Member end forces and moments

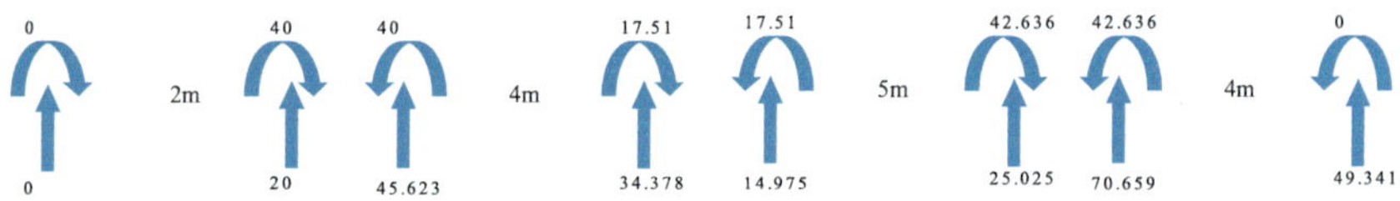

2. Final moments and reactions

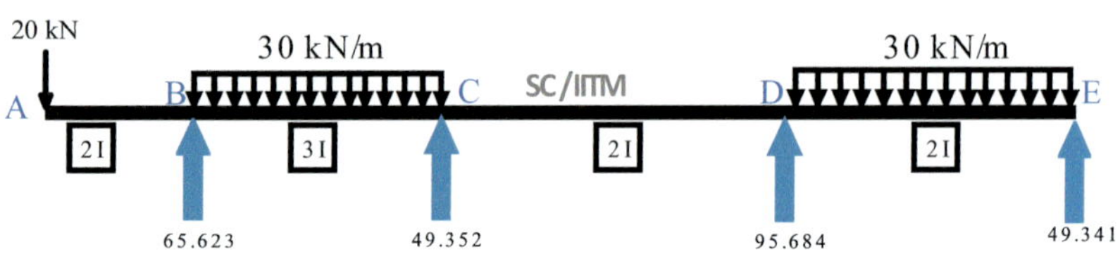

EXERCISE 12

Using the stiffness method of analysis, find the moments and reactions of the continuous beam shown in the figure.

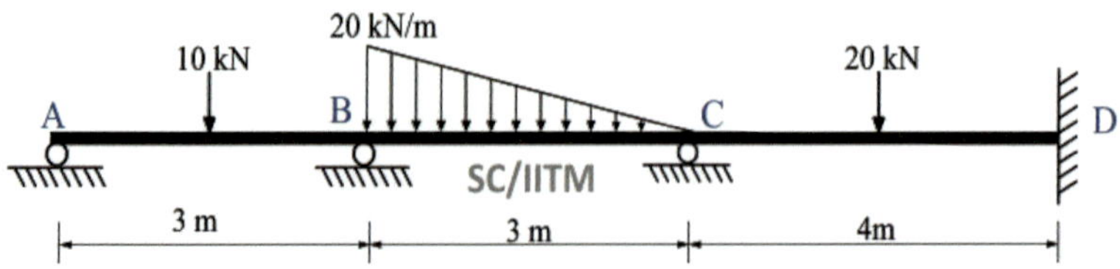

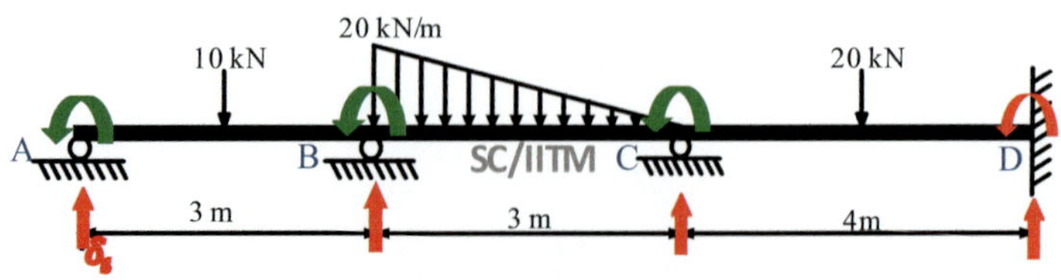

Technical Discussions

No. of members = 3 (member 1 = AB; member 2 = BC; member 3 = CD)
Length of each member = {3, 3, 4}
No. of unrestrained degrees of freedom = 3
No. of restrained degrees of freedom = 5
Axial deformation in the beam neglected, making $[K]$ 4×4 only

Solution

Details as required by the computer program are given in what follow.

Input

n = 3, l = [3 3 4], I = [1 1 1]
uu = 3, ur = 5

Global Labels

AB = [1 2 5 6]
BC = [2 3 6 7]
CD = [3 4 7 8]

Fixed End Moments

AB = [3.75–3.75 5 5]
BC = [9–6 20 10]
CD = [10–10 10 10]

Joint Load

= $[-3.75 -5.25 -4\ 10 -5 -25 -20 -10]^T$

Final Solution

1. Member end forces and moments

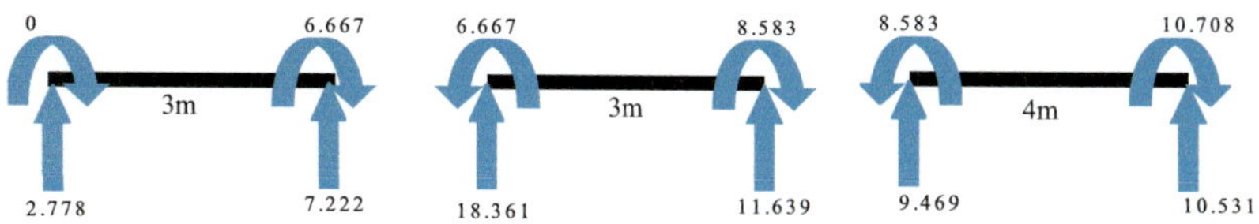

2. Final moments and reactions

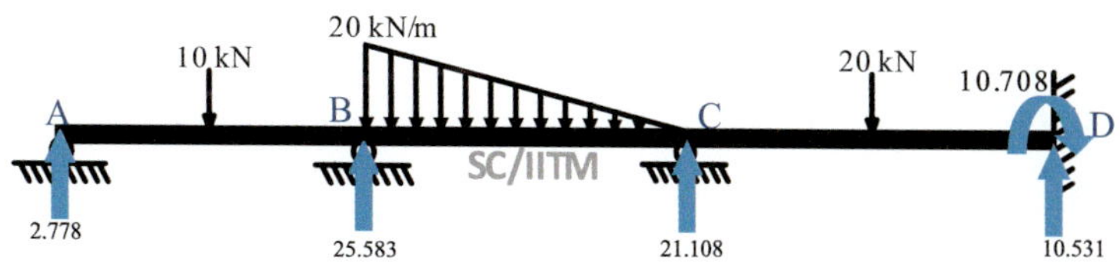

EXERCISE 13

Using the stiffness method of analysis, find the moments and reactions of the continuous beam shown in the figure.

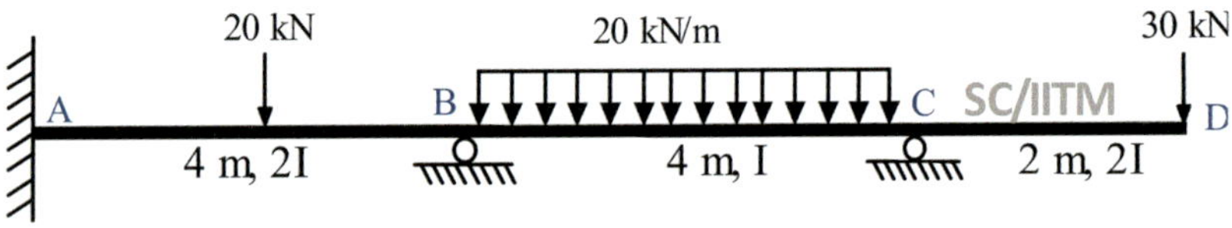

Figure 1.23.1 Continuous beam.

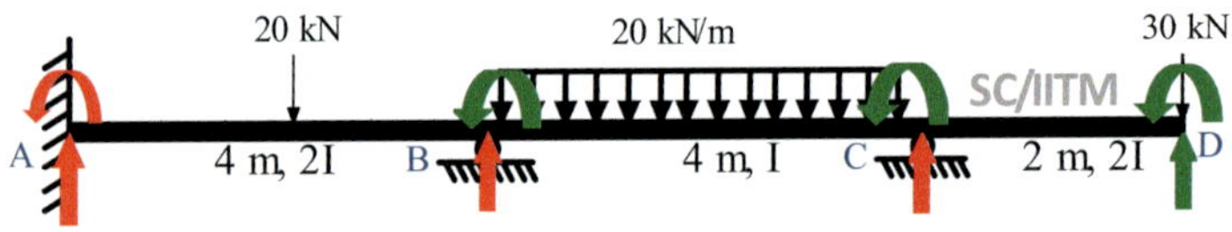

Technical Discussions

No. of members = 3 (member 1 = AB; member 2 = BC; member 3 = CD)
Length of each member = {4, 4, 2}
No. of unrestrained degrees of freedom = 4
No. of restrained degrees of freedom = 4
Axial deformation in the beam neglected, making [*K*] 4 × 4 only

Solution

Details as required by the computer program are given in what follow.

Input

n = 3, l = [4 4 2], I = [2 1 2]
uu = 4, ur = 4

Global Labels

AB = [5 1 6 7]
BC = [1 2 7 8]
CD = [2 3 8 4]

Fixed End Moments

AB = [10–10 10 10]
BC = [33.33–33.33 50 50]
CD = [0 0 0 30]

Joint Load

$= [-23.33\ 33.33\ 0{-}30{-}10{-}10{-}60{-}50]^T$

1. Member end forces and moments

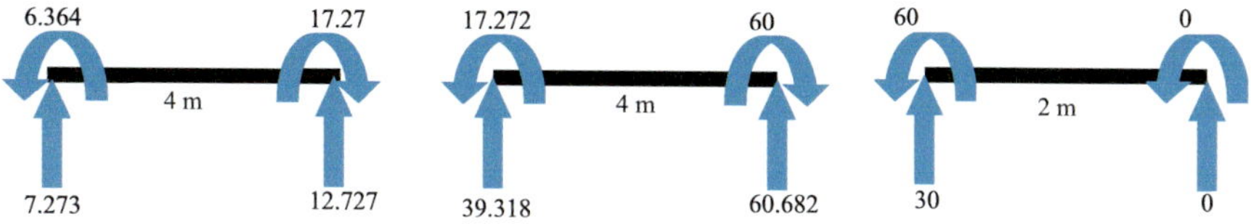

2. Final moments and reactions

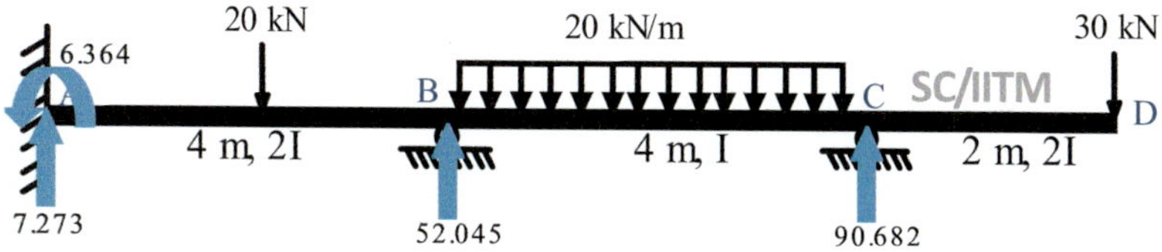

EXERCISE 14

Using the stiffness method of analysis, find the moments and reactions of the truss shown in the figure.

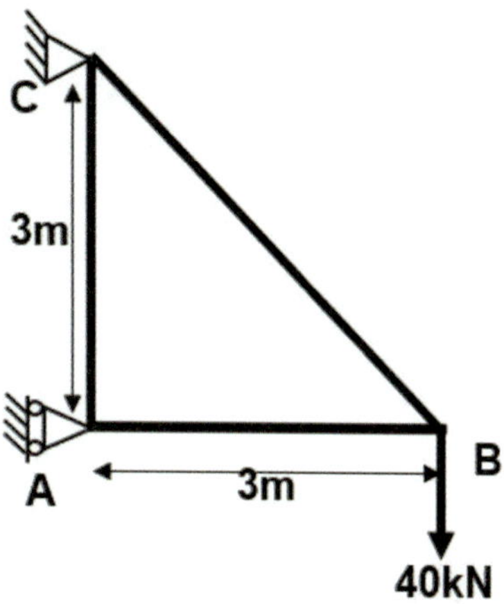

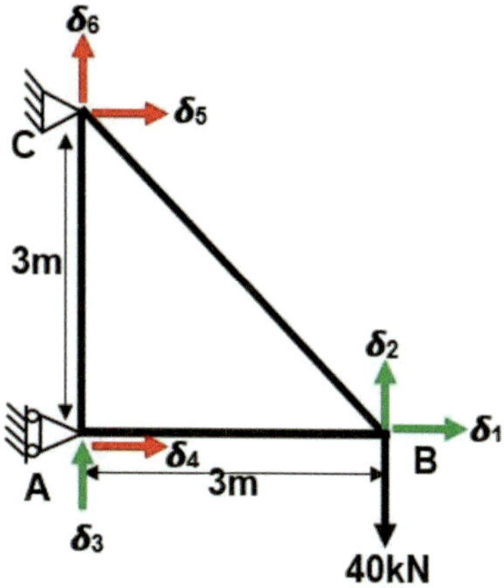

Technical Discussions

No. of members = 3
Length of each member = {3, 3, 4.243}
No. of unrestrained degrees of freedom = 3 (displacement at A, B)
No. of restrained degrees of freedom = 3 (displacement at B, C)

Solution

Details as required by the computer program are given in what follow.

Input

n = 3

Member	*j*th End	*k*th End	Length (m)	Area, *A* (m²)	Inertia (m⁴)	Angle, *θ* (Degrees)	Global Labels
AB	A	B	3	0.08	1	0°	3, 2, 4, 1
BC	B	C	3	0.04	1	90°	3, 6, 4, 5
CA	C	A	4.243	0.06	1	–45°	6, 2, 5, 1

uu = 3, ur = 3

Joint Load

$[0 -40\ 0\ 0\ 0\ 0]^T$

Output: Final Solution

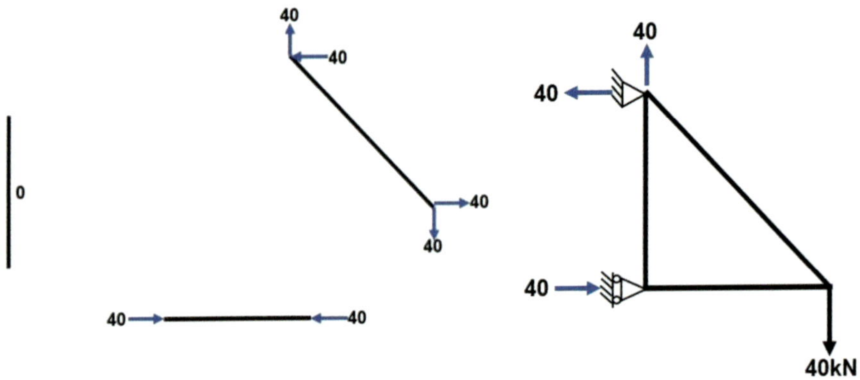

EXERCISE 15

Using the stiffness method of analysis, find the moments and reactions of the truss shown in the figure.

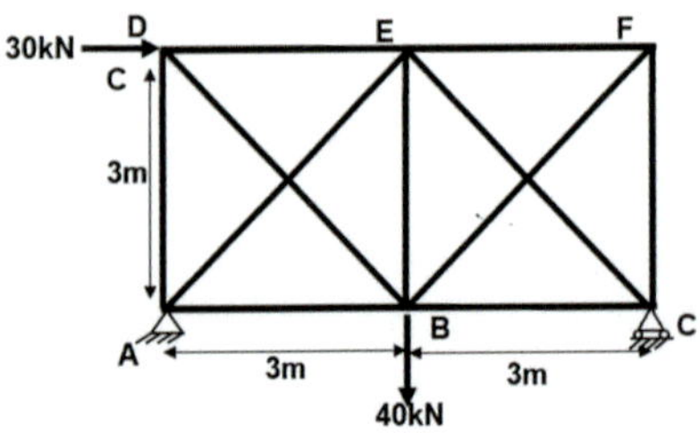

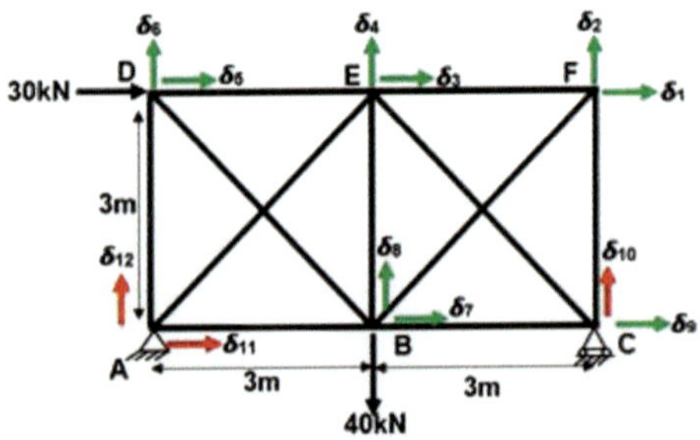

Technical Discussions

No. of members = 11
Length of each member = {3, 3, 3, 3, 3, 3, 3, 4.243, 4.243, 4.243, 4.243}
No. of unrestrained degrees of freedom = 9 (displacement at B, C, D, E, F)
No. of restrained degrees of freedom = 3 (displacement at A, C)

Solution

Details as required by the computer program are given in what follow.

Input

n = 11

Member	*j*th End	*k*th End	Length (m)	Area, *A* (m^2)	Inertia (m^4)	Angle, θ (Degrees)	Global Labels
AB	A	B	3	1	1	0°	12, 8, 11, 7
BC	B	C	3	1	1	0°	8, 10, 7, 9
DE	D	E	3	1	1	0°	6, 4, 5, 3
EF	E	F	3	1	1	0°	4, 2, 3, 1
AD	A	D	3	1	1	90°	12, 6, 11, 5
BE	B	E	3	1	1	90°	8, 4, 7, 3
CF	C	F	3	1	1	90°	10, 2, 9, 1
DB	D	B	4.243	1	1	–45°	6, 8, 5, 7
AE	A	E	4.243	1	1	45°	12, 4, 11, 3
EC	E	C	4.243	1	1	–45°	4, 10, 3, 9
BF	B	F	4.243	1	1	45°	8, 2, 7, 1

uu = 9, ur = 3

Joint Load

$[0\ 0\ 0\ 0\ 30\ 0\ 0 -40\ 0\ 0\ 0\ 0]^T$

Output: Final Solution

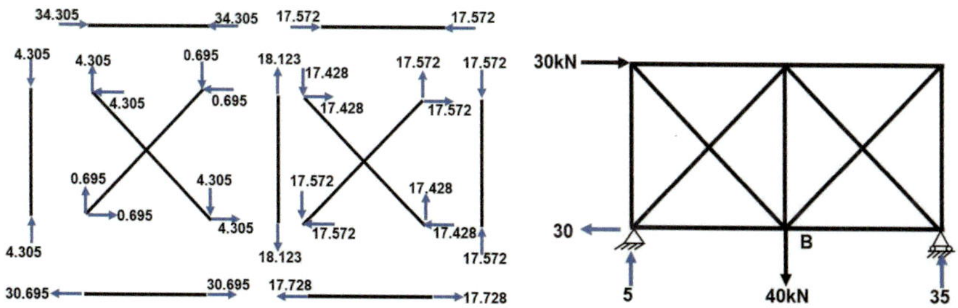

EXERCISE 16

Using the stiffness method of analysis, find the moments and reactions of the truss shown in the figure.

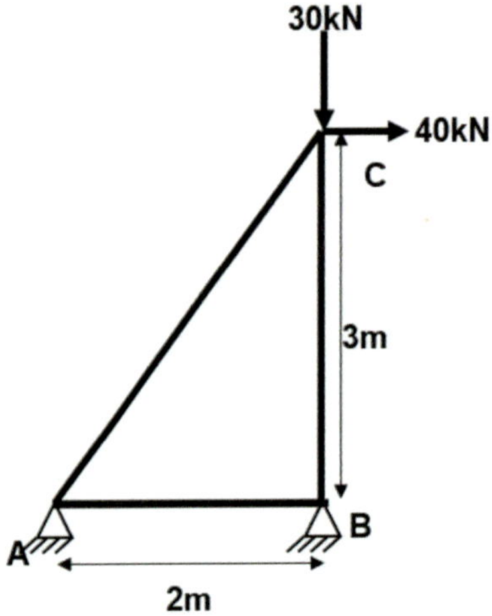

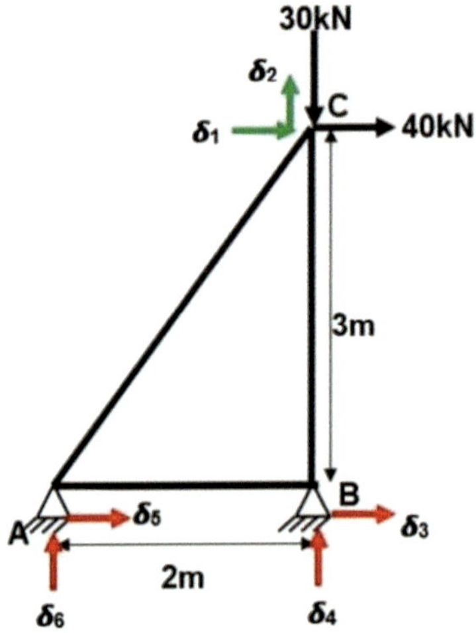

Technical Discussions

No. of members = 3
Length of each member = {2, 3.606, 3}
No. of unrestrained degrees of freedom = 2 (displacement at C)
No. of restrained degrees of freedom = 4 (displacement at A, B)

Solution

Details as required by the computer program are given in what follow.

Input

n = 3

Member	*j*th End	*k*th End	Length (m)	Area, *A* (m^2)	Inertia (m^4)	Angle, θ (Degrees)	Global Labels
AB	A	B	2	1	1	0°	6, 4, 5, 3
BC	B	C	3.606	1	1	56.31°	6, 2, 5, 1
CA	C	A	3	1	1	90°	4, 2, 3, 1

uu = 2, ur = 4

Joint Load

$[40–30\ 0\ 0\ 0\ 0]^T$

Output: Final Solution

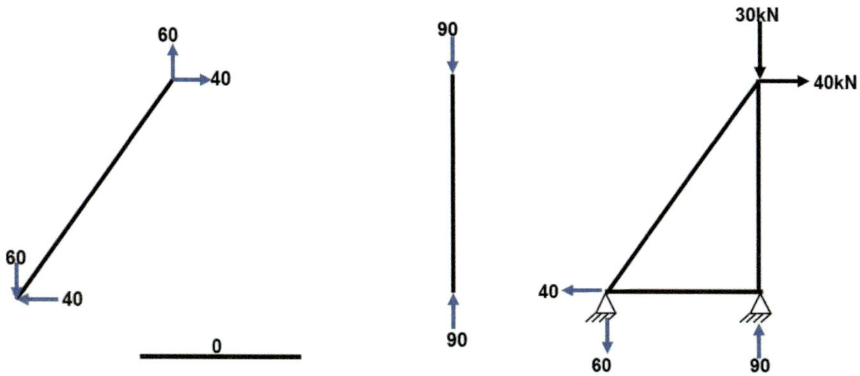

EXERCISE 17

Using the stiffness method of analysis, find the moments and reactions of the truss shown in the figure.

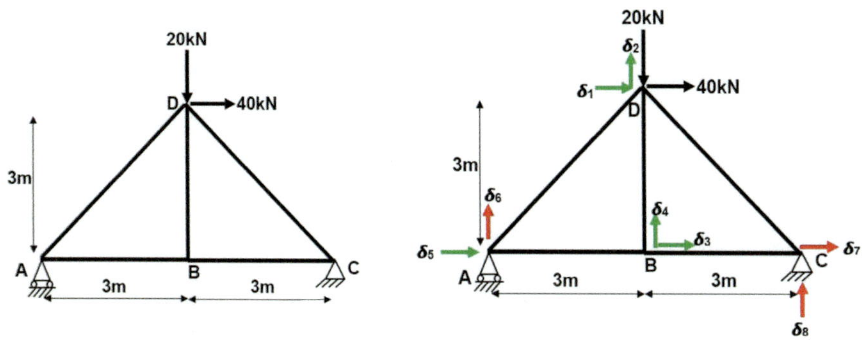

Technical Discussions

No. of members = 5
Length of each member = {3, 3, 3, 4.243, 4.243}
No. of unrestrained degrees of freedom = 5 (displacement at A, B, D)
No. of restrained degrees of freedom = 3 (displacement at A, C)

Solution

Details as required by the computer program are given in what follow.

Input

n = 5

Member	*j*th End	*k*th End	Length (m)	Area, *A* (m²)	Inertia (m⁴)	Angle, *θ* (Degrees)	Global Labels
AB	A	B	3	1	1	0°	6, 4, 5, 3
BC	B	C	3	1	1	0°	4, 8, 3, 7
BD	B	D	3	1	1	90°	4, 2, 3, 1
AD	A	D	4.243	1	1	45°	6, 2, 5, 1
DC	D	C	4.243	1	1	−45°	2, 8, 1, 7

uu = 5, ur = 3

Joint Load

$[40 -20\ 0\ 0\ 0\ 0\ 0\ 0]^T$

Final Solution

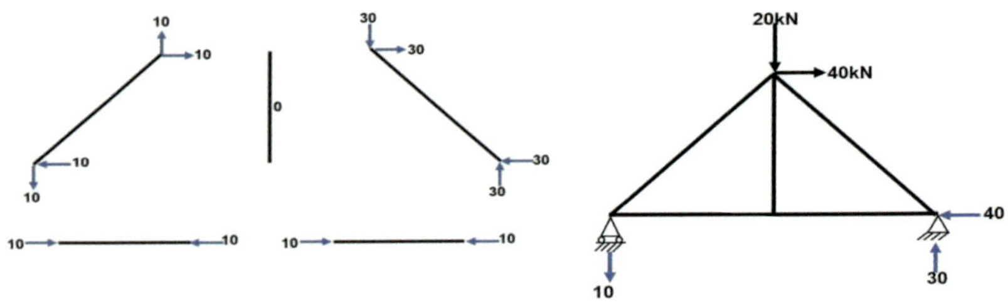

EXERCISE 18

Using the stiffness method of analysis, find the moments and reactions of the truss shown in the following figure.

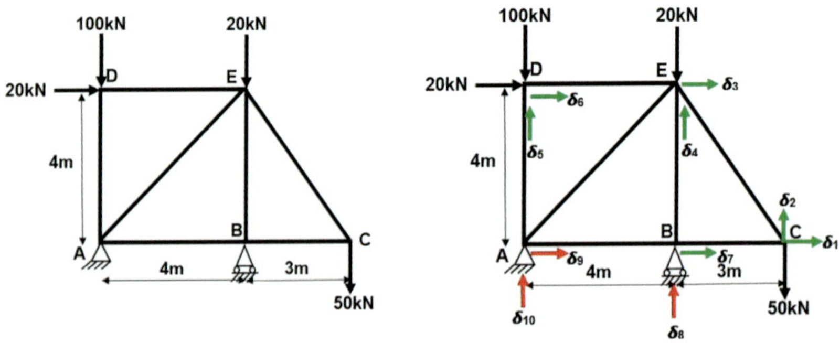

Technical Discussions

No. of members = 8
Length of each member = {4, 3, 4, 4, 4, 5.657, 5.657, 5}
No. of unrestrained degrees of freedom = 7 (displacement at B, C, D, E)
No. of restrained degrees of freedom = 3 (displacement at A, B)

Solution

Details as required by the computer program are given in what follow.

Input

n = 8

Member	*j*th End	*k*th End	Length (m)	Area, *A* (m²)	Inertia (m⁴)	Angle, θ (Degrees)	Global Labels
AB	A	B	4	0.06	1	0°	10, 8, 9, 7
BC	B	C	3	0.06	1	0°	8, 2, 7, 1
DE	D	E	4	0.06	1	0°	6, 4, 5, 3
AD	A	D	4	0.06	1	90°	10, 6, 9, 5
BE	B	E	4	0.06	1	90°	8, 4, 7, 3
DB	D	B	5.657	0.06	1	−45°	6, 8, 5, 7
AE	A	E	5.657	0.06	1	45°	10, 4, 9, 3
EC	E	C	5	0.06	1	−53.13°	4, 2, 3, 1

uu = 7, ur = 3

Joint Load

$[0 -50\ 0 -50\ 20 -100\ 0\ 0\ 0\ 0]^T$

Output

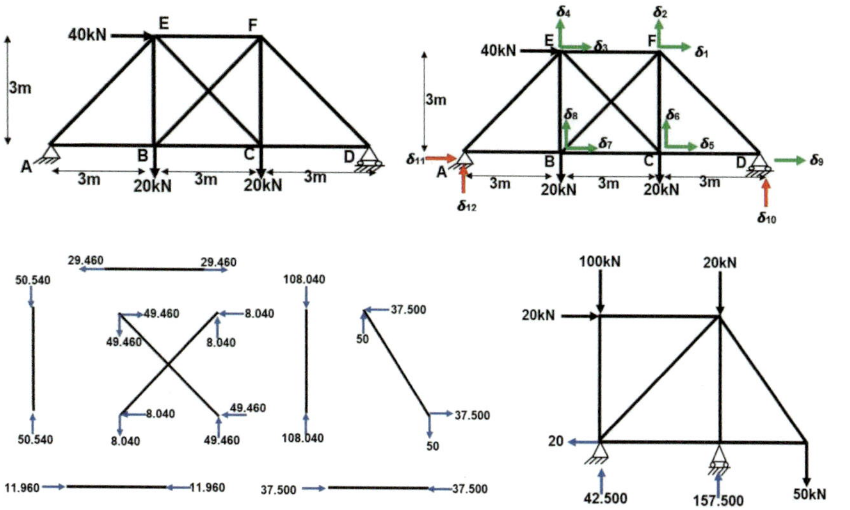

EXERCISE 19

Using the stiffness method of analysis, find the moments and reactions of the truss shown in the figure.

Technical Discussions

No. of members = 10
Length of each member = {3, 3, 3, 3, 3, 3, 4.243, 4.243, 4.243, 4.243}
No. of unrestrained degrees of freedom = 9 (displacement at B, C, D, E, F)
No. of restrained degrees of freedom = 3 (displacement at A, D)

Solution

Details as required by the computer program are given in what follow.

Input

n = 10

Member	*j*th End	*k*th End	Length (m)	Area, *A* (m^2)	Inertia (m^4)	Angle, θ (Degrees)	Global Labels
AB	A	B	3	1	1	0°	12, 8, 11, 7
BC	B	C	3	1	1	0°	8, 6, 7, 5
CD	C	D	3	1	1	0°	6, 10, 5, 9
EF	E	F	3	1	1	0°	4, 2, 3, 1
BE	B	E	3	1	1	90°	8, 4, 7, 3
CF	C	F	3	1	1	90°	6, 2, 5, 1
AE	A	E	4.243	1	1	45°	12, 4, 11, 3
EC	E	C	4.243	1	1	–45°	4, 6, 3, 5
BF	B	F	4.243	1	1	45°	8, 2, 7, 1
FD	F	D	4.243	1	1	–45°	2, 10, 1, 9

uu = 9, ur = 3

Joint Load

$[0\ 0\ 40\ 0\ 0 -20\ 0 -20\ 0\ 0\ 0\ 0]^T$

Output: Final Solution

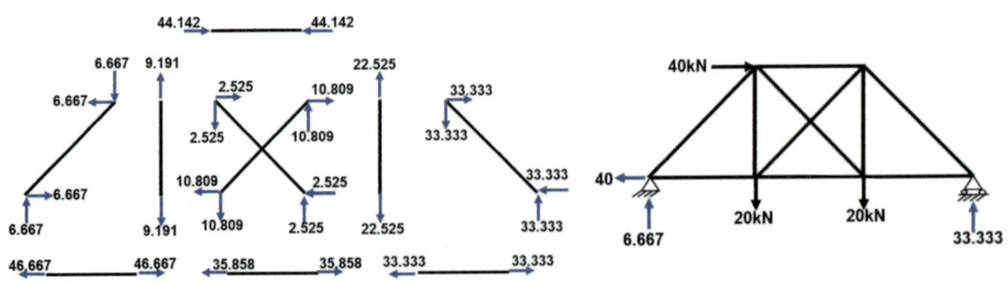

EXERCISE 20

Using the stiffness method of analysis, find the moments and reactions of the truss shown in the figure.

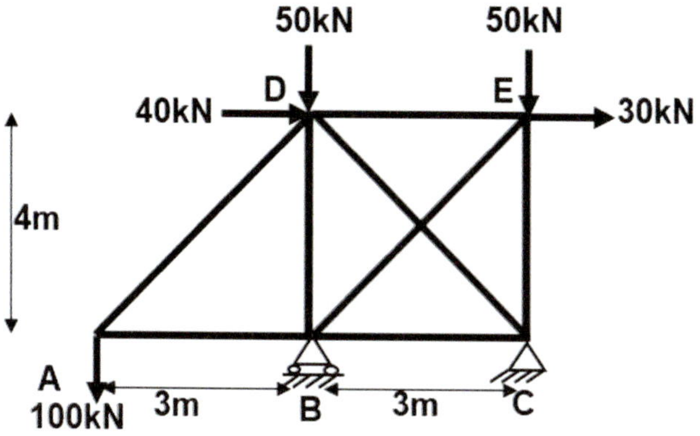

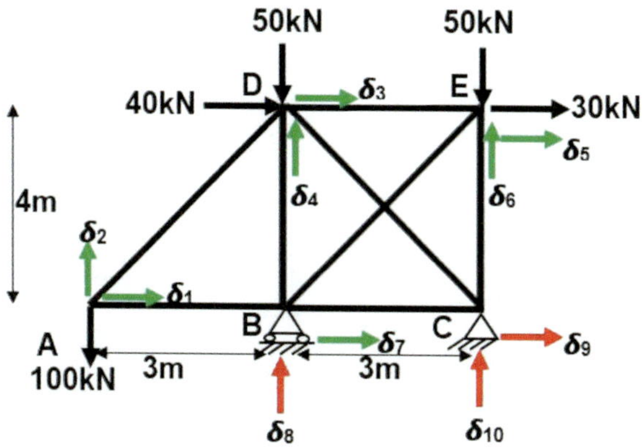

Technical Discussions

No. of members = 8
Length of each member = {4, 4, 4, 5, 5, 6.403, 6.403, 6.403}
No. of unrestrained degrees of freedom = 7 (displacement at A, B, D, E)
No. of restrained degrees of freedom = 3 (displacement at B, C)

Solution

Details as required by the computer program are given in what follow.

Input

n = 8

Member	*j*th End	*k*th End	Length (m)	Area, *A* (m²)	Inertia (m⁴)	Angle, θ (Degrees)	Global Labels
AB	A	B	4	0.06	1	0°	2, 8, 1, 7
BC	B	C	4	0.06	1	0°	8, 10, 7, 9
DE	D	E	4	0.06	1	0°	4, 6, 3, 5

(*Continued*)

Member	*j*th End	*k*th End	Length (m)	Area, *A* (m^2)	Inertia (m^4)	Angle, θ (Degrees)	Global Labels
BD	B	D	5	0.06	1	90°	8, 4, 7, 3
CE	C	E	5	0.06	1	90°	10, 6, 9, 5
AD	A	D	6.403	0.06	1	51.340°	2, 4, 1, 3
DC	D	C	6.403	0.06	1	–51.340°	4, 10, 3, 9
BE	B	E	6.403	0.06	1	51.340°	8, 6, 7, 5

uu = 7, ur = 3

Joint Load

$[0 -100\ 0 -50\ 30 -50\ 0\ 0\ 0\ 0]^T$

Output: Final Solution

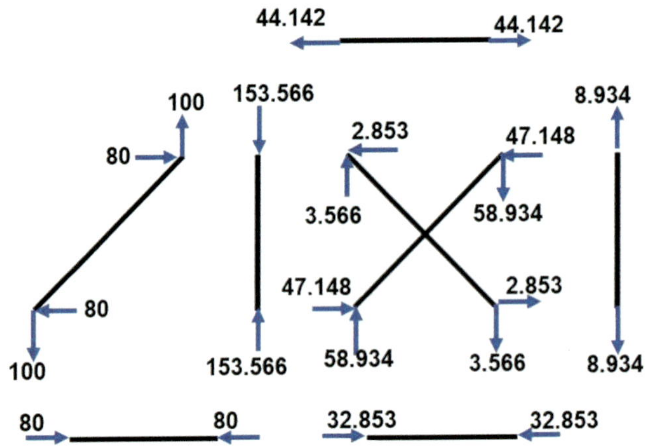

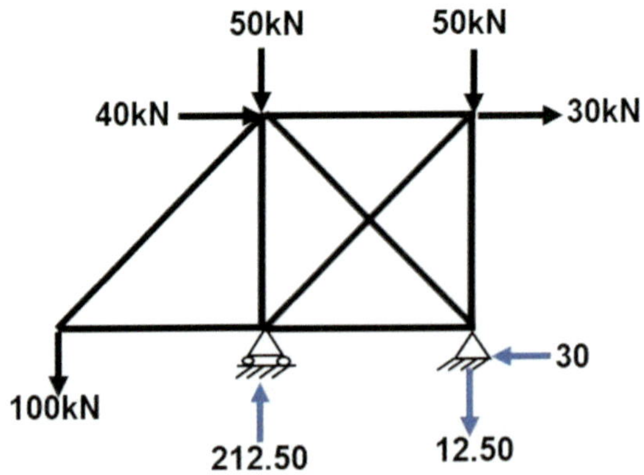

EXERCISE 21

Using the stiffness method of analysis, find the moments and reactions of the truss shown in the figure.

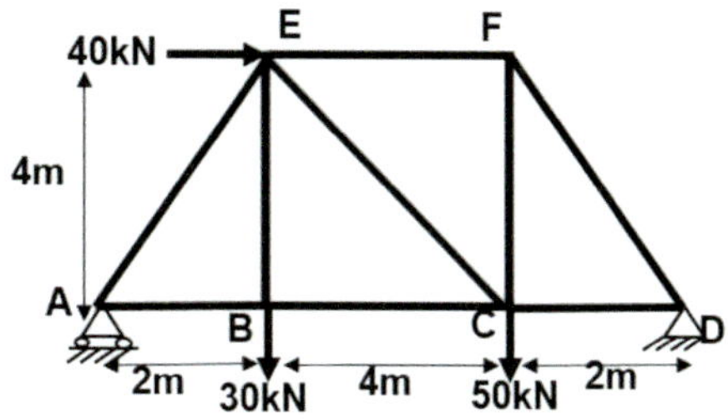

Technical Discussions

No. of members = 9
Length of each member = {2, 4, 2, 4, 4, 4.472, 5.657, 4.472, 4}
No. of unrestrained degrees of freedom = 9 (displacement at A, B, C, E, F)
No. of restrained degrees of freedom = 3 (displacement at A, D)

Solution

Details as required by the computer program are given in what follow.

Input

n = 9, l = [2 4 2 4 4 4.472 5.657 4.472 4]
A = [1 1 1 1 1 1 1 1 1]
θ = [0 0 0 90 90 63.435 – 45 – 63.435 0]

Member	*j*th End	*k*th End	Length (m)	Area, *A* (m^2)	Inertia (m^4)	Angle, θ (Degrees)	Global Labels
AB	A	B	2	1	1	0°	10, 8, 9, 7
BC	B	C	4	1	1	0°	8, 6, 7, 5
CD	C	D	2	1	1	0°	6, 12, 5, 11
BE	B	E	4	1	1	90°	8, 4, 7, 3
CF	C	F	4	1	1	90°	6, 2, 5, 1
AE	A	E	4.472	1	1	63.435°	10, 4, 9, 3
EC	E	C	5.657	1	1	–45°	4, 6, 3, 5
FD	F	D	4.472	1	1	–63.435°	2, 12, 1, 11
EF	E	F	4	1	1	0°	4, 2, 3, 1

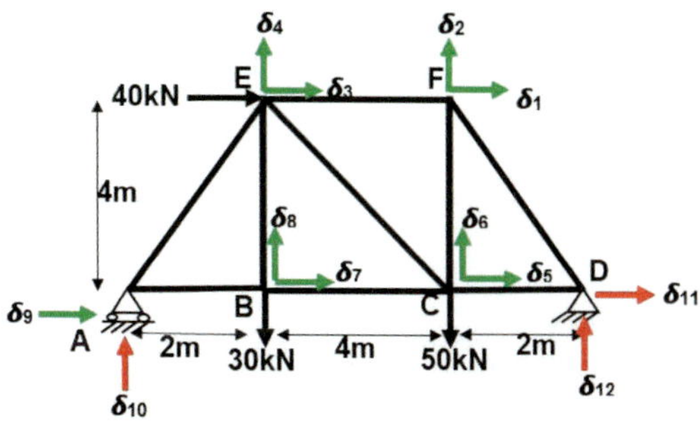

uu = 9, ur = 3

Joint Load

[0 0 40 0 0–50 0–30 0 0 0 0]^T

Output: Final Solution

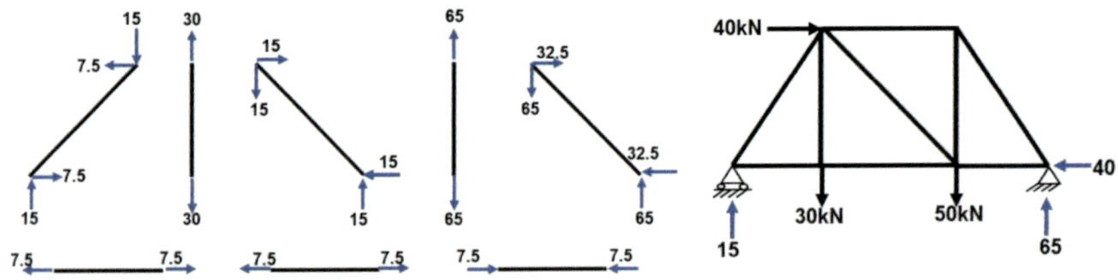

EXERCISE 22

Using the stiffness method of analysis, find the moments and reactions of the truss shown in the figure.

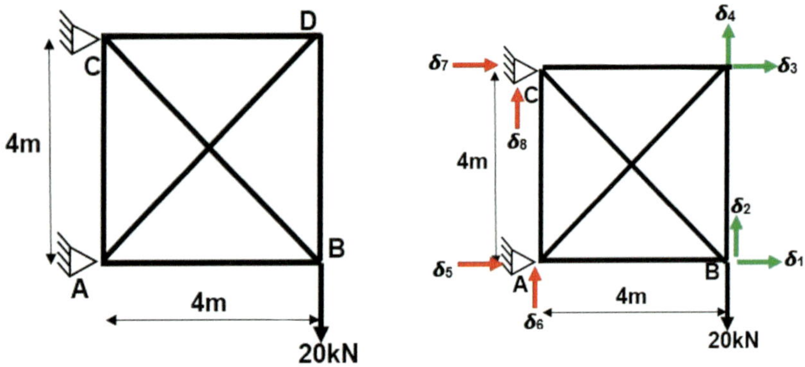

Technical Discussions

No. of members = 6
Length of each member = {4, 4, 4, 4, 5.657, 5.657}
No. of unrestrained degrees of freedom = 4 (displacement at B, D)
No. of restrained degrees of freedom = 4 (displacement at A, C)

Solution

Details as required by the computer program are given in what follow.

Input

n = 6, l = [4 4 4 4 5.657 5.657]
A = [0.05 0.05 0.05 0.05 0.05]
$\theta = [0\ 90\ 0\ 90 + 45 - 45]$

Member	*j*th End	*k*th End	Length (m)	Area, *A* (m^2)	Inertia (m^4)	Angle, θ (Degrees)	Global Labels
AB	A	B	4	0.05	1	0°	6, 2, 5, 1
BD	B	D	4	0.05	1	90°	2, 4, 1, 3
CD	C	D	4	0.05	1	0°	8, 4, 7, 3
AC	A	C	4	0.05	1	90°	6, 8, 5, 7
AD	A	D	5.657	0.05	1	45°	6, 4, 5, 3
CB	C	B	5.657	0.05	1	–45°	8, 2, 7, 1

uu = 4, ur = 4

Joint Load

$[0 -20\ 0\ 0\ 0\ 0\ 0\ 0]^T$

Final Solution

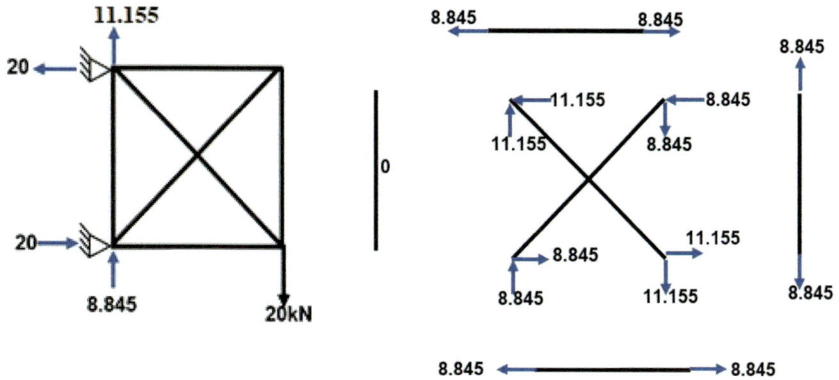

5 Design of Members and Connections

5.1 INTRODUCTION

The design of buildings consists of two parts, namely, the functional design and the structural design. While *functional design* deals with the planning of space, vertical and horizontal zoning, and lighting, *structural design* deals with arriving at the member dimensions to disburse the loads safely. The architectural concepts in design are referred to as the 3Fs, namely, feasibility, function, and form. Reinforced cement concrete (RCC) and steel are the most common construction materials, while the latter is mostly preferred for heavy industrial and tall buildings. A few common steel structures are roof brumes for factories, cinema halls, auditoriums, industrial structures, railway bridges, transmission towers, and overhead tanks and chimneys. Steel, as a construction material, possesses many advantages:

- It has a higher density (7,850 kg/m^3) and strength-to-unit-mass ratio. Hence, even for large structures, steel structural elements will become slender, saving space and improving the aesthetic view.
- It has a high strength-to-weight ratio, which means that it possesses high strength with lesser weight. Therefore, the cost of construction per unit weight is lesser, but the overall cost of steel construction is higher than for RCC.
- It has a lower life cycle cost. As rapid construction saves time and enables the building to be ready for use early, the return on investment (ROI) is higher.
- Increased ductility and toughness help resist accidental loads, earthquake loads, and blast loads, safely.
- Convenience to recycle makes it eco-friendly.
- Easy to repair and retrofit.
- Assured quality control and durability.
- Material is reusable.

A few demerits are:

1. It is susceptible to corrosion.
2. Maintenance cost is higher, and it needs periodic painting to protect it from corrosion.
3. Steel sections are costly.
4. High periodic maintenance cost.
5. High cost of dismantling the structure.
6. Less salvage value.

DOI: 10.1201/9781003581314-5

The modulus of elasticity of steel and concrete is $\left(2\times10^{5},5000\sqrt{f_{ck}}\right)$, respectively. For example, the M_{25} mix has E as 2.5×10^4. Furthermore, the design strengths of steel and concrete are $\left(\frac{250}{1.1}=227.3,\frac{25}{1.1}=16.5\right)$. This indicates that steel is one order stronger than concrete.

5.1.1 Critical Advantages of Steel

It possesses a few critical advantages: Its high strength to unit weight ensures lesser quantity required to cater to higher load, but the cost is relatively higher. Prefabrication is possible, enabling fast construction in comparison to RCC. The speed of RCC construction is mainly due to the restriction in the lift of RCC column construction (not more than 1.5 in one go), but steel has no such limitations. Steel is a highly ductile material, extending sufficient warning before failure. It also possesses reserve energy due to its high ductility ratio. The highest scrap value is an added advantage. For example, the cost of steel scrap is about 50,000/ton, while RCC is only 20,000/ton. It is also important to note the scrap value of the latter is due to reinforcement in RCC. Hence, there is about a 40% recovery of cost. The constituent alloying elements, such as manganese, sulfur, phosphorus, chrome, nickel, and copper, impart special properties to steel, namely, weldability, ductility, and malleability. An increase in the percentage of carbon and manganese helps increase the tensile strength but also causes a decrease in weldability. An increase in the percentage content of sulfur and phosphorous by more than 0.06% makes steel brittle and decreases its weldability. Chrome and nickel impart corrosion resistance to steel and increase its resistance to high temperatures, while copper addition increases corrosion resistance.

5.2 MECHANICAL PROPERTIES

Steel used for structural purposes is termed *structural steel* and classified as mild steel and high-tensile steel. They are manufactured as rolled steel sections in different shapes:

- **L**—Angle ISA
- **T**—Tee ISNT
- **I**—ISJB (Junior Beam)
- **I**—ISLB (Light Beam)
- **I**—ISMB (Medium Beam)
- **I**—ISWB (Wide Flange Beam)
- **H**—ISHB (Heavy Beam)

For example, ISA 70 × 90 indicates length and width as 70 and 90 mm, respectively. A steel section is designated by two characteristics: its shape and size and weight per unit run. For example, the Indian Standard Mild Beam is designated as ISMB 200 @ 25.4 kg/m, indicating its overall depth is 200 mm and unit weight is 25.4 kg. Figure 5.1 shows the overall shape and size of a rolled steel beam. The

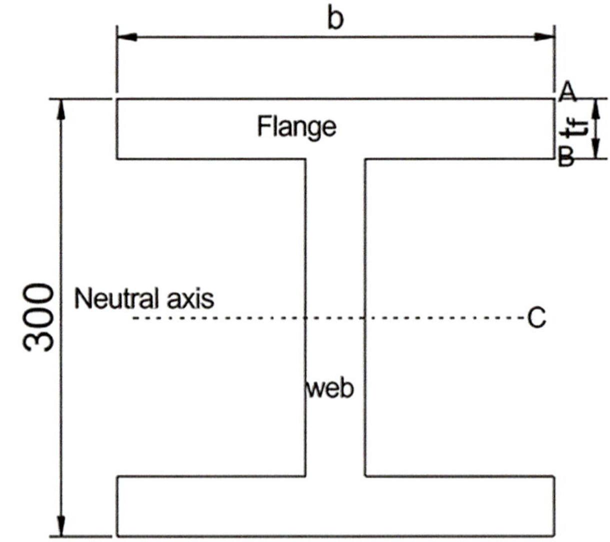

FIGURE 5.1 Cross section of a typical rolled steel beam

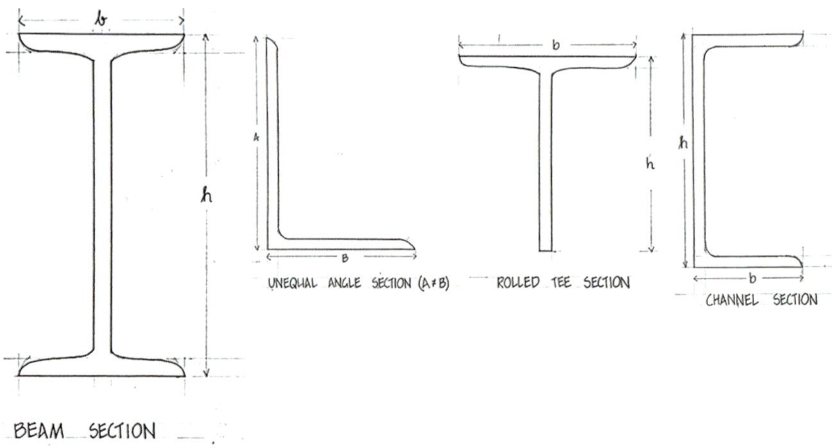

FIGURE 5.2 Rolled steel sections.

structural properties of the section can be seen in the steel tables (SP 6(1): Steel Table, BIS). Different shapes of rolled sections are shown in Figure 5.2

Properties shown in the steel table are extracted as follows:

- Overall depth $(d) = 300$ mm
- Width of flange $(b_f) = 140$ mm
- Thickness of flange $(t_f) = 12.4$ mm

- Thickness of web $(t_w) = 5.7$ mm
- Sectional area $= 56.26 \times 10^2\,mm^4$
- Radius of gyration $(r_x) = \sqrt{}(I_{xx}/A) = 123.4$ mm
- Radius of gyration $(r_y) = \sqrt{}(I_{yy}/A) = 23.4$ mm
- Moment of inertia $(I_{xx}) = 8603.6\ cm^4$
- Moment of inertia $(I_{yy}) = 453.0\ cm^4$

As the MoI about the YY-axis is lesser than that of the XX-axis, the latter is called the major axis. Hence, it is preferred to place the beam such that the bending axis becomes the major axis. For built-up beams and rolled steel sections, the following equations can be used to calculate the MoI about the respective axes.

$$\mathbf{Ixx} = \left(bd^3/12\right) + Ah^2$$

$$\mathbf{Iyy} = \left(d^3b/12\right) + bh^2$$

Other main structural properties are summarized here:

1. Unit weight $= 7850\ kg/m^3 = 7850\ N/m^3$
2. Modulus of elasticity $(E) = 2.1 \times 10^5\ N/mm^2 = 200$ GPa
3. Poisson's ratio $(\mu) = 0.3$
4. Modulus of rigidity $(E) = 0.769 \times 10^5\ N/mm^2$
5. Coefficient of thermal expansion $\alpha = 12 \times 10^{-6}/0°c$

Figure 5.3 shows the stress–strain curve of typical structural steel. The salient points of design interest are marked in the figure. The ratio of strain at the ultimate point to strain at the yield point is known as the ductility ratio. The ratio of ultimate stress to yield stress is the strength ratio.

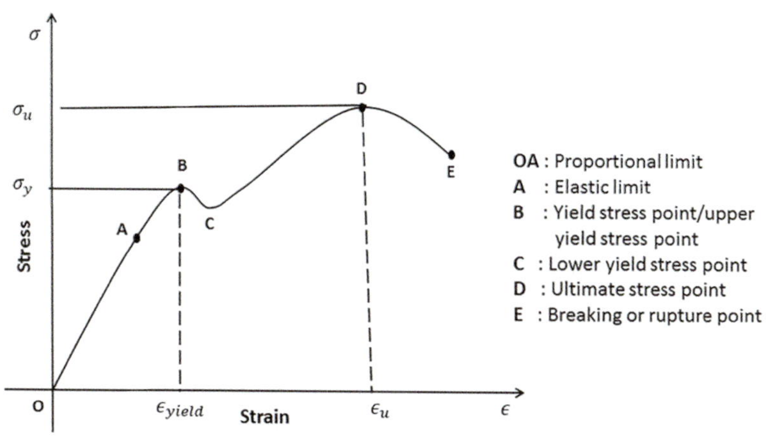

FIGURE 5.3 Stress–strain curve of structural steel.

5.3 SPECIAL DESIGN CONSIDERATIONS

The design of steel sections is mainly concerned with the thickness of the chosen section. If very thin sections are used, even a small amount of corrosion may cause large percent reductions in effective area. Hence, in the design, minimum thickness is specified.

5.3.1 Size and Shape

Size and shape need to be chosen from the steel table based on the requirement.

If two sections are available, it is preferred to choose a deeper section.

5.3.2 Buckling Considerations

If slender sections are thin, compression members may tend to buckle.

5.3.3 Minimum Thickness

Steel may get corroded, so a minimum thickness is specified. If very thin sections are used, even a small amount of corrosion may cause a large reduction in effective area. Hence, in design, minimum thickness is specified. For a fully accessible dead load, a minimum thickness of 6 mm is suggested by the design code.

5.3.4 Design Loads

Various loads act on a structure.

5.3.4.1 Dead Load

This includes (1) self-weight of the structure, (2) weight of finishes, (3) weight of partitions, (4) weight of services, and (5) weight of equipment.

TABLE 5.1
Dead Load (kN/m^2) for Various Materials

Description	Unit Weight
Asbestos sheet	0.13 kN/m^2
Mangalore tiles	0.785 kN/m^2
Terrazzo flooring	0.236 kN/m^2
Reinforced cement concrete (RCC)	25 kN/m^3
Granite masonry	24 kN/m^3
Brick masonry	20 kN/m^3
Steel	78.5 kN/m^3
Timber	6.5 kN/m^3

5.3.4.2 Live Load

The table lists the minimum live load for the design of residential, educational, commercial, office, and industrial buildings.

Description	Uniformly Distributed Load	Point Load
Living room, bedroom	2 kN/m^2	1.8 kN
Kitchen	3 kN/m^2	4.5 kN
Dining room	4 kN/m^2	2.7 kN
Office room	2.5 kN/m^2	2.7 kN
Storage room	5 kN/m^2	4.5 kN
Indoor	3 kN/m^2	1.8 kN

5.3.4.3 Wind Load

Wind exerts a horizontal force on the structure. It depends on wind speed, shape and size of the building, and location of the building.

5.3.4.3.1 Steps

Choose the basic wind pressure (V_b) from the map. It can be chosen based on the location and prevailing wind direction. The wind velocity is given as:

$$\text{Vz} = (\text{k1} * \text{k2} * \text{k3}) * \text{Vb}$$

where K_1 is the risk factor, K_2 is the terrain factor, and K_3 is the topography factor. The design wind pressure is given by:

$$p_z = 0.6 * \text{Vz}^2$$

5.3.4.4 EQ Loads

Earthquake imposes movement on the building in a horizontal (lateral) direction. Seismic force can be computed using (1) the seismic coefficient method or (2) the response spectrum method.

5.3.4.5 Erection Loads

In the case of prefabricated members, the risk of members falling increases with the increase in height of the building. Hence, erection loads, considered accident loads, are included in the design loads.

5.4 DESIGN METHOD FOR STEEL DESIGN

Structural elements should be designed using the limit state method. For design, only steel sections with the following four classes are recommended:

Class I (plastic): Used for plastic analysis.
Class II (compact): To avoid local buckling.

Class III (semi-compact): Will undergo local buckling.
Class IV (slender): Will undergo local buckling.

The maximum slenderness ratio (KL/r) is limited to 180 for tension and compression members. It depends on the effective length of the members and the support conditions.

5.4.1 Limit State Design

The design load is a product of the working load and γ_f. The partial safety factor for the loads is given in Table 4, page 29, IS 800. This is useful to take care of the various uncertainties in load calculation. The design strength is given by:

$$\text{Design strength} = \frac{\textit{Ultimate strength of the material}}{\gamma_m}$$

where γ_m is the partial safety factor for the material and is useful in addressing non-uniformities in material property. The basic design philosophy is governed by:

$$\text{Design load} \leq \text{Design strength.}$$

In the case of strength-based design, deflection controls the design check. The permissible deflection (Table 6 of IS 800) is as follows:

Maximum vertical deflection of a floor or roof = Span/300
Maximum vertical deflection of a cantilever = Span/180

5.5 DESIGN EXAMPLE

Find the direct axial tensile load capacity of a flat of cross section A_n by the (1) limit state method and the (2) working stress method.

SOLUTION

For mild steel, take f_u = ultimate stress of material = 410 N/mm^2.
Yield stress f_y = 250 N/mm^2.
Partial safety factor for material γ_m = 1.25 (Table 5).
Partial safety factor for the load γ_f = 1.5 (Table 4).
For plates: As per the equation given on page 32, Section 6.3 of IS 800:

$$T_{dn} = \frac{0.9A_n 410}{\gamma_m} = \frac{0.9A_n 410}{1.25}$$

$$\text{Load capacity} = \frac{\textit{Design load}}{\gamma_f} = \frac{0.9A_n 410}{1.25 \times 1.5} = 196.8A_n$$

Working Stress Method

Permissible strain (page 84, IS 800)

The smallest value = $0.6f_y$

Hence, load capacity = $(0.6f_y)$An

$= (0.6 \times 250)A_n$

Load capacity = $150\ A_n$

5.6 CONNECTIONS

To cater to the fabrication requirements and constraints imposed on the length of the steel section manufactured and transported, members need to be connected/spliced. Further, at the supports, structural assembly needs to be connected to the column or beam support. It is obvious to understand the fact that the connections are designed to be stronger and safer, so failure at joints/connections should be avoided. The following figure shows connections of steel members at various possible joints in a structure.

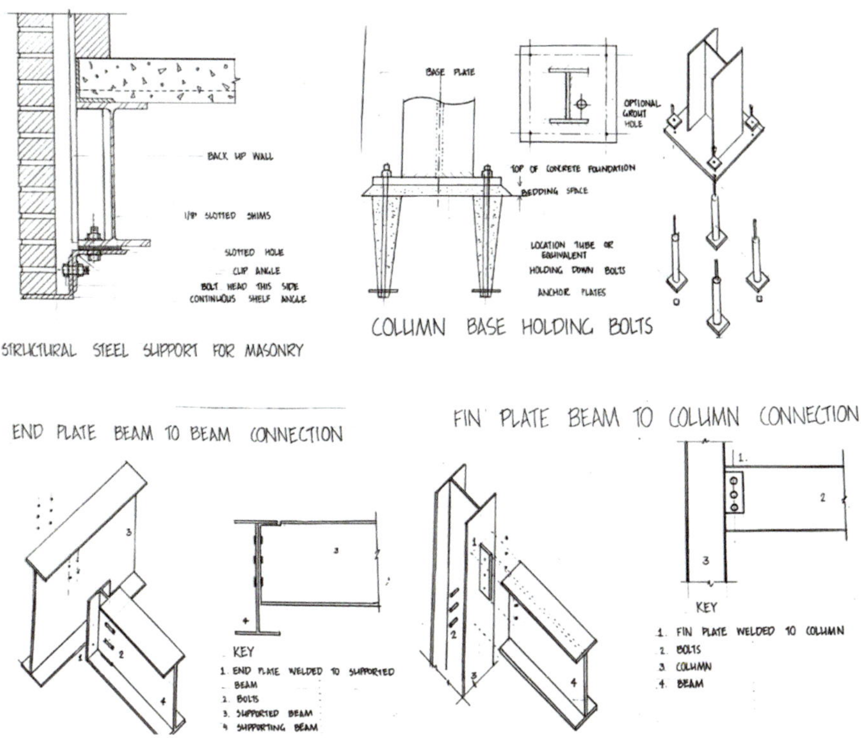

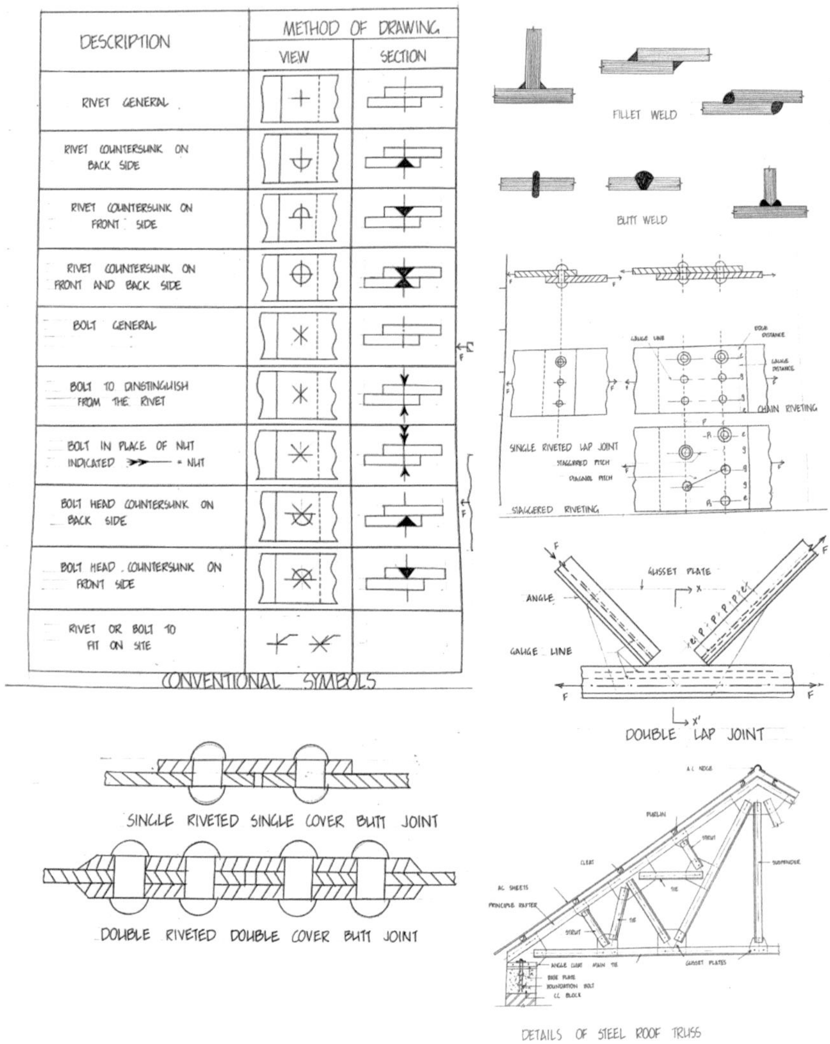

There are three types of connections to join steel members, namely, riveted, bolted, and welded. India code shares more details on the connections (IS 800, 2007). A *bolt* is a metal pin with a head at one end and a shank at the other. Using holes, one can connect the steel members by driving holes into the plates or members. Bolts are then driven into these holes and tightened with nuts/washers. Bolts are classified as unfinished (black), finished (turned), and high-strength friction grip (HSFG) bolts. Unfinished bolts are manufactured from mild steel, and they have a square or a hexagonal head. The nominal diameters available in the market are 12, 16, 20, 22, 24, 27, 30, and 36 mm, but commonly used are 16, 20, 24, and 30 mm; their diameter helps designate them. For example, a 16 mm diameter bolt is designated as $M_{16,}$ etc., whereas 16 mm is the nominal diameter of the bolt. As they are manufactured from mild steel, their yield and ultimate strength are 240 and 400 MPa, respectively. They are used in light structures.

5.6.1 Bolted Connections

A *bolt* is a metal pin with a head at one end and a shank at the other (Figure 5.4).

Holes are drilled on plates or members. Bolts are inserted, and nuts are tightened. Washers used in the bolted connection can be of three types, namely, plain washer, spring washer, and high-tensile washer (HSFG). Bolts are classified as black bolts and high-strength friction grip bolts (HSFG). Black bolts are made from mild steel and have designated threads. The actual diameter of the thread is less than the nominal diameter. HSFG bolts are made of high-strength steel. They have close tolerance threads. The actual diameter of the thread is slightly more significant than the nominal diameter. HSFG bolts are tightened to a specified tension using special tools. A few advantages of HSFG bolts are (1) a smaller number of bolts required to transfer the load, (2) enabling faster construction, and (3) providing uniformity in the load transfer process. A few disadvantages are that (1) special tools are required for connection, (2) skilled labor is required, and (3) they are expensive.

5.6.1.1 Terminology

5.6.1.1.1 Pitch (p)

Pitch is the distance between the centers of the bolts in the same row.

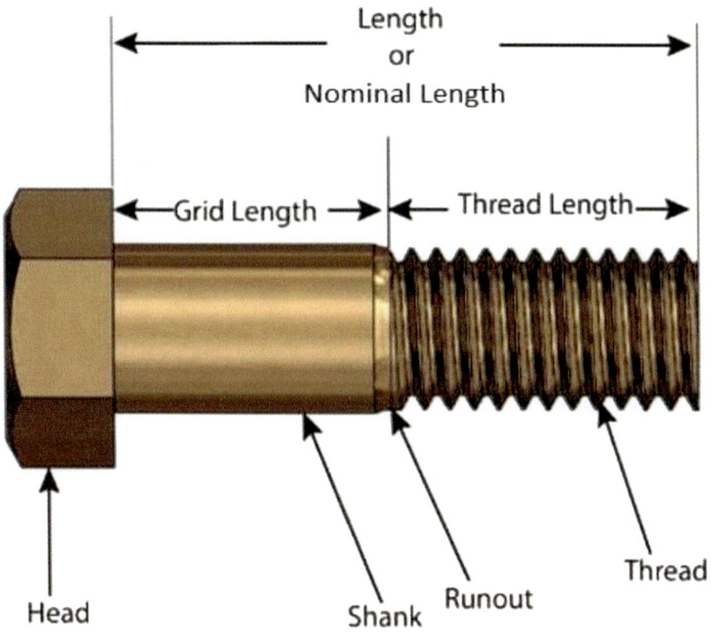

FIGURE 5.4 Bolts used in construction.

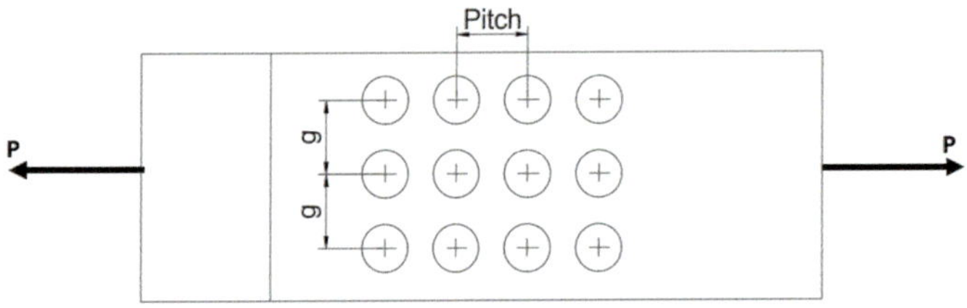

- Pitch should not be less than 2.5 d (clause 10.2.2 P 73 IS 800–2007), where *d* is the nominal diameter of the bolt.
- Pitch should not be greater than 16 t or 200 mm, whichever is less for tension members.

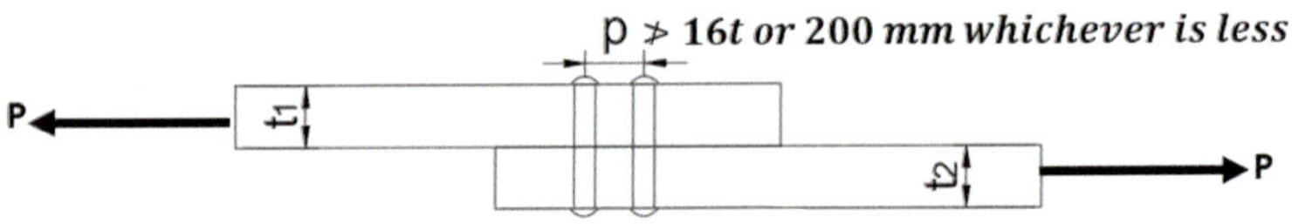

- Pitch should not be greater than 12 t or 200 mm, whichever is less for compression members.

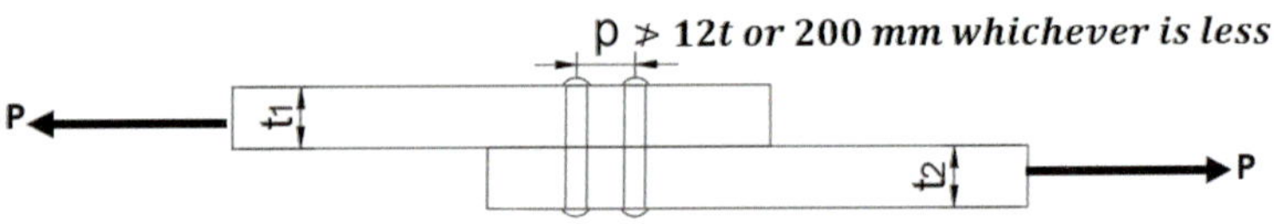

where *t* is the thickness of the plate.

Minimum pitch = 2.5 d.

5.6.1.2 Gauge Distance (*g*)

It is the center-to-center distance of the bolts in adjacent rows (Clause 10.2.3.3, p. 74).

- $g \ngtr 100$ mm + 4 t or 200 mm, whichever is less for both in tension and compression, where *t* is the thickness of the thinner plate being connected.

5.6.1.3 Edge Distance (e)

Edge distance is the distance of the nearest bolt hole from the adjacent end of the plate.

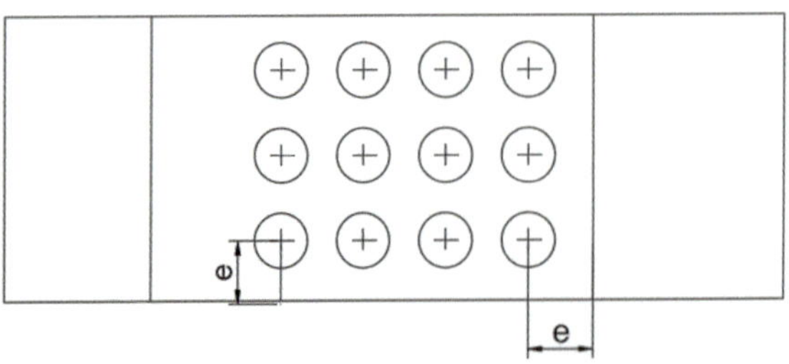

MINIMUM EDGE DISTANCE

- e = 1.7 × hole diameter (for hand-tightened bolts)
- e = 1.5 × hole diameter (for power-driven bolts)

(Clause 10.2.4, p. 74.)

5.6.1.4 Maximum Edge Distance

$$e \ngtr 12t\epsilon$$

where t is the thickness of the thinner plate being connected.

$$\epsilon = \sqrt{\frac{250}{f_y}}$$

where f_y is the yield strength of the plate.

If the plates are exposed to extensive corrosion:

$$e_{max} \ngtr 40\text{ mm} + 4t$$

5.6.1.5 Staggered Distance (p_s)

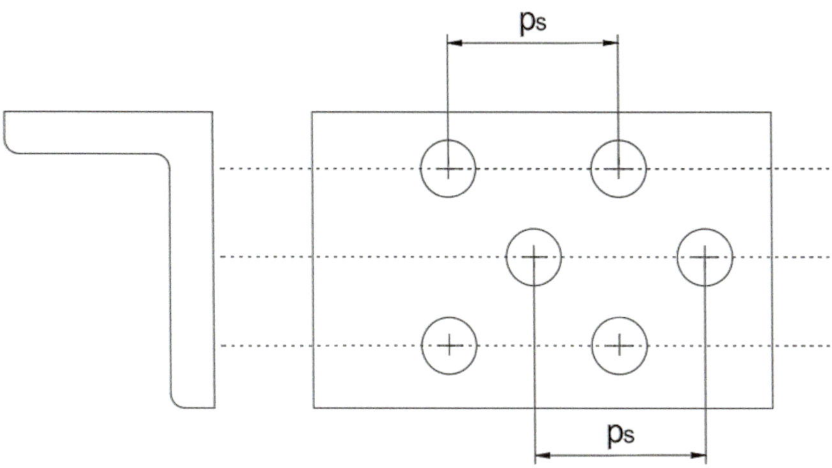

5.6.1.6 Tacking Rivets

(Clause 10.2.5, p. 74, IS 800.)

If e_{max} is exceeded, then the tacking of rivets/bolts should be provided as follows:

- Spacing of rivets/bolts $\ngtr$ 32 t or 300 mm (under standard conditions)
- Spacing of rivets/bolts $\ngtr$ 16 t or 200 mm (for very heavy loads)

Bolted connections can be grouped as lap joint and butt joint.

5.6.1.7 Lap Joint (With Multiple Bolts)

- Two plates are overlapped and bolted together.
- The number of bolts, and their spacing, depends on the load and the size of the plates.

5.6.1.8 Butt Joint (With Cover Plates)

- Two plates are joined end-to-end using cover plates.
- Bolts are placed in the cover plates to connect the main plates.

5.6.1.9 Butt Joint (With Double Cover Plates)

- Two plates are joined end-to-end using two cover plates.
- Bolts are placed in both cover plates to connect the main plates.

5.6.1.10 Butt Joint

- It is used to connect two plates of the same thickness.
- Cover plates are used to connect the plates.

5.6.1.11 Single-Cover Butt Joint

- One cover plate is used to connect the two plates.

5.6.1.12 Double-Cover Butt Joint

- Two cover plates are used to connect the two plates.

5.6.1.13 Zig-Zag Bolt Arrangement

- Bolts are arranged in a zig-zag pattern to provide better load distribution.

5.6.2 Forces on Fasteners

Based on the types of connections, the bolts are subjected to the following types of loads:

1. Single shear
    - One plane of shear
    - Example: Single-cover butt joint
2. Double shear
    - Two planes of shear
    - Example: Double-cover butt joint
3. Pure tension
    - Bolt subjected to tensile force only
    - Example: Tension member
4. Pure moment
    - Bolt subjected to bending moment only
    - Example: Cantilever beam
5. Shear and moment in the plane of connection:
    - Bolt subjected to both shear and bending moments
    - Example: Angle connection

PRINCIPLES OBSERVED IN DESIGN

1. *The center of gravity of the bolts should coincide with the CG of the connected parts.* This ensures uniform stress distribution on the bolts.
2. *Length of connection should be kept as small as possible.* This reduces the bending moment on the bolts.

A few assumptions are made in the analysis of joints:

- Friction between the plates is neglected.
- Shear is assumed to be uniformly distributed over the net area of the bolt.
- The bolt hole is assumed to be circular.
- Bolts are assumed to be subjected to direct shear and no bending moment.
- Bending stress developed in bolts is negligible.

5.6.3 Bolts in Shear and Moment in the Plane of Connection

PRINCIPLES OBSERVED IN DESIGN

1. The center of gravity of bolts should coincide with the CG of the connected parts.
2. The connection length should be kept as small as possible.

ASSUMPTIONS MADE

- Friction between the plates is neglected.
- Shear is assumed to be uniformly distributed over the net area of the bolt.
- The bolt hole is assumed to be circular.
- Bolts are assumed to be subjected to direct shear and no bending moment.
- Bending stress developed in bolts is negligible.

A bolted connection may fail in shear failure of plate or tension failure of bolt.

1. **Shear Failure of Plate:** The plate may fail in shear due to excessive shear stress.
2. **Tensile Failure of Bolt:** The bolt may fail in tension due to excessive tensile stress.

To ensure the strength of the connection, minimum distances are ensured between the bolt holes to prevent the plate from tearing.

5.6.3.1 Design Checks Are as Follows

1. **Bearing Stress:** The bearing stress in the plate should be less than the allowable bearing stress.
2. **Shear Stress in the Bolt:** The shear stress in the bolt should be less than the allowable shear stress.
3. **Tensile Stress in the Bolt:** The tensile stress in the bolt should be less than the allowable tensile stress.

5.6.3.2 Assumptions for Net Sectional Area

- In the case of shear in the plane of the bolt, P = 0.
- $A_n = (b - nt0) \times t$.

5.6.3.3 Design Strengths of Bearing Bolts

- Design shear strength of bearing bolts = $0.78\ f_{yd}\ A_n$
- Shear strength = $V_{nsb} = 0.78\ f_{yd}\ A_n$

$$\mathbf{V_{nsb} = V}$$

Where:

V = factored design shear force
f_{yd} = design yield stress of the bolt material
An = net area of bolt
$\gamma m0$ = partial safety factor for materials

5.6.3.4 Reduction in Shear Strength

- For long-edge distances, the shear capacity of the bolts is reduced by a reduction factor. (Clause 10.3.2.4, P.73, IS 800.)

5.6.3.5 Reduction Factor (*fr*)

$$f_r = 1.075 - 0.005(a/d) \text{ for } a/d <= 10$$

$$f_r = 1.075 - 0.075 \text{ for } a/d > 10$$

Reduction factor is applied if the distance between the first and the last rows of bolts measured in the direction of load exceeds 15 d.

5.6.3.6 Bearing Capacity (V_{pb})

(Clause 10.3.4, p. 75, IS 800–2007.)

$$V_{pb} = 2.5\ k_b\ d\ t\ f_{yd}$$

Where:

V_{pb} = bearing capacity of one bolt
k_b = bearing stress factor
d = diameter of the bolt hole
t = thickness of the plate
f_{yd} = design yield stress of the bolt material

Bearing stress is reduced for bigger holes other than the standard clearance by a factor:

- 0.7 for oversized holes
- 0.85 for short-slotted holes
- 0.9 for long-slotted holes

5.6.3.7 Design Procedure for Bolted Connections

Step 1: Check the existing forces (factored forces) acting on the joint.
Step 2: Select a connection with a suitable bolt size and a number of bolts.
Step 3: Determine the shear strength of the connection.
Step 4: Check if the shear strength is greater than or equal to the factored shear force.
Step 5: Check for bearing stress and slip resistance.

5.6.3.8 Bolt Hole Diameter

- Nominal bolt diameter (d): 12, 14, 16, 20, 22, 24, 30, 36 mm
- Hole diameter ($d0$): 13, 15, 18, 22, 24, 26, 37, 42 mm

5.6.3.9 Example

Find the shear capacity of the joint.

5.6.3.9.1 Given Data

- Plate thickness = 20 mm
- Bolt diameter = 20 mm
- Grade of bolt = 4.6
- Grade of plate = 250

5.6.3.9.2 Solution

1. **Number of Bolts**
    - Total number of bolts = 6
2. **Net Area of Bolt**
    - Diameter of bolt hole = 20 + 2 = 22 mm
    - Net area of bolt (An) = $(\pi/4) \times (20)^2 = 314.16$ mm^2
3. **Shear Capacity of One Bolt**
    - Design shear stress (fvd) = $0.7 \times 0.6 \times$ fu = $0.42 \times 410 = 172.2$ MPa
    - Shear capacity of one bolt (Vnsb) = fvd $\times$ An = 172.2×314.16 = 54,157.55 N
4. **Total Shear Capacity of the Joint**
    - Total shear capacity = Vnsb $\times$ number of bolts = 54,157.55 $\times$ 6 = 324,945.3 N ≈ 325 kN

5.6.3.11 Design Shear Strength (*Vnsb*)

$$\text{Vnsb} = 0.78 \text{ fyd An}$$

Where:
fyd = design yield stress of the bolt material
An = net area of the bolt

5.6.3.12 Calculation

$$\text{fyd} = 0.7 \times 410 = 287 \text{ MPa}$$
$$\text{An} = (\pi/4) \times 20^2 = 314 \text{ mm}^2$$
$$\text{Vnsb} = 0.78 \times 287 \times 314 = 69{,}757.96 \text{ N} \approx 70 \text{ kN}$$

5.6.3.13 Design Bearing Strength (Vpb)

$$Vpb = 2.5\ kb\ d\ t\ fyd$$

Where:

kb = bearing stress factor
d = diameter of the bolt hole
t = thickness of the plate
fyd = design yield stress of the bolt material

5.6.3.14 Calculation

$$kb = 1.2 \text{ (from table)}$$
$$d = 22 \text{ mm}$$
$$t = 20 \text{ mm}$$
$$fyd = 287 \text{ MPa}$$
$$Vpb = 2.5 \times 1.2 \times 22 \times 20 \times 287 = 379{,}200 \text{ N} \approx 379 \text{ kN}$$

5.6.3.15 Efficiency of the Joint

$$Efficiency = \frac{V_{nsb}}{V_{pb}} x100 = \frac{70}{379} x100 = 18.47\%$$

5.6.3.16 Calculation for a Single Plate

$$An = 12 \times 20 = 240 \text{ mm}^2$$
$$Vnsb = 0.78 \times 287 \times 240 = 51{,}019.2 \text{ N} \approx 51 \text{ kN}$$

5.6.3.17 The Efficiency of the Joint

$$Efficiency = \frac{51}{379} x100 = 13.46\%$$

5.7 DESIGN OF RCC SLAB

A *slab* is a flat, two-dimensional structural element with a small thickness relative to its other two dimensions. It provides a flat working surface or serves as a building covering or shelter. Slabs mainly transfer loads through bending in one or two directions. Reinforced concrete slabs are commonly used in floors, roofs, walls, and bridge decks. The floor system in a building can take various forms, such as in situ solid slabs, ribbed slabs, or precast units. Slabs may be supported by monolithic concrete beams, steel beams, walls, or even directly on columns. Concrete slabs primarily function as flexural members, and their design approach is similar to that of beams.

5.7.1 Slab Classification Based on Several Aspects

1. **Shape:** Slabs can be square, rectangular, circular, or polygonal.
2. **Type of Support:**
    - Slabs supported on walls
    - Slabs supported on beams
    - Slabs supported on columns (flat slabs)

3. **Support or Boundary Condition:**
    - Simply supported slabs
    - Cantilever slabs
    - Overhanging slabs
    - Fixed or continuous slabs
4. **Usage:**
    - Roof slabs
    - Floor slabs
    - Foundation slabs
    - Water tank slabs
5. **Cross Section or Configuration:**
    - Ribbed or grid slabs
    - Solid slabs
    - Filler slabs
    - Folded plates
6. **Spanning Direction:**
    - One-way slabs (spanning in one direction)
    - Two-way slabs (spanning in two directions)

5.7.2 Design Steps

5.7.2.1 Effective Span of Slab

The *effective span* of a slab is determined as the smaller of the following two values:

1. L = clear span + d (where *d* is the effective depth of the slab)
2. l = center-to-center distance between the supports

5.7.2.2 Depth of Slab

The slab depth is influenced by the bending moment and deflection criteria. The trial depth can be estimated using:

- Effective depth d = span/[(l/d) × basic modification factor]
- The modification factor can be obtained based on the percentage of steel reinforcement, typically assumed between 0.2% and 0.5% for slabs.
- For two-way slabs, the effective depth *d* may also be estimated using Clause 24.1 of IS 456, provided the short span is less than 3.5 m and the loading class is less than 3.5 kN/m^2.

Type of Support	Fe-250	Fe-415
Simply supported	l/35	l/28
Continuous	l/40	l/32

Or the following thumb rules can be used:

- One-way slab d = (l/22) to (l/28).
- Two-way simply supported slab d = (l/20) to (l/30).
- Two-way restrained slab d = (l/30) to (l/32).

5.7.2.3 Load on Slab

The total load on a slab includes dead load, floor finish load, and live load. These loads are calculated per unit area (kN/m^2).

- **Dead Load** = D $\times$ 25 kN/m^2 (where *D* is the slab thickness in meters)
- **Floor Finish Load** (typically assumed) = 1 to 2 kN/m^2
- **Live Load** (typically assumed) = 3 to 5 kN/m^2, depending on the building's occupancy type

5.7.2.4 Detailing Requirements as Per IS 456:2000

1. **Nominal Cover:**
    - For mild exposure: 20 mm
    - For moderate exposure: 30 mm

If the bar diameter does not exceed 12 mm, the cover may be reduced by 5 mm. Therefore, for main reinforcement up to a 12 mm diameter bar under mild exposure, the nominal cover is 15 mm.

2. **Minimum Reinforcement:**
    - The reinforcement in each direction of the slab should not be less than:
        - 0.15% of the total cross-sectional area for Fe-250-grade steel
        - 0.12% of the total cross-sectional area for Fe-415- and Fe-500-grade steel
3. **Spacing of Bars:**
    - The maximum spacing of bars should not exceed:
        - Main reinforcement: 3 d, or 300 mm, whichever is smaller
        - Distribution reinforcement: 5 d or 450 mm, whichever is smaller
    - Here, *d* represents the effective depth of the slab.
    - Note: Although the code does not specify a minimum clear spacing, it is generally kept above 75 mm, preferably 100 mm.
4. **Maximum Diameter of Bar:**
    - The maximum diameter of bars used in a slab should not exceed D/8D/8D/8, where *DDD* is the total thickness of the slab.

EXAMPLE PROBLEM

Design a simply supported one-way slab with a clear span of 3.5 m, subjected to a live load of 4 kN/m^2 and a floor finish load of 1.5 kN/m^2. The supporting walls have a width of 230 mm. Use M20-grade concrete and Fe-415-grade steel.

5.7.2.4.1 Trail Depth and Effective Span

Assume approximate depth d = L/26 3500/26 = 134 mm.

Assume overall depth D = 160 mm and clear cover 15 mm for mild exposure.

$$d = 160 - 15\ \text{(cover)} - 10/2\ \text{(dia of bar/2)} = 140\ \text{mm}$$

Effective span is lesser of the two:

1. l = 3.5 + 0.23 (width of support) = 3.73 m
2. l = 3.5 + 0.14 (effective depth) = 3.64 m

Effective span = 3.64 m.

5.7.2.4.2 Load on Slab

1. Self-weight of slab = 0.16 × 25 = 4.00
2. Floor finish = 1.50
3. Live load = 4.00 = 9.5 kN/m2
4. Ultimate load Wu = 9.5 × 1.5 = 14.25 kN/m2

5.7.2.4.3 Design Bending Moment and Check for Depth

$$M_u = \frac{W_u l^2}{8}$$

$$M_u = \frac{14.25 \times 3.64^2}{8} = 23.6\ kNm$$

Minimum depth required from BM consideration:

$$d = \sqrt{\frac{M_u}{0.138 f_{ck} b}}$$

$$d = \sqrt{\frac{23.6 \times 10^6}{0.138 \times 20 \times 1000}} = 92.4 < 140(oK)$$

5.7.2.4.4 Area of Reinforcement

The area of steel is obtained using the following equation:

$$M_u = 0.87 f_y A_{st} d \left(1 - \frac{f_y A_{st}}{f_{ck} b}\right)$$

$$23.6 \times 10^6 = 0.87 \times 415 \times A_{st} \times 140 \left(1 - \frac{415 \times A_{st}}{20 \times 1000 \times 140}\right)$$

$$23.6 \times 10^6 = 50547 A_{st} - 749 A_{st}^{\ 2}$$

5.7.2.4.5 On Solving

$$A_{st} = 504\, mm^2$$

OR

$$A_{st} = \frac{0.5 f_{ck}}{f_y}\left[1 - \sqrt{1 - \frac{4.6 M_u}{f_{ck} b d^2}}\right] bd$$

$$A_{st} = \frac{0.5 \times 20}{415}\left[1 - \sqrt{1 - \frac{4.6 \times 23.6 \times 10^6}{20 \times 1000 \times 140^2}}\right]1000 \times 140$$

$$A_{st} = 505\,mm^2$$

5.7.2.4.6 Spacing of 10 mm

$$S_v = \frac{a_{st}}{A_{st}} \times 1000$$

$$S_v = \frac{78}{505} \times 1000 = 154\,mm$$

Provide 10 mm at 150 C/C (<3 d, or 300).
(420 or 300) OK
Provided steel (Ast = 524 mm^2, $P_t = 0.37\%$)
Distribution steel at 0.12% of the gross area

$$= 0.12 \times 10^{-2} \times 1000 \times 160$$
$$= 192 \text{ mm}^2$$

Spacing of 8 mm:

$$S_v = \frac{a_{st}}{A_{st}} \times 1000$$

$$S_v = \frac{50}{192} \times 1000 = 260\,mm$$

Provide 8 mm at 260 C/C (<5 d, or 450)
(700 or 450) OK.

5.7.2.4.7 Check for Shear

$$\text{Design shear } V_u = \frac{W_u l}{2}$$

$$V_u = \frac{14.25 \times 3.64}{2}$$

$$= 25.93 \text{ kN}$$

$$\tau_v = \frac{25.93 \times 10^3}{1000 \times 140} = 0.18\frac{N}{mm^2} \quad \left(\tau_{cmax} = 2.8\frac{N}{mm^2}\right)$$

Shear resisted by concrete $\tau_c = 0.42\frac{N}{mm^2}$ for $P_t = 0.37$ (Table 19, IS 456–2000).

However, for solid slab design, shear strength is given as:

$$= \tau_c k$$

where K is obtained from Cl.40.2.1.1, IS 456–2000.

$$\tau_{cd} = 0.42 \times 1.28 = 0.53 \frac{N}{mm^2}$$

$$\tau_{cd} > \tau_v \text{ OK}$$

5.7.2.4.8 Check for Deflection

$$\left(\frac{l}{d}\right)_{Actual} < \left(\frac{l}{d}\right)_{Allowable}$$

$$\left(\frac{l}{d}\right)_{Actual} = \left(\frac{l}{d}\right)_{Basic} \times k_1 k_2 k_3 k_4$$

Where:

k1 = modification factor for tension steel
k2 = modification factor for compression steel
k3 = modification factor for T-sections
k4 = only if span exceeds 10 m (10/span)
$k_1 = 1.38$ for $P_t = 0.37$ (Figure 4, cl. 32.2.1)

$$\left(\frac{l}{d}\right)_{Allowable} = 20 \times 1.38 = 27.6$$

$$\left(\frac{l}{d}\right)_{Actual} = \frac{3630}{140} = 25.92$$

$$\left(\frac{l}{d}\right)_{Actual} < \left(\frac{l}{d}\right)_{Allowable} \quad \text{OK}$$

5.7.2.4.9 Check for Development Length

Development length:

$$L_d = \frac{\phi \sigma_s}{4 \tau_{bd}}$$

$$L_d = \frac{0.87 \times 415 \times 10}{4 \times 1.2 \times 1.6} = 470\ mm$$

At simple support, where compressive reaction confines the bars, to limit the diameter of the bar:

$$L_d \le 1.3\left(\frac{M_1}{V}\right) + L_0$$

Since alternate bars are cranked M1 = Mu/2 = 23.2/2 = 11.8 kNm.

V1 = 5.93 kN, providing 90° bend and 25 mm end cover
Lo = 230/2 – 25 + 3 (diameter of bar) = 120
470 < (1.3 × 11.8 × 106)/(25.9 × 103) + 120 = 711 mm OK

However, from the end anchorage requirement, extend the bars for a length equal to ld/3 = 156 mm from the inner face of the support.

5.7.2.4.10 Check for Cracking

Steel is more than 0.12% of the gross area, with spacing <3 d and the diameter of the bar used < 160/8 = 20 mm. The check for cracking is satisfied.

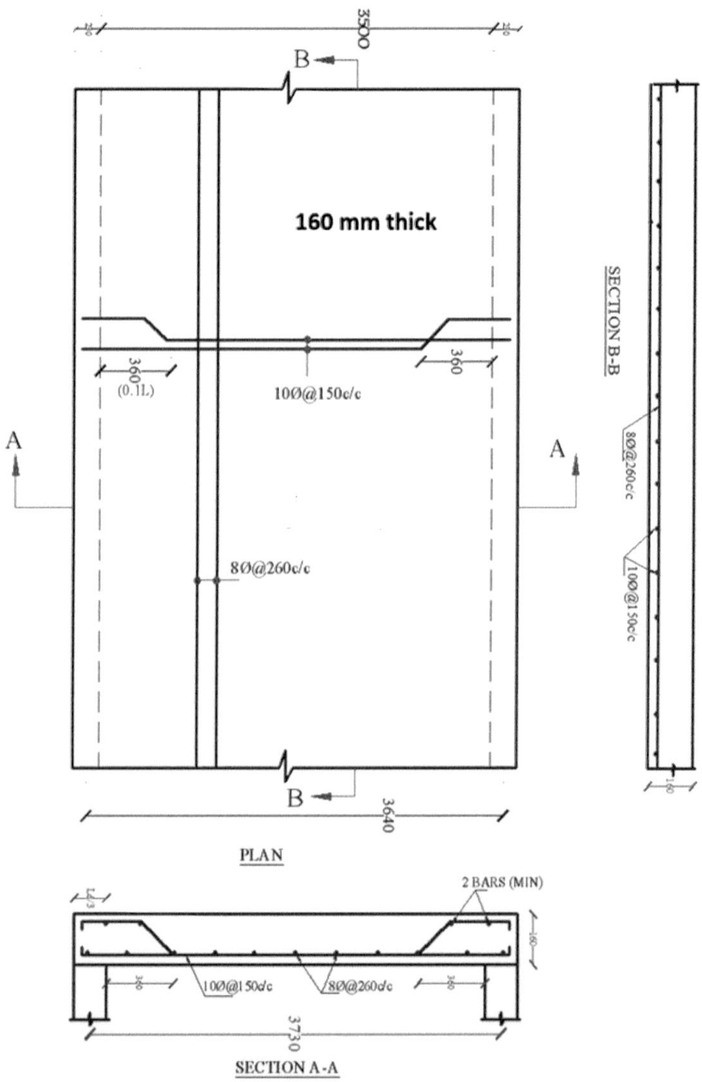

EXAMPLE

Design an RCC slab for a room measuring 6.5 m × 5 m, cast monolithically over supporting beams with the corners restrained. The supporting beams are 230 mm in width. The slab is subjected to a superimposed load of 4.5 kN/m^2. Use M20-grade concrete and Fe-500-grade steel.

SOLUTION

Since the length-to-width ratio of the slab is less than 2.0 and it is supported on beams, the slab will be designed as a two-way restrained slab (Case-9).

5.7.2.4.10.1 Trail Depth and Effective Span

Assume approximate depth d = 1/30 = 5,000/30 = 166 mm.

Assume D = 180 mm and clear cover 15 mm for mild exposure.
D = 180 – 15 – 10/2 = 160 mm.

Effective span is a minimum of the two:

1. ly = 6.5 + 0.23 = 6.73 m, lx = 5 + 0.23 = 5.23 m
2. ly = 6.5 + 0.16 = 6.66 m, lx = 5 + 0.16 = 5.16 m

ly = 6.66 m, lx = 5.16 m

$$\alpha = \frac{l_y}{l_x} = \frac{6.66}{5.16} = 1.3$$

5.7.2.4.10.2 Load on Slab

1. Self-weight of slab = 0.18 × 25 = 4.50 kN/m^2.
2. Superimposed load = 4.50

Total = 9.0 kN/m2
Ultimate load w_u = 9 × 1.5 = 13.5 kN/m2

5.7.2.4.10.3 Design Bending Moment, and Check for Depth

The boundary condition of the slab in all four edges is discontinuous (Case 9, Table 9.5.2)

$$M_x = \alpha_x w_u l_x^2$$
$$M_y = \alpha_y w_u l_x^2$$

For

$$\frac{l_y}{l_x} = 1.3$$

$$\alpha_x = 0.079$$

$$\alpha_y = 0.056$$

Positive moment at mid-span of short span = Mx = $0.079 \times 13.5 \times 5.162 = 28.40$ kN.m

Positive moment at mid-span of longer span = My = 0.056 × 13.5 × 5.162 = 20.13 kN.m

Minimum depth required from the maximum BM consideration

Minimum depth required from BM consideration

$$d = \sqrt{\frac{M_u}{0.133 f_{ck} b}}$$

$$d = \sqrt{\frac{28.40 \times 10^6}{0.133 \times 20 \times 1000}} = 103\ mm$$

However, provide d = 160 mm.

5.7.2.4.10.4 Area of Reinforcement

$$M_u = 0.87 f_y A_{st} d \left(1 - \frac{f_y A_{st}}{f_{ck} b}\right)$$

Steel along the shorter direction (Mx):

$$28.17 \times 10^6 = 0.87 \times 500 \times A_{st} \times 160 \left(1 - \frac{500 \times A_{st}}{20 \times 1000 \times 160}\right)$$

$$28.40 \times 10^6 = 69600 A_{st} - 10.875 A_{st}^{\ 2}$$

On solving $A_{st} = 438\, mm^2$:

Provide 10 mm at 175 C/C (Pt = 0.27%).

Steel along the longer direction (My).

Since long-span bars are placed above short-span bars, d = 160 – 10 = 150.

$$20.13 \times 10^6 = 0.87 \times 500 \times A_{st} \times 150 \left(1 - \frac{500 \times A_{st}}{20 \times 1000 \times 150}\right)$$

$$20.13 \times 10^6 = 65250 A_{st} - 10.875 A_{st}^{\ 2}$$

On solving $A_{st} = 327\, mm^2$:

Spacing at 10 mm:

$$S_v = \frac{79}{372} \times 1000 = 241\, mm$$

Provide 10 mm at 240 mm c/c (<3d = 450).

5.7.2.4.10.5 Check for Shear and Development

Check for shear and development length, which are generally satisfied in the slab case, so they are not checked.

5.7.2.4.10.6 Check for Deflection

$$\left(\frac{l}{d}\right)_{Actual} < \left(\frac{l}{d}\right)_{Allowable}$$

$$\left(\frac{l}{d}\right)_{Actual} = \left(\frac{l}{d}\right)_{Basic} \times k_1$$

$k_1 = 1.5$ for $P_t = 0.27\%$ and $f_s = 0.58 \times f_y = 240$ (Figure 4, cl. 32.2.1, IS 450 2000)

$$\left(\frac{l}{d}\right)_{Allowable} = 26 \times 1.5 = 39$$

$$\left(\frac{l}{d}\right)_{Actual} = \frac{5.16}{0.16} = 32$$

$$\left(\frac{l}{d}\right)_{Actual} < \left(\frac{l}{d}\right)_{Allowable} \quad \text{OK}$$

5.7.2.4.10.7 Detailing

Torsion Steel

Area of torsion steel = 0.75 × Ast = 0.75 × 438 = 328 mm^2
Provide 8 mm bars at spacing (50/328) × 1,000 = 152 mm.
Size of mesh = (lx/5) = 5,160/5 = 1,032 mm

Provide 8 mm at 150 c/c in both directions for a length of 1,035 mm mesh at the top and bottom. The calculated steel in shorter and longer directions will only be provided in the middle strip. The steel in the edge strip contains only 0.12% of the gross area.

Steel in the edge strip = (0.12/100) × 1,000 × 180 = 216 mm^2
Spacing of 8 mm (50/216) × 1,000 = 230 mm c/c.

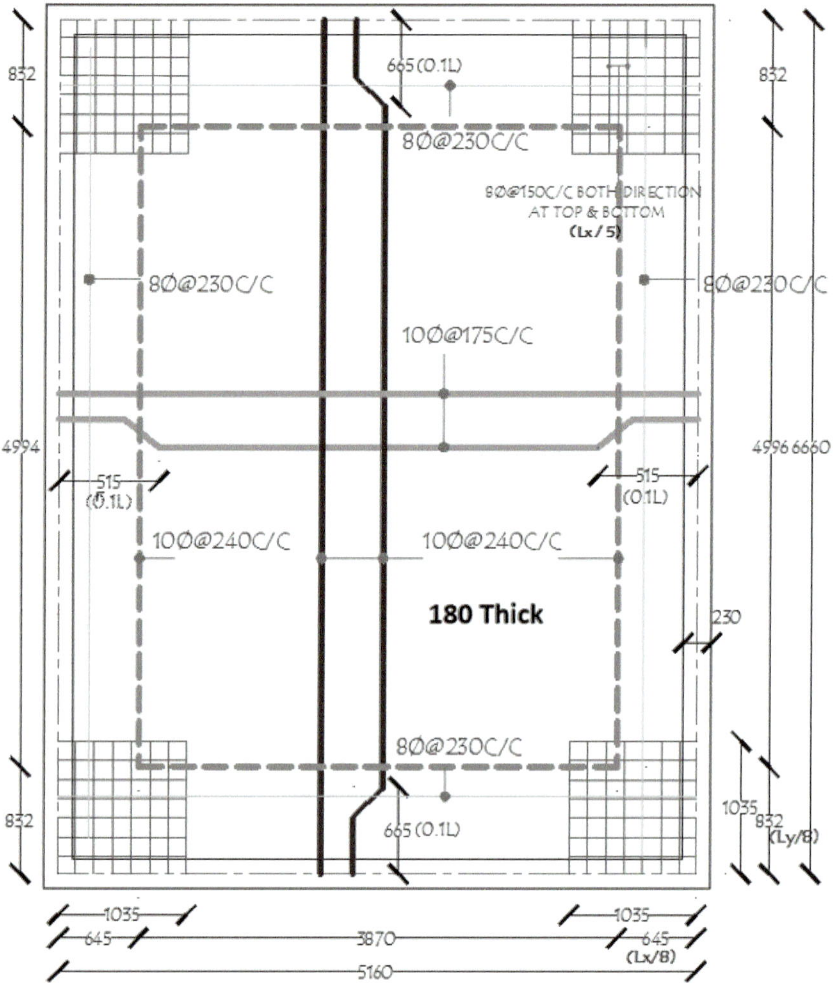

5.8 DESIGN OF RCC BEAM

Reinforced concrete (*RC*) *beams* are horizontal structural elements that carry transverse loads, causing bending and shear. They consist of concrete for compressive strength and steel reinforcement to resist tensile stresses, which concrete alone cannot handle effectively. In an RC beam, steel reinforcement bars (rebar) are placed mainly near the bottom of the beam, where tensile stresses occur due to bending. Additional shear reinforcement, typically in the form of stirrups, is provided to prevent shear failure and to support the longitudinal bars. The design of RC beams is guided by codes, such as IS 456, which specifies criteria for strength, serviceability, and durability. Factors like beam depth, effective span, load conditions, and reinforcement placement are carefully considered to ensure the beam can resist bending moments, shear forces, and deflection within acceptable limits, making RC beams essential in building frames, bridges, and other structures.

DESIGN STEPS FOR A SINGLY REINFORCED BEAM AS PER IS 456:2000

Step 1: Determine Design Parameters

- *Material grades.* Choose the concrete grade (e.g., M20) and steel grade (e.g., Fe-415 or Fe-500).
- *Load calculations.* Identify the loads applied on the beam, including dead load, live load, and any additional loads (measured in kN/m or kN/m^2).
- *Span and dimensions.* Note the clear span of the beam and any preliminary section dimensions, if available.

Step 2: Calculate Factored Loads

- Calculate the factored load (*Wu*) using load combinations as per IS 456: Wu = 1.5 × (dead load + live load)

Step 3: Effective Span (l_{eff})

- Effective span (l_{eff}) is taken as:
  - Lesser of clear span + effective depth (*d*) or center-to-center distance between supports.

Step 4: Moment Calculation

- Calculate the factored bending moment (Mu) using the standard bending moment formulas:
  - For a simply supported beam:

$$M_u = \frac{W_u l_{eff}^2}{8}$$

Step 5: Determine Effective Depth (*d*)

- Estimate the effective depth (*d*) of the beam for deflection control using the following guidelines from IS 456: d = span/[(l/d) × basic modification factor].
- Use Table 19 in IS 456 to obtain the modification factor based on the percentage of tension reinforcement.

Step 6: Area of Steel Required (*Ast*)

- Use the limiting moment of resistance equation to calculate the required area of steel (*Ast*):

$$M_u = 0.87 f_y A_{st} d \left(1 - \frac{f_y A_{st}}{f_{ck} b}\right)$$

Or using

$$A_{st} = \frac{0.5f_{ck}}{f_y}\left[1-\sqrt{1-\frac{4.6M_u}{f_{ck}bd^2}}\right]bd$$

Where:

fy = yield strength of steel
fck = characteristic compressive strength of concrete
b = width of the beam

- Rearrange this equation to solve for *Ast*, and select the appropriate bar diameters.

Step 7: Minimum Reinforcement Requirement

- Verify that the area of steel provided Ast meets the minimum reinforcement requirements as per IS 456:

$$A_{stmin} = \frac{0.85bd}{f_y}$$

Step 8: Check for Shear Reinforcement

- Calculate the factored shear force (*Vu*), and check for the shear capacity of the section using the shear provisions from IS 456.
- Provide shear reinforcement (stirrups) if necessary, based on calculated shear force and permissible shear stress limits.

Step 9: Detailing Requirements

- Ensure compliance with IS 456 for spacing of bars, clear cover, and anchorage length.
- For Fe-415 and Fe-500 steel, ensure adequate development length for tension bars as per IS 456 requirements.

Step 10: Verify Deflection Control

- Check deflection using the effective depth and modification factors as per Table 19 of IS 456. Ensure the section complies with deflection limits.

EXAMPLE

A reinforced concrete beam is supported on two walls 250 mm thick, spaced at a clear distance of 6 m. The beam carries a super-imposed load of 9.8 KN/m. Design the beam using M20 concrete and HYSD bars of Fe-415-grade.

Solution

The minimum depth of the beam is based on the limited state of serviceability requirements.

As per IS 456:2000 cl no 23.2.1(a), for simply supported beam, l/d = 20, and for the balanced section, P_{tlimit} 0.96%.

Also, for Fe 415 steel, fs = 0.58 × 415 = 240 N/mm^2.

As per IS 456:2000 cl no 23.2.1(c) and Figure 4 of IS 456:2000, modification factor = 1.

Therefore, l/d = 20 × 1.

$$d = l/20 = 6{,}000/20 = 300 \text{ mm}$$

This is the minimum value of *d*. The actual value of *d*, based on bending, may be more than this.

Now, from experience, assume d = l/15 = 400 mm.

Therefore, overall depth = effective depth + clear cover + diameter of stirrup + 0.5 (diameter of main reinforcement)

$$= 400 + 25 + 8 + 0.5 \times 20 = 443 \text{ mm} = 450 \text{ mm}$$

Assume b = 250 mm

Therefore, try a trial section of dimensions 250 × 450.

5.8.1 Load Calculation

Self-weight of beam (DL) = 0.25 × 0.45 × 1 × 25 = 2.8125 KN/m
Super-imposed load (LL) = 9.8 KN/m
Therefore, total load w = (DL + LL) = (2.8125 + 9.8) = 12.6125 KN/m
Design load, w_u = 1.5 × w = 18.9187 KN/m

5.8.2 Calculation of Effective Span

As per IS 456:2000, cl no 22.2(a), the effective span of a simply supported beam is lesser of the following two:

Clear span + the effective depth of beam or slab
Or center-to-center distance between supports
Clear span = 6 m
The effective depth of the beam, d = 450 – 25 – 8 – 0.5 × 20 = 407 mm
Therefore, clear span + effective depth of beam = (6 + 0.407) m = 6.407 m
Centre-to-center distance between support = (6 + 0.25/2 + 0.25/2) m = 6.25 m
Minimum of two = 6.25 m
Therefore, effective span = 6.25 m

5.8.3 Calculation of BM and SF

Maximum BM:

$$M_u = \frac{W_u l^2}{8}$$

$$M_u = \frac{18.9187 \times 6.25^2}{8} = 92.376\ kNm$$

Maximum SF:

$$V_u = \frac{W_u l}{2}$$

$$V_u = \frac{18.9187 \times 6.25}{2} = 59.12\ kN$$

5.8.4 Computation of Effective Depth

For M20-grade concrete and Fe-415-grade steel

$$d = \sqrt{\frac{M_u}{0.138 f_{ck} b}}$$

$$d = \sqrt{\frac{92.376 \times 10^6}{0.138 \times 20 \times 250}} = 365.89\ mm$$

Now, assumed depth was = 407 mm.

So the section assumed it is safe from a bending moment point of view. Since the available depth (407 mm) is greater than required (365.89 mm), the section is under-reinforced.

5.8.5 Calculation of Steel Reinforcement

The reinforcement for an under-reinforced section is given by:

$$A_{st} = \frac{0.5 f_{ck}}{f_y}\left[1 - \sqrt{1 - \frac{4.6 M_u}{f_{ck} b d^2}}\right] bd$$

$$A_{st} = \frac{0.5 \times 20}{415}\left[1 - \sqrt{1 - \frac{4.6 \times 92.376 \times 10^6}{20 \times 250 \times 407^2}}\right] 250 \times 407$$

$$A_{st} = 740.88\ mm^2$$

$$\%\text{pt} = 100 \frac{740.88}{250 \times 407} = 0.728\% < \text{p}_{\text{tlim}}$$

Therefore, using 20 mm diameter, the number of bars required $= \dfrac{740.88}{\dfrac{\pi}{4}20^2} = 2.35 = 3.$

Provide three numbers 20 diameter bar.

5.8.6 Shear Reinforcement

As per IS 456:2000 Cl. No. 22.6.2, the critical section for shear is at a distance of *d* from the face of the support.

So shear force at that distance Vu = 59.12 − 18.9187 (0.25/2 + 0.407) = 49.05 kN.

Nominal shear stress, $\tau_v = \dfrac{V_u}{bd}$

$$\tau_v = \frac{49.05 \times 10^3}{250 \times 407} = 0.482 \frac{N}{mm^2}$$

$$\% \text{ pt at support} = 100 \frac{3\frac{\pi}{4}20^2}{250 \times 407} = 0.926\%$$

As per IS 456:2000, Table 19, the design shear strength of concrete τ_c for %pt = 0.926, and M20 grade of concrete, $\tau_c = 0.61 \dfrac{N}{mm^2}$.

Since $\tau_v < \tau_c$, no shear reinforcement is necessary. However, minimum shear reinforcement as per cl no 26.5.1.6 of IS 456:2000 should be provided.

$$\frac{A_{sv}}{bS_v} \geq \frac{0.4}{0.87 \times f_y} = S_v = \frac{2.172\ A_{sv} f_y}{b}$$

As per cl no 26.5.1.5 of IS 456:2000, the maximum spacing of shear reinforcement is at least one of the following: (1) 0.75 d, or (2) 300 mm

Hence, provide two-legged 8 mm diameter at 300 mm c/c throughout the length of the beam.

5.8.7 Check for Development Length

As per cl no 26.2.1 of IS 456:2000, the development length *Ld* is given by:

$$L_d = \frac{\phi \sigma_s}{4\tau_{bd}}$$

$\tau_{bd} = 1.2$ N/mm^2 for M20 grade of concrete.

For deformed bars conforming to IS 1786, these values shall be increased by 60%.

$$L_d = \frac{0.87 \times 415 \times 20}{4 \times 1.2 \times 1.6} = 940.23\ mm$$

Now, as per cl no 26.2.3.3 (c) of IS 456:2000, at a simple support and at points of inflection, positive moment tension reinforcement shall be limited to a diameter such that Ld computed for fd does not exceed $\left(\frac{M_1}{V}\right)+L_0$.

The value of $\left(\frac{M_1}{V}\right)+L_0$ in the preceding expression may be increased by 30% when the ends of the reinforcement are confined by a compressive reaction.

$$L_d \leq 1.3\left(\frac{M_1}{V}\right)+L_0$$

$$x_u = \frac{0.87\times f_y \times A_{st}}{0.36\times f_{ck}b} = \frac{0.87\times 4150\times 3\times 314.15}{0.36\times 20\times 250} = 189\ mm$$

$$\begin{aligned} M_1 &= 0.87\times f_y \times A_{st}\left(d-0.416x_u\right) \\ &= 0.87\times 415\times(3\times 314.15)\times(407-0.416\times 189) \\ &= 111.73\ \text{kNm} \\ \text{V} &= 59.12\ \text{kN.} \end{aligned}$$

Lo = greatest of effective depth of member or 12∅ = 407 mm

$$1.3\left(\frac{M_1}{V}\right)+L_0 = 2.863\ m$$

Therefore:

$$L_d \leq 1.3\left(\frac{M_1}{V}\right)+L_0$$

5.8.8 Reinforcement Details

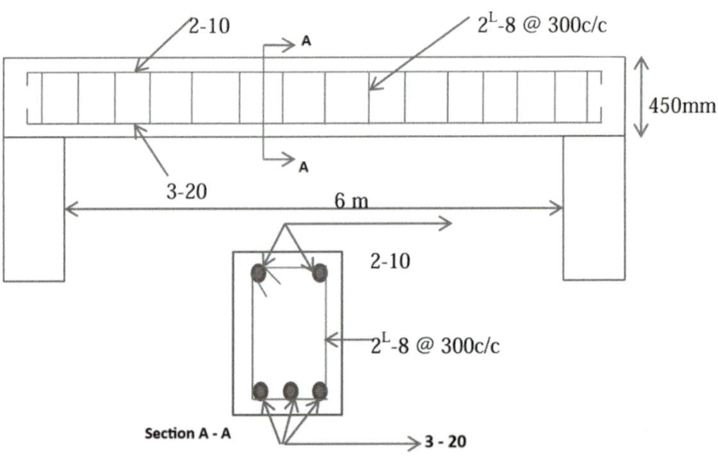

5.9 DESIGN OF RCC COLUMN

Column members are structural elements that primarily carry **axial compressive forces**. These members are found in many structures, including buildings, bridges, and towers, and are crucial for resisting vertical and sometimes lateral loads. Due to the nature of compressive loading, the design of compression members is governed by considerations of **strength** and **stability** (specifically **buckling**).

5.9.1 Design Steps

Step 1: Determine the Design Loads

- Calculate the factored axial load *Pu* on the column using the load combinations provided in IS 875 Part 5: Code of Practice for Design Loads.

Pu = 1.5 × (Dead Load + Live Load)

Step 2: Choose the Column Type and Size

- Decide whether the column is **short** or **long** based on its slenderness ratio.
  - **Short Column:** Slenderness ratio (effective length to least lateral dimension) is less than 12.
  - **Long Column:** Slenderness ratio is greater than 12.
- If it is a rectangular column, assume an initial section size. For a circular column, assume an initial diameter.

Step 3: Calculate the Area of Steel (*Ast*)

1. *Assume concrete grade (f_{ck}).* M20 or M25 concrete is typically used in columns. Higher grades can also be used depending on design requirements.
2. *Assume steel grade (f_y).* Fe415 or Fe500 is generally used for reinforcement.
3. Calculate the required area of steel using the formula:

$$P_u = 0.4 \times f_{ck} \times A_c + 0.67 \times f_y \times A_{st}$$

where

Ac = area of concrete (gross area minus area of steel)

Ast = area of steel

Rearrange the formula to solve for *Ast* and find the required area of steel.

Step 4: Provide Longitudinal Reinforcement

- The minimum reinforcement in a column should not be less than **0.8%** and not more than **6%** of the gross cross-sectional area, as per IS 456.
- Distribute the calculated area of steel into a suitable number of bars, ensuring that they are symmetrically placed around the column.
- Typically, the minimum diameter of longitudinal bars should be 12 mm.

Step 5: Provide Lateral Ties (Transverse Reinforcement)

1. **Spacing of Ties:** As per IS 456:2000, the spacing of lateral ties should be the **least of**:
   - 300 mm
   - 16 times the diameter of the longitudinal bar
   - Least lateral dimension of the column
2. **Diameter of Ties:** The minimum diameter of lateral ties should be 8 mm or not less than **one-fourth** of the largest diameter of the longitudinal bar.

Step 6: Check the Slenderness Ratio

- Calculate the slenderness ratio (effective length/least lateral dimension).
  - For a short column, the slenderness ratio should be less than 12.
  - For a long column, if the slenderness ratio exceeds 12, additional checks and design adjustments are required as per IS 456.

Step 7: Check Load-Carrying Capacity

- Confirm that the provided section meets the load-carrying requirement by rechecking the axial load capacity *Pu* using the chosen dimensions and reinforcement area. Ensure it satisfies:

$$P_u \leq 0.4 \times f_{ck} \times A_c + 0.67 \times f_y \times A_{st}$$

EXAMPLE

Design the reinforcement in a 400 mm × 600 mm column subjected to an axial load of 2,000 kN under service dead load and live load. The column has an unsupported length of 4.0 m and is effectively held in position and restrained against rotation in both ends. Use M 25 concrete and Fe-415 steel.

Solution

Step 1: Check If the Column Is Short or Slender

Given: $l = 4{,}000$ mm, $b = 400$ mm, and $D = 600$ mm. Table 28 of IS 456 = lex = ley = 0.65(l) = 2,600 mm. So we have:

lex/D = 2,600/600 = 4.33 < 12
ley/b = 2,600/400 = 6.5 < 12

Hence, it is a short column.

Step 2: Minimum Eccentricity

$$e_{xmin} = Greater\,of\left(\frac{l_{ex}}{500} + \frac{D}{30}\right) and\,20\ mm = 25.2\ mm$$

$$e_{ymin} = Greater\,of\left(\frac{l_{ey}}{500} + \frac{b}{30}\right) and\,20\ mm = 20\ mm$$

$$0.05\ \text{D} = 0.05(600) = 30\ \text{mm} > 25.2\ \text{mm}\ (= e_{xmin})$$

$$0.05\ b = 0.05(400) = 20\ \text{mm} = 20\ \text{mm}\ (= e_{ymin})$$

Hence, the equation given in cl. 39.3 of IS 456 (Eq. (1)) is applicable for the design here.

Step 3: Area of Steel

From equation 10.4, we have:

$$P_u = 0.4 \times f_{ck} \times A_c + 0.67 \times f_y \times A_{st}$$

$$3000 \times 10^3 = 0.4 \times 25 \times 400 \times 600 + 0.67 \times 415 \times A_{st}$$

$$A_{st} = 2238.39\ mm^2$$

Provide 6–20 mm diameter and 2–16 mm diameter rods, giving 2,287 mm (> 2238.39), and p = 0.953%, more than the minimum percentage of 0.8 and less than the maximum percentage of 4.0. Hence, OK.

Step 4: Lateral Ties

The diameter of transverse reinforcement (lateral ties) is determined from cl.26.5.3.2 C-2 of IS 456 as not less than (1) $\varnothing / 4$ and (2) 6 mm. Here, $\varnothing$ = largest bar diameter used as longitudinal reinforcement = 20 mm. So the diameter of bars used as lateral ties = 6 mm.

The pitch of lateral ties, as per cl.26.5.3.2 C-1 of IS 456, should not be more than the least of the following:

1. The least lateral dimension of the column = 400 mm
2. Sixteen times the smallest diameter of the longitudinal reinforcement bar to be tied = 16(16) = 256 mm
3. 300 mm

Let us use p = pitch of lateral ties = 250 mm.

5.9.2 Reinforcement Details

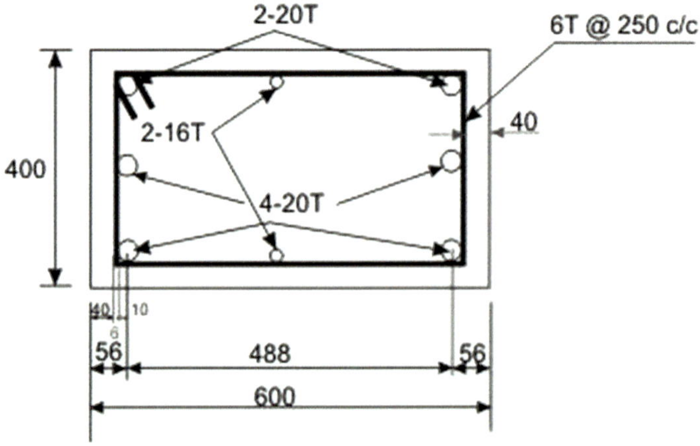

References

Arturson, G. 1987. The tragedy of San Juanico—the most severe LPG disaster in history. *Burns Incl Therm Inj*. 13: 87–102.

ASCE. 2005. *Standard calculation methods for structural fire protection*, Standard ASCE 29. USA.

Assael, M. J. and K. E. Kakosimos. 2010. *Fire, explosion and toxic gas dispersions: Effects, calculation and risk analysis*. CRC Press, NY.

ASTM. 2020. *Standard test methods for fire tests of building construction and materials*, E119. American Society for Testing and Materials.

Aven, T. and J. E. Vinnem. 2007. *Risk management with applications to offshore petroleum industry*. Springer, London, p. 212. ISBN: 978-1846286520.

Babrauskas, V. 1981. A closed form approximation for post flashover compartment fires. *Fire Saf J*. 4: 67–73.

Babrauskas, V. and S. J. Grayson (Eds.). 1992. *Heat release in fires*. Elsevier Applied Science, Amsterdam.

Barrell, D. J. A., H. Jack, M. Gadsby, D. T. Strong, R. J. Van Dissen, N. J. Litchfield, D. B. Townsend and J. G. Begg. 2013. Characterization of active faults in the Canterbury region: A tool for risk-based minimization of fault surface-rupture hazards to communities and infrastructure. *Proceedings, 19th NZGS Geotechnical Symposium*.

Barrell, D. J. A., N. J. Litchfield, D. B. Townsend, M. Quigley, R. J. Van Dissen, R. Cosgrove, S. C. Cox, K. Furlong, P. Villamor, J. G. Begg and S. Hemmings-Sykes. 2011. Strike-slip ground-surface rupture (Greendale Fault) associated with the 4 September 2010 Darfield earthquake, Canterbury, New Zealand. *Q J Eng Geol Hydrogeol*. 44(3): 283–291.

Behnam, B. 2019. Fire structural response of the plasco building: A preliminary investigation report. *Int J Civil Eng*. 17: 563–580.

Bergman, T. L., A. S. Lavine, D. P. DeWitt and F. P. Incropera. 2011. *Introduction to heat and mass transfer*, 6th edition. John Wiley & Sons, USA.

BIS. 1979. *IS 3809: Fire resistance tests for structures*. Bureau of Indian Standards, India.

Brode, H. L. 1959. Blast wave from a spherical charge. *Phys Fluids*. 2(2): 217–229.

BS EN 10025–2. 2019. *Hot rolled products of structural steels. Technical delivery conditions for non-alloy structural steels*. BSI, UK.

BS EN 1993–1–1. 2005. *UK national annex to Eurocode 3: Design of steel structures general rules and rules for buildings*. BSI, UK.

Buchanan, A. H. (Ed.). 2001. *Fire engineering design guide*, 2nd edition. Centre for Advanced Engineering, University of Canterbury, New Zealand.

Buchanan, A. H. and A. K. Abu. 2017. *Structural design for fire safety*. Hoboken, NJ, Wiley.

Budnick, E. K., D. D. Evans and H. E. Nelson. 1997. Simple fire growth calculations, Chapter 10, *NFPA Fire Protection Handbook*, 18th edition. National Fire Protection Agency, Quincy, Massachusetts, USA.

Buonocunto, V., G. Serino and F. Parisi. 2023. Large-scale vulnerability analysis of historical masonry buildings. *Proceedings of 9th ECCOMAS Thematic Conference on Computational Methods in Structural Dynamics and Earthquake Engineering*, Athens, Greece, 12–14 June, pp. 2762–2776.

Bureau of Indian Standards. 1987. *Code of practice for deign loads- part 2*. Imposed loads, New Delhi, India.

Bureau of Indian Standards. 2015. *Criteria for earthquake resistant design of structures: Part 4 Industrial structures including stack-like structures*, IS 1893: Part 4: 2015. Bureau of Indian Standards.

Bureau of Indian Standards. 2016. *Criteria for earthquake resistant design of structures: Part 1 General provisions and buildings*, IS 1893: Part 1: 2016. Bureau of Indian Standards.

CEN. 2002. *Eurocode 1: Basis of design and design actions on structures*, Part 2–2. EN 1991–2–2, Brussels. European Committee for Standardization, Brussels.

CEN. 2004. *Eurocode 2: Design of concrete structures*, Part 1–2: General rules—structural fire design. EN 1992–1–2, Brussels. European Committee for Standardization, Brussels.

CEN. 2005. *Eurocode 3: Design of steel structures*, Part 1–2: General rules—structural fire design. EN 1993–1–2, Brussels. European Committee for Standardization, Brussels.

Chandrasekaran, S. 2013a. *Dynamics of ocean structures*. Web-based course on the national program on technology-enhanced learning, IIT Madras. nptel.ac.in/courses/114106036

Chandrasekaran, S. 2013b. *Ocean structures and materials*. Web-based course on the national program on technology-enhanced learning, IIT Madras. nptel.ac.in/courses/114106035

Chandrasekaran, S. 2014. *Advanced theory on offshore plant FEED engineering*. Changwon National University Press, Republic of South Korea, p. 237. ISBN:978-89-969-7928-9.

Chandrasekaran, S. 2015a. *Advanced marine structures*. CRC Press, Florida, USA.

Chandrasekaran, S. 2015b. *Dynamic analysis and design of ocean structures*, 1st edition. Springer, India.

Chandrasekaran, S. 2015c. *Dynamic analysis of offshore structures*. Video course on MOOC, NPTEL portal. https://onlinecourses.nptel.ac.in/noc15_oe01/preview

Chandrasekaran, S. 2016a. *Health, safety and environmental management for offshore and petroleum engineers*. John Wiley & Sons, UK. ISBN: 978-11-192-2184-5.

Chandrasekaran, S. 2016b. *Offshore structural engineering: Reliability and risk assessment*. CRC Press, Florida.

Chandrasekaran, S. 2017. *Dynamic analysis and design of ocean structures*, 2nd edition. Springer, Singapore.

Chandrasekaran, S. 2018. *Advanced structural analysis with MATLAB*. CRC Press, Florida, USA.

Chandrasekaran, S. 2019a. *Advanced steel design of structures*. CRC Press, Florida.

Chandrasekaran, S. 2019b. *Structural health monitoring with application to offshore structures*. World Scientific, Singapore.

Chandrasekaran, S. 2024. *Structural analysis and design for fire and blast loads*. Web-based course offered on Digital Academy Thailand portal. Kasetsart University, Bangkok, Thailand.

Chandrasekaran, S. and S. K. Bhattacharyya. 2012. Analysis and design of offshore structures with illustrated examples. In *Human resource development center for offshore and plant engineering (HOPE center)*. Changwon National University Press, Republic of Korea, p. 285. ISBN: 978-89-963-9155-5.

Chandrasekaran, S., M. Fraldi, L. Nunziante, F. Carannante and M. C. Pernice. 2009a. Mechanics of distributed fibre optic sensors for strain measurements on rods. *Struct Eng*. 35(5): 323–333.

Chandrasekaran, S. and G. Gaurav. 2008. Offshore triangular tension leg platform earthquake motion analysis under distinctly high sea waves. *J Ships Offshore Struct*. 3(3): 173–184.

Chandrasekaran, S., S. Hari and A. Murugaiyan. 2020. Wire arc additive manufacturing of functionally graded material for marine risers. *J Mat Sci Eng A*. 792: 139530.

Chandrasekaran, S., S. Hari and A. Murugaiyan. 2022. Functionally graded materials for marine risers by additive manufacturing for high-temperature applications: Experimental investigations. *Structures*. 35: 931–938.

Chandrasekaran, S. and A. K. Jain. 2016. *Ocean structures: Construction, materials and operations*. CRC Press, Florida. ISBN 978-14-987-9742-9.

Chandrasekaran, S., A. K. Jain and N. R. Chandak. 2006. Seismic analysis of offshore triangular tension leg platforms. *Int J Struct Stab Dyn*. 6(1): 1–24.

Chandrasekaran, S., A. K. Jain, N. Shafiq and M. A. Wahab. 2021. *Design aids for offshore platforms under special loads*. CRC Press, Florida, p. 280.

Chandrasekaran, S. and A. Kiran. 2014a. Accident modelling and risk assessment of oil and gas industries. *Proceedings of 9th structural engineering convention*, IIT Delhi, India, 22–24 December, pp. 2533–2543.

Chandrasekaran, S. and A. Kiran. 2014b. Consequence analysis and risk assessment of oil and gas industries. *Proceedings of international conference on safety & reliability of ship, offshore and subsea structures*, Glasgow, UK, 18–20 August.

Chandrasekaran, S. and A. Kiran. 2015. Quantified risk assessment of LPG filling station. *Prof Saf J Am Soc Saf Eng*. 60(9): 44–51.

Chandrasekaran, S., L. Nunziante, V. Gupta and F. Carannante. 2008a. Nonlinear seismic analyses of high-rise reinforced concrete buildings. *ICFAI J Struct Design*. 1(1): 7–24. ICFAI Press.

Chandrasekaran, S., L. Nunziante, G. Serino and F. Carannante. 2008b. Axial force-bending moment failure interaction and deformation of reinforced concrete beams using Eurocode. *J Struc Eng*. 35(1): 181–190.

Chandrasekaran, S., L. Nunziante, G. Serino and F. Carannante. 2009b. *Seismic design aids for nonlinear analysis of reinforced concrete structures*. CRC Press, Florida, USA.

Chandrasekaran, S., L. Nunziante, G. Serino and F. Carannante. 2010. Axial force-bending moment limit domain and flow rule for reinforced concrete elements using Euro Code. *Int J Damage Mech*. 19: 523–558. Q1 category, Impact factor 1.17.

Chandrasekaran, S., L. Nunziante, G. Serino and F. Carannante. 2011. Curvature ductility of RC sections based on Eurocode: Analytical procedure. *J Korean Soc Civil Eng*. 15: 131–144.

Chandrasekaran, S. and A. Roy. 2006. Seismic evaluation of multi-storey RC frames using modal pushover analysis. *Int J Nonlinear Dyn*. 43(4): 329–342.

Chandrasekaran, S., G. Serino and M. Spizzuoco. 2024. *Earthquake engineering and structural control: Theory and applications*. CRC Press, Florida, p. 432.

Chandrasekaran, S. and G. Srivastava. 2017. *Design aids for offshore structures under special environmental loads including fire resistance*. Springer Nature, Singapore. ISBN 978-981-10-7608-4.

Chandrasekaran, S. and G. Srivastava. 2022. *Fire-resistant design of structures*. CRC Press, Florida, USA, p. 212.

Chandrasekaran, S. and U. K. Tripati. 2005. Geometric irregularity effects on seismic vulnerability of buildings. *J Vib Eng Tech*. 4(2): 115–123.

Chang, Y. F., Y. H. Chen, M. S. Sheu and G. C. Yao. 2006. Residual stress-strain relationship for concrete after exposure to high temperatures. *Cem Concr Res*. 36(10): 1999–2005. https://doi.org/10.1016/j.cemconres.2006.05.029

Chen, M. C., J. I. Restrepo, C. Blandon and J. F. Velásquez. 2021. Performance-based seismic design framework for inertia-sensitive nonstructural components in base-isolated buildings. *J Build Eng*. 43: 103073.

Chen, X., J. Li, X. Cheng, B. He, H. Wang and Z. Huang. 2017. Microstructure and mechanical properties of the austenitic stainless steel 316L fabricated by gas metal arc additive manufacturing. *Mater Sci Eng A*. 703: 567–577. https://doi.org/10.1016/j.msea.2017.05.024

Chopra, A. K. 2003. *Dynamics of structures: Theory and applications to earthquake engineering*, 2nd edition. Pearson Education, Singapore.

Constantinou, M. C. and M. D. Symans. 1992. Experimental and analytical investigation of seismic response of structures with supplemental fluid viscous dampers. In *Technical report, NCEER-92-0032*. Multidisciplinary Center for Earthquake Engineering Research, University at Buffalo, State University of New York, Buffalo, NY, USA.

Crowl, D. A. and J. F. Louvar. 2002. *Chemical process safety: Fundamentals with applications*. Prentice Hall, New Jersey. ISBN: 0-13-0181765.

Dadashzadeh, M., F. Khan, K. Hawboldt and P. Amyotte. 2013. An integrated approach for fire and explosion consequence modelling. *Fire Saf J*. 61: 324–337.

Dellchatsios, M. A. 1984. Flame height of turbulent wall fire with significant flame radiation. *Combust Sci Technol*. 39: 195–214.

Denel, B. 1981. *Basic design and creativity*. O.D.T.U. Faculty of Architecture Publications, Ankara, Turkey.

Divanlıoglu, D. 1997. *Basic design: Elements and principles of design*. Birsen Publishing House, Istanbul, Turkey.

Drysdale, D. 1998. *An introduction to fire dynamics*, 2nd edition. John Wiley & Sons, Chichester, UK.

Dubnikova, F., R. Kosloff, J. Almog, Y. Zeiri, R. Boese, H. Itzhaky, A. Alt and E. Keinan. 2005. Decomposition of triacetone triperoxide is an entropic explosion. *J Am Chem Soc*. 127(4): 1146–1159.

Ealangi, I. 2010a. Earthquake protection of buildings by seismic isolation. Devices and concepts. *Proceedings of young researchers conference*. Technical University of Bucharest.

Ealangi, I. 2010b. Earthquake protection of buildings by seismic isolation. Devices and concepts. *Eng Struct*. 14(1): 35–48.

Eem, S. and D. Hahm. 2019. Large strain nonlinear model of lead rubber bearings for beyond design basis earthquakes. *Nuclear Eng Technol*. 51(2): 600–606.

El Ouni, M. H., M. Abdeddaim, S. Elias and N. B. Kahla. 2022. Review of vibration control strategies of high-rise buildings. *Sensors*. 22(21): 8581. https://doi.org/10.3390/s22218581

Evans, R. 1995. *The projective cast: Architecture and its three geometries*. The MIT Press, Cambridge, MA.

Faisal, I. K. and S. A. Abbassi. 1999. Major accidents in process industries and analysis of causes and consequences. *J Loss Prev Process Ind*. 12: 361–378.

Feasey, R. and A. H. Buchanan. 2002. Post-flashover fires for structural design. *Fire Saf J*. 37(1): 83–105.

Fenz, D. M. and M. C. Constantinou. 2008. Spherical sliding isolation bearings with adaptive behavior: Theory. *Earthq Eng Struct Dyn*. 37: 163–183. https://doi.org/10.1002/eqe.751

Fergusson, J. 2020. History of Indian and Eastern architecture. *Hist Indian Eastern Architect*. II: 1–615. https://doi.org/10.1017/cbo9781139814638

Gambhir, M. L. 2004. *Stability analysis and design of structures*. Springer, New York. ISBN: 3-540-20784-8.

Gelernter, M. 1988. Reconciling lectures and studios. *JAE*. 41(2): 46–52.

Guarini, G. 1937. Architettura civile, del padre d. Guarino Guarini cherico regolare: Opera postuma. In *Torino: Appresso Gianfrancesco Mairesse all'insegna di Santa Teresa di Gesù*. Getty Research Institute.

Gungor, I. H. 1972. *Basic design*. Celtut Printing, Istanbul, Turkey.

Gungor, I. H. 2005. *Basic design: For visual arts and architecture*. Potato Printing, Istanbul.

Guzman Pujols, J. C. and K. L. Ryan. 2020. Slab vibration and horizontal–vertical coupling in the seismic response of low-rise irregular base-isolated and conventional buildings. *J Earthq Eng*. 24(1): 1–36.

Hall, J. R. 2014. *The total cost of fire in the United States*. National Fire Protection Association, Quincy, MA.

Hancerlioglu, O. 2008. *Glossary of philosophy*. Remzi Bookstore, Istanbul, Turkey.

Hemmatian, B., E. Planas and J. Casal. 2015. Fire as a primary event of accident domino sequences: The case of BLEVE. *Reliabil Eng Syst Saf*. 139: 141–148.

Hemmatian, B., E. Planas and J. Casal. 2017. Comparative analysis of BLEVE mechanical energy and overpressure modelling. *Process Saf Environ Protect*. 106: 138–149.

Hernried, A. G. and K. M. Lei. 1995. Semi-analytical techniques for the determination of the dynamic response of equipment in structures supported on coulomb friction elements. *Earthq Eng Struct Dynam*. 24(6): 801–810.

IBC. 2015. *International building code*. International Codes Council, NJ, USA.

Inaudi, J. A. and J. M. Kelly. 1993. Optimum damping in linear isolation systems. *Earthq Eng Struct Dynam*. 22(7): 583–598.

ISO. 2012. *Fire resistance tests—elements of building construction*, ISO 834–2012. International Organization for Standardization.

James, G. Q. 2016. *Principles of fire behaviour*, 2nd edition. Taylor and Francis. ISBN: 9781498735629.

Jasieńko, J., K. Raszczuk, K. Kleszcz, and P. Frąckiewicz. 2021. Numerical analysis of historical masonry domes: A study of St. Peter's Basilica dome. *Structures*. 31: 80–86. ISSN: 2352-0124. https://doi.org/10.1016/j.istruc.2021.01.082

Jia, J. 2016. *Modern earthquake engineering: Offshore and land-based structures*. Springer.

Jin, Y. and B. S. Jang. 2015. Probabilistic fire risk analysis and structural safety assessment of FPSO topside module. *Ocean Eng*. 104: 725–737. https://doi.org/10.1016/j.oceaneng.2015.04.019

Johnston, J. F. and S. R. White. 2018. *Understanding the Meckering earthquake: Western Australia*. 14 October 1968. Geological Survey of Western Australia, 26p.

Joye, Y. 2007. Fractal architecture could be good for you. *Nexus Netw J*. 9(2): 159–382.

Juhn, G., G. D. Manolis, M. C. Constantinou and A. M. Reinhorn. 1992. Experimental study of secondary systems in base-isolated structure. *J Struct Eng*. 118(8): 2204–2221.

Kawagoe, K. 1958. *Fire behavior in rooms*, Report No. 27. Building Research Institute, Tokyo, Japan.

Khan, N. and G. Srivastava. 2018. Enhanced fire severity in modern Indian dwellings. *Curr Sci*. 115(2): 320–325.

Kodur, V. 2014. Properties of concrete at elevated temperatures. *ISRN Civil Eng*. https://doi.org/10.1155/2014/468510

Koffa, K. 1963. *Principles of gestalt psychology*. Harbinger Books, New York.

Lees, F. P. 1996. *Loss prevention in chemical process industries*. Butterworth, London.

Li, H.-N., G. Li and S.-Y. Wang. 2014. Study and application of metallic yielding energy dissipation devices in buildings. *NCEE 2014–10th U.S. National Conference on Earthquake Engineering: Frontiers of Earthquake Engineering*. https://doi.org/10.4231/D3FQ9Q593

Lie, T. T. 1995. Fire temperature-time relations. In *SFPE handbook of fire protection engineering*, 2nd edition. Society of Fire Protection Engineers, USA.

Lin, Y.-S. 2005. Estimations of the probability of fire occurrences in buildings. *Fire Saf J*. 40(8): 728–735.

Lorenzi, M. G. and M. Francaviglia. 2010. Art & mathematics in antoni gaudí's architecture: "la sagrada família. *J App Math*. 1.

Losanno, D., H. A. Hadad and G. Serino. 2019. Design charts for Eurocode-based design of elastomeric seismic isolation systems. *Soil Dyn Earthq Eng*. 119: 488–498.

Manco, M. R., M. A. Vaz and A. Landesmann. 2014. Analysis of oil tanker deck under hydrocarbon fire. *Int J Model Simul Pet Ind*. 8(2): 17–24.

Mandelbrot, B. B. 1982. *The fractal geometry of nature*, Rev. edition. Freeman Publications, San Francisco.

Mittal, A. 2017. *Handbook on construction of earthquake resistant buildings*, CAMPTECH, Gwalior, INDIA, p. 172.

Murty, C. V. R., R. Goswami, A. R. Vijayanarayanan and V. Mehta. 2012. Earthquake behaviour of buildings. *Gujarat State Disaster Management Authority, Gandhinagar*. 53: 268.

NBC. 2016. *National building code*, Part 4—Fire & Life Safety. Bureau of Indian Standards, New Delhi, India.

NDMA. 2016. *National disaster mitigation authority: Simplified guidelines for earthquake safety of buildings*. Bureau of Indian Standards, New Delhi.

NFPA. 1997. *Fire protection handbook*, 18th edition. National fire Protection Association, Quincy, MA, USA.

NFPA Report. 2020. *Fire loss in the U.S. National fire Protection Association*. Quincy, MA, USA.

Nuzzo, I., D. Losanno, N. Caterino, G. Serino and L. M. B. Rotondo. 2018. Experimental and analytical characterization of steel shear links for seismic energy dissipation. *Eng Struct*. 172(1): 405–418.

Pan, J., R. Zou and F. Jin. 2017. Experimental study on specific heat of concrete at high temperatures and its influence on thermal energy storage. *Energies*. 10(1). https://doi.org/10.3390/en10010033

Polatoglu, C. 2009. *Building evaluation*. Y.T.U. Publishing House, Istanbul.

Politopoulos, I. 2008. A review of adverse effects of damping in seismic isolation. *Earthq Eng Struct Dynam*. 37(3): 447–465.

Punmia, B. C. and K. K. Khandelwal. 2002. *Project planning control with PERT and CPM*. Laxmi Publications (P) Ltd.

Purandare, A. 2016, Jan–Mar. The escalating fire hazards of modern homes. *Fire Eng*. 41.

Purser, D. 2002. *Society of fire protection engineers (SPFE) guide to human behavior*, 3rd edition. Springer, Maryland, USA.

Quintiere, J. G. 2006. *Fundamentals of fire phenomena*. John Wiley & Sons, pp. 1–439. https://doi.org/10.1002/0470091150

Renzi, E. and G. Serino. 2004. Testing and modelling of a semi-actively controlled steel frame equipped with MR dampers. *J Struct Control Health Monitor*. 11(3): 189–221.

Sakamoto, M. 1993. Practical applications of active structural response control and earthquake & strong wind observation systems. In *Planning workshop for the Hong Kong international full-scale control test facility*. Hong Kong University of Science & Technology, Hong Kong, China.

Skinner, R. I., W. H. Robinson and G. H. McVerry. 1993. *An introduction to seismic isolation*. John Wiley & Sons.

Sucuoğlu, H. and S. Akkar. 2014. *Basic earthquake engineering*. Spring International Publishing, Switzerland.

Thomas, P. H. and A. J. M. Heselden. 1972. *Fully developed fires in single compartments*, CIB Report No. 20. Fire Research Note 923, Fire Research Station, UK.

Til, B., R. Abbassi, V. Garaniya, F. Khan and M. Dadashzadeh. 2018. Review and analysis of fire and explosion accidents in maritime transportation. *Ocean Eng*. 158: 350–366.

Torunbalci, N. and G. Ozpalanlar. 2008. Earthquake response analysis of mid-story buildings isolated with various seismic isolation techniques. *Proceedings of 14th World Conference*.

Verderame, G. M., P. Ricci, F. De Luca, C. Del Gaudio and M. T. De Risi. 2014. Damage scenarios for RC buildings during the 2012 Emilia (Italy) earthquake. *Soil Dyn Earthq Eng*. 66: 385–400.

Wickstrom, U. 1986. A very simple method for estimating temperature in fire exposed concrete structures. In S. J. Grayson and D. A. Smith (Eds.), *Proceedings of new technology to reduce fire losses and costs*. Elsevier, NY, USA.

Wickstrom, U. 2016. *Temperature calculation in fire safety engineering*. Springer-Nature.

Wielgosz, E., T. Kargul and J. Falkus. 2014. Comparison of experimental and numerically calculated thermal properties of steels. *Proceedings of 23rd International Conference Metallurgy and Materials Conference*, May, pp. 1528–1533.

Yeo, J. G., Y. G. Jung and S. C. Choi. 1998. Design and microstructure of ZrO2/SUS316 functionally graded materials by tape casting. *Mater Lett*. 37(6): 304–311. https://doi.org10.1016/S0167–577X(98)00111–6.

Zhang, C. and A. Ali. 2021. The advancement of seismic isolation and energy dissipation mechanisms based on friction. *Soil Dyn Earthq Eng*. 146: 106746.

cityfire.co.uk (accessed June 2022).

Index

Note: Page numbers in *italic* indicate a figure and page numbers in **bold** indicate a table on the corresponding page.

D

E

G

H

I

J

K

L

N

O

P

Q

R

Z

For Product Safety Concerns and Information please contact our EU representative GPSR@taylorandfrancis.com Taylor & Francis Verlag GmbH, Kaufingerstraße 24, 80331 München, Germany